Computational Mineral Physics

Computational mineralogy is fast becoming the most effective and quantitatively accurate method for successfully determining structures, properties, and processes at the extreme pressure and temperature conditions that exist within the Earth's deep interior. It is now possible to simulate complex mineral phases using a variety of theoretical computational techniques that probe the microscopic nature of matter at both the atomic and subatomic levels. This introductory guide is for geoscientists and researchers performing measurements and experiments in a lab, those seeking to identify minerals remotely or in the field, and those seeking specific numerical values of particular physical properties. Written in a user- and property-oriented way and illustrated with calculation examples for different mineral properties, it explains how property values are produced, how to tell if they are meaningful or not, and how they can be used alongside experimental results to unlock the secrets of the Earth.

Razvan Caracas is a computational mineral physicist with a background in both geology and materials sciences. He was awarded a PhD from the Catholic University of Louvain prior to holding a postdoctoral position at the University of Minnesota, a Carnegie Fellowship at the Carnegie Institution of Washington, and a Humboldt Fellowship at the University of Bayreuth. He is a Fellow of the Mineralogical Society of America, a recipient of the Dana Medal of the same society, and a member of the Academia Europaea. He was a CNRS researcher at the Ecole Normale Supérieure de Lyon, and adjunct professor at the Center for Planetary Habitability of the University of Oslo. He is presently a senior researcher at the Institute de Physique du Globe de Paris, working on a wide range of topics in planetary mineralogy, going from the supercritical state that dominated the protolunar disk to the internal structure of exoplanets. With the help of atomistic simulations, his work explores the early Earth's evolution – helping to decipher the condensation of the Earth and the Moon, the formation of the primordial atmosphere, and exploring what conditions planets must fulfil to make prebiotic chemistry thrive.

Computational Mineral Physics

A Practical Guide for Earth Scientists

RAZVAN CARACAS

Institut de Physique du Globe de Paris, CNRS, Université Paris Cité

CAMBRIDGE
UNIVERSITY PRESS

Shaftesbury Road, Cambridge CB2 8EA, United Kingdom

One Liberty Plaza, 20th Floor, New York, NY 10006, USA

477 Williamstown Road, Port Melbourne, VIC 3207, Australia

314–321, 3rd Floor, Plot 3, Splendor Forum, Jasola District Centre, New Delhi – 110025, India

Cambridge University Press is part of Cambridge University Press & Assessment, a department of the University of Cambridge.

We share the University's mission to contribute to society through the pursuit of education, learning and research at the highest international levels of excellence.

www.cambridge.org
Information on this title: www.cambridge.org/9781108416771
DOI: 10.1017/9781108241908

First published 2026

A catalogue record for this publication is available from the British Library

A Cataloging-in-Publication data record for this book is available from the Library of Congress

ISBN 978-1-108-41677-1 Hardback

In memory of my father.

Contents

Boxes

Preface

If you would have the courage to walk in a session on theoretical mineralogy at one of the big conferences in earth sciences (American Geophysical Union Conference, European Geosciences Union Conference, Goldschmidt, etc.), you would probably be *rained down* with the following abbreviations: DFT, PAW, GGA, PBE, XC, WC90, B3LYP, +U, and so on. Unless you were an expert active in the field for years or a highly achieving student familiar with all this foreign language, you would be lost and stop paying attention pretty soon. Alas, you would probably lose an entirely beautiful piece of science, or you wouldn't be able to say if it's the correct choice or not, if LDA is *good enough*, or if one would need GGA or meta-GGA, or if highly accurate, but computationally expensive QMC is really worth the effort.

This book tells the story of what happens during the long hours a simulation runs on a (super)computer. It tries to teach you how to perform such a simulation. Which number is important, why and how much, and what are all those abbreviations? It tries to show you what you could learn from that simulation and how to judge the quality of that particular theoretical result.

The large majority of the examples in this book are realized using the abinit package (https://www.abinit.org/).

This book is for geoscientists in the broadest possible sense: those who want to have a first guide into performing first-principles calculations, those who do measurements and experiments in a lab, those who try to identify minerals remotely or in the field, and those who need specific numerical values of some physical properties. Advanced master students, PhD students, and postdoctoral and young researchers who want to explore the numerical world of mineralogy will have a first taste in the pages of this book. Confirmed researchers in complementary fields, like experimental mineralogy and petrology, geochemistry, or geodynamics, will find explanations and seek understanding for the origin and meaningfulness of particular numerical results provided by their theoretician colleagues. The book spares the reader the unnecessary theoretical details and lengthy equations and instead focuses on the practical aspects of producing and interpreting useful numbers, particularly in mineralogy.

In-depth reviews of various sorts have been written about almost every imaginable technique and aspect of earth sciences, except for atomistic computational mineralogy. This book tries to fill this gap in the geoscience literature today.

Chapter 1 introduces the reader to the basis of crystallography necessary to set up the atomic arrangements for the input files of the simulations. Chapter 2 briefly discusses the topic of phenomenological interatomic potentials, their use in large-scale simulations, and also their limitations. Chapters 3, 9, 10, and 16 form the theoretical core of the entire computational mineral physics. Chapter 3 presents the foundations of the density functional theory and describes the numerical implementations and the common approximations employed. Computations of electronic properties are covered in Chapters 4 and 5, ferroic properties in Chapters 6 and 7, and mechanical properties in Chapter 8. Chapter 9 introduces the dynamical properties, the treatment of small perturbations in terms of the development of energy as Taylor series, and the quasi-harmonic approximation. Their actual computation, in the framework of the perturbation theory, is explained in Chapter 10. The calculations of various dynamical properties are discussed in Chapters 11–15. Chapter 15 is of particular interest to geochemists, who will find here the foundations of isotope fractionation. This is followed by illustrative examples of simulations of melts and fluids in Chapters 16 and 17. The book ends with a few words about supercomputers and large-scale simulations.

At the end of reading the book, the interested reader should hopefully be able to perform first-principles calculations on their own. But the main purpose is to cover as much as possible practical aspects, rather than pure theory. The entire computational machinery will be put into motion, and the steps of how to compute specific physical properties and how to judge the quality of the results will be thoroughly discussed. Reading the book should come after advanced classes in solid-state and statistical physics, typically from a strong master's or doctoral curriculum.

And a final, personal note. My voyage to computational mineral physics was meandrous, with many bricks accumulating over time. Oscillating between physics and biology, computer science and geology, this path was marked by many people's presence and knowledge. After a short physics and computational period, geology offered me a much-needed window into the natural world. From the beginning of my geology years, a unique role in my career and life played Prof. Gyuri Ilinca, who made me discover the beauty of crystallography and mineralogy and whose correspondence was highly inspiring in many ways. Then the geology professors at the University of Bucarest, Romania, like Professors Nicolae Anastasiu, Gheorghe C. Popescu, Marin Seclaman, Emil Constantinescu, and others, the colleagues, and the assistant professors chiselled my geological view of the world. After my geology years, Professor Xavier Gonze, in the ABINIT group at Louvain-la-Neuve, Belgium, saved me from the entropy of the university and accepted me in his group, where he made me fully appreciate my return to computational materials science; I am highly grateful to him for all those years. During this time, Professor Philippe Sonnet's understanding smoothed my transition from crystallography and descriptive mineralogy to physics. Later on, in the cold winter of Minnesota, Professor Renata Wentzcovitch introduced me to the fascinating world of the deep Earth, and I thank her sincerely for this. The following years in the Geophysical Laboratory at the Carnegie Institution of Washington, USA, where I learnt so much from the staff scientists, and especially Drs Ron Cohen and Russ Hemley, completed my association with the extreme conditions. Many pages of this book were written in Oslo, and I warmly thank Carmen Gaina, Trond Torsvik, Stephanie Werner, and the entire Earth Evolution and Dynamics (CEED)/Center for Planetary HABitability (PHAB) team for hosting me so many times and offering many insightful discussions about earth sciences. Special thanks go to Chris Mohn, Mandy Bethkenhagen, Tim Bögels, and Wendy Panero for reading many parts of this manuscript and providing valuable insight. I would also like to thank the staff of the supercomputing centres in France, such as Centre Informatique National de l'Enseignement Supérieur (CINES), Centre de calcul pour la recherche et la technologie (CCRT), and the Institut du développement et des ressources en informatique scientifique (IDRIS), for their longstanding computational and technical support, as well as the Laboratory of Geology of Lyon, the Centre Blaise Pascal, and the PSMN supercomputing centre of the Ecole Normale Superieure de Lyon, France, for their continuous support and for hosting me and my group. Over the years, countless interactions and immeasurable advice from so many other colleagues and friends, younger and older, as well as students and professors, whose names are too many to list here, helped me gain a physical and materials-oriented understanding of the natural world, which you might retrieve as well in this book. Last but not least, I deeply thank my parents, Gabriela and Virgiliu, my daughter Sarah, and my partner Cécile, for their support that long precedes and outlasts the writing of this book.

PART I

GETTING STARTED

1 Introduction

1.1 The Internal Structure of the Earth

From the entire radius of the Earth of almost 6,400 km, the direct sampling performed by humans is only a bare scratch at the surface. The deepest well took two decades of drilling, at the dusk of the last century by soviet scientists in the Kola Peninsula (present-day Russia, close to Murmansk). The bottom of the well reached a little more than 12 km depth and was stopped due to high temperature and pressure. At the base of the well, there was a level of gneisses with epidote, biotite and plagioclase, and amphibolites (Figure 1.1), forming the old crystalline basement. Even so, it was a tremendous achievement and still holds tight to its world record. As for the remaining thousands of kilometres of the Earth's interior, we have to look for alternative ways of sampling.

One of the most reliable ways to investigate the Earth's mantle is to use xenoliths brought to the surface by various volcanic eruptions or trapped in magmatic bodies. These xenoliths sample large parts of the lithosphere and carry information from the upper mantle. Solid inclusions in diamonds provide further information from even greater depths [203]. These different inclusions and xenoliths are brought to the surface by volcanoes, some having roots deeper than 1,000 km below the surface of the Earth. Many minerals trapped in these inclusions undergo retrograde phase transitions during their uplift. The lavas and gases erupting from the mid-ocean ridges or on top of mantle plumes also offer direct information about the Earth's mantle, usually in the form of isotopic signatures. But the rest of the deep Earth's interior is only indirectly sampled by seismology (propagation of seismic waves) or inferred from experimental and computational studies in petrology and mineralogy.

The present-day internal structure of the Earth is the result of processes spanning the entire life of our very active planet. Chemical (amount and ratio between the major elements, presence of volatiles, etc.), physical (temperature, pressure, and redox state), and hazardous (fall of meteorites) factors, as well as the presence and development of life, have all contributed to this mineralogical and geological outcome. The mineralogical diversity at the Earth's surface spans today about 4,000 mineral species. The diversity of minerals increased immensely over geological time, as their crystallography and chemistry became more and more complex and diverse [130]. Imagine that you start with a relatively simple pile

Fig. 1.1   A sample of basement gneiss with epidote, biotite, and plagioclase, coming from a depth of 11,100 m, close to the bottom of the Kola well hole, the deepest well ever executed. Image courtesy of Professor Gh. C. Popescu, University of Bucharest.

of oxides and silicates, add water, a few other volatiles, heat, and pressure, and let this chemical system evolve/react for more than 4 billion years. With time, crystal structures and chemistry became more complex. Many of the minerals we know today are stable only under the narrow conditions of ambient pressures and temperatures. Others span a larger stability field or are found at the surface in a metastable state – those brought from the deep or those that survive or form during the shocks of the meteorite falls. This new concept of increasing mineral diversity finally makes justice to the image of an evolving mineral world with time.

Bodies in the solar system that reach about 100 km in size have accumulated enough heat to melt and differentiate [186]. The heat comes notably from the radioactive decay of the short-lived Al^{26} and Fe^{60} isotopes, from impacts during accretion, and from gravitation during compaction. This means that based on immiscibility relations (i.e. minerals unable to mix as a solid solution), the heavy minerals, usually Fe-based alloys for the rocky bodies or oxide and silicate rocks for the icy bodies, would separate and fall gravitationally to form a denser core. At the same time, the lighter fraction would float on top. At the end of this chemical separation, the remaining boundaries of the geological layers correspond to the stability fields of the major minerals constituting these global layers.

The outermost solid layer of the Earth, the crust, is the most mineralogically diverse. Being in contact with the atmosphere, the hydrosphere, and the biosphere, it is also the most active layer. The width of the crust ranges from 0 km at the mid-ocean ridges, where it is continuously formed, up to 6–8 km in the middle of the oceanic plates and several tens of kilometre in the middle of the continental crust, where old pieces of accumulated rocks form the stable cratons. The crust is separated by the mantle by the Mohorovicic discontinuity, which was discovered in 1909 as an abrupt change in seismic wave velocity. This marks a sharp seismological transition that might correspond petrologically to the basalt-to-eclogite transition and/or mineralogically to the feldspathic-garnet transition in peridotites. Despite its depth, there was an international effort for drilling through the Mohorovicic discontinuity [265, 83].

The outermost part of the mantle, directly below the Mohorovicic discontinuity, is a layer of solid hard rock. The crust and this layer form the lithosphere. This is broken into several parts, forming the tectonic plates. The tectonic plates behave like rigid solid bodies at short timescales, that is, on the order of seconds, and like viscoelastic plates at long timescales, that is, on the order of years; they plunge into the mantle at the subduction zones, contributing to the mantle convection. Consequently, the term 'lithosphere' reflects rheological properties, and the crust versus mantle distinction is related to seismic observations.

The tectonic plates float and move on top of a low-viscosity layer, containing partially molten rocks, called the asthenosphere. Its origin is probably related to the presence of large amounts of volatiles, like H_2O and CO_2, which decrease the melting point of the silicates below the geotherm (the thermal profile of the interior of the Earth), hence the occurrence of partial melting.

The upper mantle is dominated by the presence of $(Mg,Fe)_2SiO_4$ olivine, alongside other low-pressure light silicates, like pyroxenes and garnets. The phase transformation of olivine into wadsleyite, taking place at a depth of about 440 km, marks the discontinuity in the TZ. Its actual depth depends on the amount of Fe in olivine and the presence of water. Wadsleyite transforms into ringwoodite inside the transition zone. This transition is also highly dependent on the amount of iron and is observed in seismic data.

At even higher pressures, the decomposition of ringwoodite into a mixture of bridgmanite $(Mg,Fe)SiO_3$, with an orthorhombic distorted perovskite structure, and $(Mg,Fe)O$ magnesiowüstite, with a rock salt structure, marks the beginning of the lower mantle. The phase decomposition takes place at a depth of 660 km. The lower mantle is the largest part by volume of our planet. It extends down to the core-mantle boundary (CMB), at a depth of about 2,890 km. At the planetary scale, most of the lower mantle is largely homogeneous. But at its base, things become complicated, with several local and discontinuous structures. In contact with the core lies a widespread but discontinuous layer called the D" layer. Its occurrence is due to the further transition of bridgmanite to a post-perovskite structure. Considerable amounts of iron can be accumulated in the D" layer. Chemical and/or thermal lateral variations with variable heights lead to the formation of other zones, the large low shear-wave velocity provinces (LLSVPs). Finally, regions of very low shear velocity have been identified lying on the mantle side of the CMB, on top of the core, many in the vicinity of the D" layer: these are the ultra low velocity zones (ULVZs). The LLSVPs can be related to narrow mantle upwellings whose tops may arrive at the surface in places like Hawaii or Iceland.

It is predicted that in super-Earths, extrasolar rocky planets presumably similar in composition to our own but of much larger size and mass, various other layers will be differentiated inside the mantle. Further higher-pressure phase transitions in post-perovskite and in its mixtures with magnesiowüstite [91, 252, 92] would lead to other distinct layers. Depending on the size and age of the planet and on the temperature of its surface, which in turn depends on the presence and characteristics of an atmosphere, a molten silicate layer can lie on top of the core in these super-Earths. Changes in the ratio of major elements can affect the internal layers, as minerals rich in alkalis or calc-alkalis can dominate the mantle of some of these super-Earths [254].

Coming back to the Earth, the core, lying below the CMB, is formed in its large majority by iron, with nickel coming second, together with an uncertain number and amount of light elements. The major candidates for these elements are volatiles soluble in molten iron, like Si, O, S, C, or H. Because of the remoteness of the core and its inaccessibility, actual mineralogical and petrological data are missing. So the problem of solving its composition based solely on seismic data is under-constrained [82]. While geochemical and geophysical [50, 94, 61] considerations can help further reduce the spectrum of possible chemical solutions, atomistic simulations, like the ones discussed in this book, have the potential to bring the actual answers geoscientists are looking for.

From seismic measurements, we know that the core has two layers: a liquid outer core, shrinking in size as the Earth cools down, and a solid inner core, whose size increases. The Earth's magnetic field is generated by convection in the outer core. The outer core is laterally homogeneous in terms of both chemistry and temperature. Any possible heterogeneity would be quickly erased by convection. The inner core might crystallise radially or, alternatively, only on one side and melt at a lower rate on the opposite side, resulting in a net translation effect [2]. The inner core started to differentiate about 1 billion years after the core-formation event [112], which raises questions about the mechanism of the origin of the magnetic field prior to its formation. A current explanation is that the driving force of the convection was the chemical exsolution of light elements from the cooling liquid, like Mg, Si or O, into the mantle [17, 140].

1.2 A Few Words about Thermodynamics

We will outline only a few basic thermodynamic concepts in the following lines. For in-depth lectures on thermodynamics, the reader is advised to consult specialised books.

All thermodynamic systems evolve towards minimising their free energy and maximising their entropy. The Gibbs free energy, G, has three terms:

$$G = U + PV - ST, \tag{1.1}$$

where U is the internal energy, P is the pressure, V is the volume, S is the entropy, and T is the temperature.

The volume, V, density, ρ, and the temperature, T, are the intensive variables that determine the state of the system. The other variables are extensive and stem from the behaviour of the system as a function of ρ and T. The internal energy is the ground-state energy – the energy of the ensemble of nuclei and electrons at 0 K, with no electronic excitations and no perturbations. This is the typical result of the first-principles simulations and will be described in detail in Chapter 3. At finite positive temperatures, the internal energy gains a component dependent on the thermal agitation of the atoms. (Internal) pressure is a thermodynamic potential, computed as the energy derivative with respect to volume. Entropy is a measure of the disorder of the ensemble of particles. It has various components: electronic entropy due to the Fermi distribution of electrons at non-null temperature, for example electronic excitations in metals; magnetic entropy due to the partial ordering of the local magnetic moments, for example local spins in Fe-bearing bridgmanite at mantel conditions; dynamical entropy due to the vibrations or the thermal agitation of the atoms, for example atomic vibrations in all minerals; configurational entropy due to the partial ordering of the atoms on a lattice, for example Mg/Fe distribution on the M1 and M2 sites in the olivine solid solution; rotational and librational entropy due to the free rotations or librations of molecules, for example rigid rotations of water molecules in the plastic phase of ice, OH librations in Brucite, $Mg(OH)_2$ around the c axis; and so on.

Often in first-principles calculations, the temperature is neglected in the first step. These calculations are called 'static', and temperature is absolute zero; all the terms multiplied by temperature are 0. At these conditions, the Helmholtz free energy, or the enthalpy, is defined as:

$$H = U + PV. \tag{1.2}$$

Computing the enthalpy for a mineral system is straightforward. This has the clear advantage of a quick calculation that offers good and powerful insight into the phase relations. In calculations, we can approximate the temperature as 0 K because most experiments are conducted at room temperature. For most natural systems, the thermal effects between 0 K and ambient temperature are not large enough to change the relative stabilities of different phases. Of course, this assumption should not be taken as an absolute rule, and caution must be paid before jumping to conclusions; for example, many H-bearing systems would undergo order–disorder phase transitions below the ambient temperature that would be missed entirely out if only the enthalpy is computed.

Most natural systems are solid solutions, not pure phases. In solid solutions, the chemistry can continuously vary between two or more ENS-member terms. For example, forsterite, Mg_2SiO_4, and fayalite, Fe_2SiO_4, are, respectively, the Mg- and Fe-pure end-member terms of the olivine, $(Mg,Fe)_2SiO_4$, solid solution.

For the solid solution, the free energy is the chemical potential μ, which, for each component i, of concentration X_i, is the change in free energy corresponding to a change in the amount of the component i.

$$\mu_i = \Delta G / \Delta X_i dX_i. \tag{1.3}$$

This can be rewritten in terms of free energy and additional terms related to the configurational disordering of the atoms:

$$\mu_i = \mu_i^0 + R T ln(a_i), \tag{1.4}$$

where R is the gas constant, and a_i is the activity of the component i. The first term on the right-hand side of the equation is:

$$\mu_i^0 = \Delta G_i / \Delta X_i. \tag{1.5}$$

For an ideal solid solution, the activity $a_i = X_i$, but for a real solid solution, the two are different. In practice, especially when you want to compute solubilities, you can separate the activity into an ideal contribution ($R T ln(X_i)$) and a non-ideal contribution.

Once we compute the free energies as a function of pressure, temperature, and composition, we can decipher the relations between the different polymorphs and the equilibrium conditions of the different mineral reactions and mineral assemblages. These are represented graphically in phase diagrams, where the phase stability and phase relations are shown as a function of the intensive parameters. The topology of the diagrams is such that they need to respect a few rules, like the phase rule:

$$F = C - P + 2, \tag{1.6}$$

where F is the number of phases, C is the number of components, and P is the number of degrees of freedom. Here, the *phases* are the different polymorphs of various minerals, the *components* are the atomic types or groups of atomic types that form the building blocks of the phases, and the *degrees of freedom* are the intensive variables.

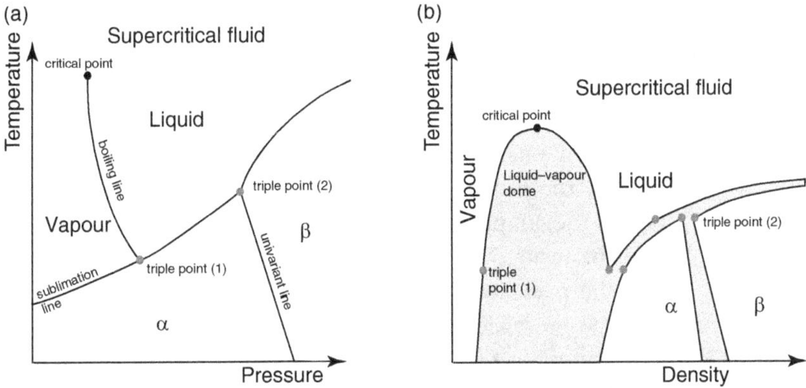

Fig. 1.2 Schematic phase diagram of a single-component system represented in the pressure–temperature (a) and temperature–density reference (b). α, β, and γ are solid polymorphs. Only a mechanical mixture of two or three phases is stable in the grey areas in (b); there is no single phase at those temperatures with that specific density.

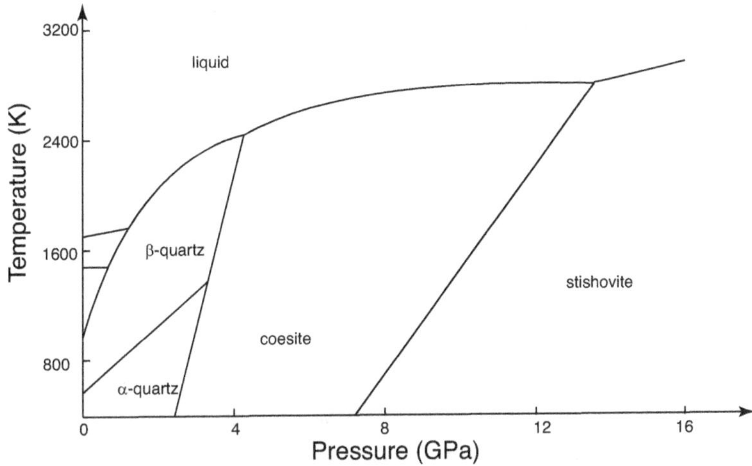

Fig. 1.3 Schematic phase diagram of silica represented in the temperature–pressure space. The divariant fields define the stability region of single polymorphs delimited by the univariant lines. The intersection between lines is triple points – invariant places at which three phases are in equilibrium.

Figure 1.2 shows a model phase diagram for a single-component system. The pressure–temperature reference is the most widely used in petrology. The equilibrium lines are called univariant, as only one thermodynamic parameter can change independently to keep the system in equilibrium between two phases. The lines intersect in fixed invariant points, where three phases are in equilibrium at one specific set of values for pressure and temperature. The univariant lines delimit the divariant stability fields of the individual phases, where both pressure and temperature can vary independently within those boundaries.

The phase diagram of SiO_2 (C = 1) is represented in Figure 1.3. The divariant stability fields correspond to various silica minerals, like quartz, cristobalite, and tridymite. The univariant lines define the conditions at which one phase is in equilibrium with another. The relation between pressure and temperature at which the transition between phases occurs corresponds to the conditions at which the free energies of the two corresponding phases are equal. The triple points are fixed P and T conditions at which three silica phases coexist.

In phase diagrams with two components, like MgO and SiO_2 (C = 2), the divariant fields (F = 2) correspond to two mineral assemblages. For example, periclase, MgO, may coexist with forsterite, Mg_2SiO_4; then forsterite, Mg_2SiO_4, may coexist with enstatite, $MgSiO_3$; and so on. But the presence of one of the two silicates, forsterite or enstatite, excludes the presence of periclase and quartz. This is the reason why you cannot find quartz in basalts that contain forsterite and enstatite. The univariant lines correspond either to phase transitions, like ortho- to clino-enstatite, or to mineral reactions, like (at mantle pressures) ringwoodite, Mg_2SiO_4, transforming to bridgmanite, $MgSiO_3$, and periclase, MgO.

Various other rules stem from the phase rule: for single-component systems, the triple points are found at the intersection of three univariant lines, each univariant line arriving at a triple point has to bisect the other two lines, the melting lines exhibit a change in slope at the solid–solid phase transitions. The topology of the phase diagram is a useful exercise in itself, and mastering its rules allows us to quickly understand phase relations in natural systems.

One special point in the diagram is the end of the liquid–vapour equilibrium line in Figure 1.2b. This is called the critical point, with P_c as the critical pressure, ρ_c as the critical density, and T_c as the critical temperature. At temperatures higher than T_c, there is no clear distinction between gas and liquid – the transition between the two is continuous. One implication is that a gas can be transformed into a liquid by heating above T_c, compression, and cooling below T_c. Similarly, a liquid can be transformed into a gas without boiling, following a path of heating, isothermal decompression, and cooling. Most silicate minerals have critical points at temperatures below 10,000 K and pressures on the order of a few kilobars.

At temperatures considerably higher than T_c, physicists place the domain of the *warm dense matter*. At these conditions, a large fraction of the electrons occupy excited states and are at least partly delocalised. This domain largely corresponds to the interior of giant planets and exoplanets. The electrons become fully delocalised at even higher temperatures, above 100,000 K. This domain corresponds to the interior of various stars.

In a standard phase diagram in the P–T space, the slope of the phase boundaries is given by the Clapeyron relation:

$$dP/dT = -dS/dV = dH/TdV. \qquad (1.7)$$

The order of the phase transition is given by the smallest order in which the derivatives of the thermodynamic potentials are discontinuous. If there is an abrupt change in specific volume, entropy, heat capacity, and so on, then ΔV or, respectively ΔS, ΔCv, and so on, are non-zero, this dX changes through the transition, and the transition is first order. If the changes in thermodynamic properties are continuous, ΔX are null, but then the d^2X are discontinuous, and the transitions are second order. There are cases where the first two derivatives are continuous, and the third derivative has a discontinuity – these are the third-order phase transitions, like the Bose–Einstein condensate. Theoretically, according to this scheme, one can imagine phase transitions of any order; this certainly is an interesting exercise of mathematical physics, but it has little to no practical application, or at least not in mineralogy.

The first-order phase transitions are oftentimes associated with structural reconstructions. Because of kinetic reasons, some reactions are more sluggish than others, and a certain amount of energy needs to be put into the system to overcome the energy barriers and allow for the transformation to proceed. The result is the existence of hysteresis. In particular, the hysteresis is the only visible sign of iso-symmetrical phase transitions [55], a particular case of transition where the two phases have exactly the same symmetry, but they exhibit a discontinuity in dV, dS, and so on. Nature abounds in examples of first-order phase transitions: melting is one of the most common transitions, then decomposition, or the transitions at high pressure inside the Earth's mantle, and so on.

The second-order phase transitions exhibit a continuous change of the crystal structure during the transition. The change usually happens via a gradual collapse of an ordering parameter, like a lattice distortion due to a vibrational mode or an elastic instability. The symmetries of the two phases, before and after the transition, are related by a group–subgroup relation. The phase transition between quartz α and quartz β is a good example of a second-order phase transition, even though the reality is a bit more complicated [131], but we will come back later on this. Magnetic transitions, where the magnetic spin vanishes during compression, are other examples of such transitions.

There is one more thermodynamic property that is worth mentioning: the heat capacity, which is defined as the amount of heat, or energy, necessary to raise the temperature of a unit of mass by one degree. Reversely, this can also give us the amount of heat that is released during cooling down by one unit of mass of a geological body by one degree. Imagine the cooling of a magmatic batholith: in order to estimate its cooling history, one of the things you need to know is the heat capacities of its constituting magma and of the surrounding rocks to be able to estimate the total amount of energy that will be released during this process. At the planetary scale, knowing the heat capacity of the magma forming the magma ocean allows us to estimate the cooling amount of the magma ocean; knowing the heat capacity of the liquid iron and of the mantle allows us to estimate the cooling of the Earth's core, and so on. There are various ways to compute the heat capacity detailed later in the book.

We talked so far mainly about thermodynamics at equilibrium. But phase transitions, chemical reactions, and heat release all take time; they do not happen instantaneously. For this, we need to know material transport properties, of which one of the most important is thermal conductivity. Because once we know the amount of heat that is required for a certain geological process to take place or the amount of heat that is released during a certain geological process, the thermal conductivity allows us to estimate time scales: how long would it take for a phase transition to complete, how long will it actually take for a geological body to cool or to heat up, and how long would it take for the magma ocean or the liquid core to crystallise.

There are several contributions to thermal conductivity: an electronic contribution, a vibrational contribution, a mixed electron–phonon coupling contribution, and then a whole series of contributions coming from lattice defects, impurities, and so on. For metals, like iron or liquid metallic hydrogen (inside Jupiter!), the electron–phonon contribution dominates the thermal conductivity, with the electronic contribution coming second. The electronic conductivity and the thermal conductivity can be actually related by an empirical proportionality law, the Wiedemann–Franz law [12]. For insulators like quartz and feldspars and most of the mantle minerals, the lattice effects dominate the thermal conductivity. Here, the most important effects are vibrational, like phonon–phonon interactions and phonon lifetimes.

Further kinetic effects influence the first-order phase transitions and the chemical reactions that require a certain amount of activation energy in order to proceed. If this energy is not available, we talk about a sluggish transition, which takes place over a long time, sometimes even long on a geological scale. This lapse of time is also translated in the presence of a hysteresis, which is oftentimes found in the first-order phase transition, as we mentioned earlier. In this context, the hysteresis also represents the departure of the observed phase boundary from its position as obtained from the theoretical difference in free energy or chemical potential between the two phases. Diamond is an excellent example of sluggish phase transition: at ambient conditions of pressure and temperature, the thermodynamically stable phase of carbon is graphite. However, the diamond that forms at high pressures and temperatures deep in the mantle survives the surface conditions because of the lack of this activation energy. The hysteresis of the graphite–diamond phase transition can be very large. But heating up a diamond may provide enough energy to overcome the kinetic barriers to promote the transition and then transform to a block of graphite. Determining the rate of a chemical reaction is a very tedious process and occurs at scales that most of the time go beyond the scales encountered in typical *ab initio* or even classical molecular dynamics atomistic simulations, which at most can cover microseconds.

1.3 Notions of Crystallography. Symmetry, and the Direct Space

All the atomistic simulations we perform require, before anything else, placing the atoms in space. The ability to correctly identify the geometrical relations between atoms requires a basic understanding of crystallography.

A symmetry operation is a geometrical transformation of the space that leaves its metrics invariant. In other words, the distances between any two points and the angles between any three points are preserved upon the application of the symmetry operation.

Symmetries operate on space and objects. In an n-dimensional space, the positions of points are given by n-dimensional vectors. The symmetry operations are represented as $n \times n$ matrices that act on these vectors. In the real-life mineralogical world, the one-dimensional space corresponds for example to stacking sequences usually found in polytypes, the two-dimensional space to surfaces and interfaces, and, finally, the three-dimensional (3D) space to the standard crystals.

Examples of such operations are the rotations around an axis, the reflections on mirror planes, and the inversion, whose names are self-explanatory. Another symmetry operation is the translation, represented by a vector. Its effect is the rigid displacement of the entire space. The symmetry operations can be combined, such as to obtain, for example, glide planes, where a translation is applied after a reflection, screw axis, where a translation is applied after a rotation, or rotoinversions, where inversion is applied after a rotation.

The standard notation adopted today by the International Union of Crystallography [125] for the rotation axis is A_n, where n denotes the order of the rotation, that is the number of times the same image of the space is retrieved during a complete rotation by 360 degrees. The notation for rotoinversions is $\bar{A}_n$, but as most text editors do not allow the bar on top of characters, more and more people today prefer the notation with the negative sign in front of the symbol: $-A_n$. As such, inversion is then denoted by -1. The notation for mirror planes is m. Some of the old crystallography books and some physics texts still use the old notation based on the German classification, but we will adopt here only the International Notation, as explained earlier.

Figure 1.4a illustrates the symmetry axis of a square prism. By convention, the highest-order symmetry axis, here the A_4 axis, is positioned vertically. Note that there are two sets of A_2 axes and that each axis is superimposed onto itself by applying twice the A_4 axis. Two of the A_2 axes are parallel to the Cartesian axes, x and y, and the other two axes are diagonal. Figure 1.4a illustrates the symmetry planes of the prism. In this particular case, each plane is perpendicular to one axis, though this is not a general rule. Whenever a symmetry plane is perpendicular to an odd-order axis, that is, A_2, A_4, or A_6, that body also contains a centre of symmetry.

Figure 1.5 shows the symmetry elements of the Fm3m unit cell of NaCl. The cubic structure has $3A_4$, $4A_3$, and $6A_2$ symmetry axes, $3m$ and $6m\prime$ mirror planes, perpendicular respectively to the A_4 and A_2 axes. There are two families of mirror planes denoted with m and $m\prime$. The structure also has an inversion centre.

In a Cartesian 3D reference system, the matrix notation for a rotation axis of order n around the z axis is:

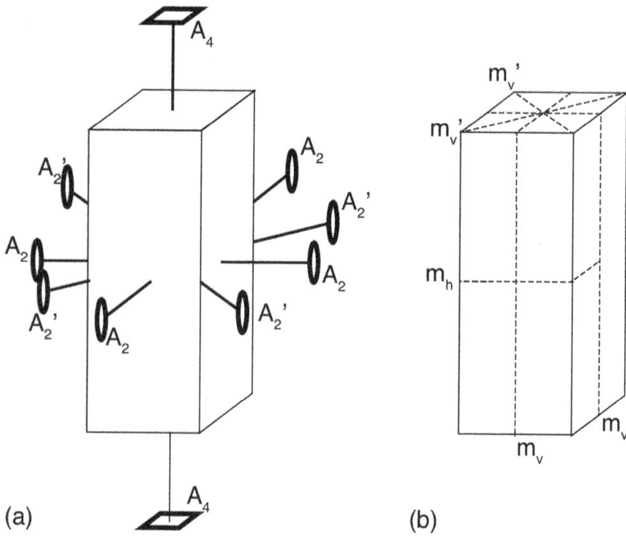

Fig. 1.4 A prism with a square base has a fourfold symmetry axis passing through its centre. By convention, this major axis is positioned vertically, along the c axis. (a) One set of a twofold symmetry axis in the horizontal plane perpendicular to the A_4 crosses the prism in the middle of the lateral faces (the A_2 axis) and a second set crosses the middle of the lateral edges (the A_2' axis). (b) The prism has two families of vertical mirror planes m_v, parallel to the vertical faces, and m_v', bisecting the horizontal square basis. There is one additional horizontal mirror plane cutting the prism into two equal parts. The prism also contains a symmetry centre in its middle.

$$\begin{pmatrix} cos(2\pi/n) & -sin(2\pi/n) & 0 \\ sin(2\pi/n) & cos(2\pi/n) & 0 \\ 0 & 0 & 1 \end{pmatrix} \tag{1.8}$$

The matrix for A_1 is the identity matrix, with 1 on the main diagonal and 0 in the rest; the one for inversion contains -1 on the diagonal and 0 in the rest. The corresponding matrix for an A_2 axis along z is:

$$\begin{matrix} -1 & 0 & 0 \\ 0 & -1 & 0 \\ 0 & 0 & 1 \end{matrix} \tag{1.9}$$

The mirror plane, m, perpendicular to the z axis leaves invariant the x and y directions, and reflects the objects perpendicular to the plane, along the z axis:

$$\begin{matrix} 1 & 0 & 0 \\ 0 & 1 & 0 \\ 0 & 0 & -1 \end{matrix} \tag{1.10}$$

The determinant of the pure rotation axis is always $+1$, while the determinant of the inversions and of the reflection planes is -1.

Figure 1.4 shows the symmetry operations as applied to a square prism. Because the symmetry operations act on the entire space and on all the objects that this space contains, they also act on each other, leading to strict constraints about their possible mutual coexistence. For example, the

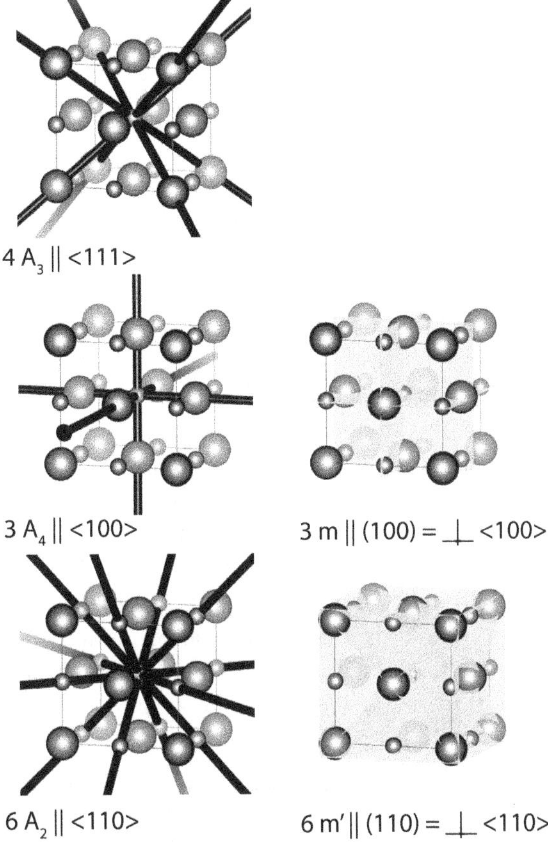

Fig. 1.5 Symmetry elements of the cubic unit cell of NaCl organised by relative relations. The face-centred cubic structure of NaCl is characterised by four three-fold axes along the diagonals of the cube, three four-fold axes passing through the middle of the faces, and six two-fold axes passing through the middle of the edges of the cube. The symmetry planes from the right column are perpendicular to the rotational axis on the left column. The structure has an inversion centre and supplementary translations towards the middle of the faces: $(\mathbf{a}+\mathbf{b})/2$, $(\mathbf{b}+\mathbf{c})/2$, $(\mathbf{c}+\mathbf{a})/2$. The Na and Cl atoms are respectively shown by large and small grey spheres.

vertical A$_4$ axis acts on the two sets of horizontal A$_2$ axes and relates them. It is enough to identify the first A$_2$ axis and then apply the vertical A$_4$ axis and find the other. The same happens with the vertical mirror planes.

The ensemble of symmetry operations thus related forms a group, G, of matrix multiplication, x, in the mathematical sense. They have the four basic properties of groups:

* Closure (the application of symmetry operations onto each other yields symmetry operations belonging to the same group): any A, B from the group G, A × B = C, and C is in G.

* Associativity (the order of operation is interchangeable): any A, B, C from the group G, (AB)C = A(BC).

* The group contains the identity operation, I: any A from the group G, AI = A.

* Within the group, every symmetry operation has its inverse (the application of the two results in the identity matrix): Any A from the group G, there is a B such as $AB = BA = I$.

The symmetry groups that are formed only with rotations and/or mirror planes and leave invariant one point of the space – the origin – are called point groups. There are 10 point groups in two-dimensional space and 32 point groups in three-dimensional space. The latter are classified in 7 crystallographic classes (Table 1.1), based on the symmetry elements they contain. Proper identification and knowledge of the 7 classes is extremely important in mineralogy, as it facilitates understanding certain mineral names, some of the sequences of phase transitions, particular physical properties, and simply allows us to better communicate with each other. With the convention that the symmetry axes are characterised by their direction and the mirror planes by their normal directions, there is a set of simple rules to quickly identify each of the 7 crystallographic classes based on the symmetry elements, as listed in Table 1.1.

Triclinic groups contain at most the inversion. Monoclinic groups have no more than one 2-fold axis and/or a mirror plane. By convention, the 2-fold axis and the normal of the mirror plane are oriented along the y axis. The orthorhombic groups have one 2-fold axis along each of the three basis vectors of the space and/or a mirror plane perpendicular to

Table 1.1 Notation scheme for symmetry groups and their characteristic symmetry elements.

Symmetry class	Minimum elements	Notation	Bravais centring	Relations between lattice parameters
Triclinic	1 and/or -1	1 or -1	P	a, b, c $\alpha, \beta, \gamma \alpha$
Monoclinic	A_2 and/or m	2 or m or 2/m	P, C	a, b, c $\alpha, \beta = 90°, \gamma$
Ortho-rhombic	$3\,A_2$ and/or m	ZZZ	P, C, I, F	a, b, c $\alpha = \beta = \gamma = 90°$
Trigonal	A_3	3XX	P	a = b, c $\alpha = \beta = 90°, \gamma = 120°$
			R	a = b = c, $\alpha = \beta = \gamma$
Hexagonal	A_6	6XX or 6/mXX	P	a = b, c $\alpha = \beta = 90°, \gamma = 120°$
Tetragonal	A_4	4XX or 4/mXX	P, I	a = b, c $\alpha = \beta = \gamma = 90°$
Cubic	$4\,A_3$	Y3X	P, I, F	a = b = c $\beta = \beta = \gamma = 90°$

X = 2 or m or nothing, Y = 2, 4, or m, Z = 2 or m or 2/m. Rotoinversions are described by replacing the symbols 3, 4, or 6, respectively by -3, -4, or -6.

the three basis vectors. Trigonal, tetragonal, and hexagonal groups have one maximum symmetry axis of 3, 4, and 6 order, respectively, always oriented vertically, along the z axis. The first digit after the characteristic rotation describes the elements along or parallel to the cartesian x and y axis, and the second symbol along or parallel to their bisectors. Finally, cubic groups have 4 A_3 axis, marked on the second entry in the notation. The first symbol describes operations along the cartesian axis, and the last digit along the diagonals, that is, the bisectors.

Finally, in the 3D crystallographic space, there are three independent translations. The smallest volume of space delimited by these translations is called the primitive unit cell. Its dimensions define the unit cell parameters, a, b, and c, with their corresponding angles, α, between b, and c axis (opposing the a axis), β, between c, and a axis (opposing the b axis), and γ, between b, and a axis (opposing the c axis). The entire lattice results from integer translations of the unit cell along these parameters. But the orientation and shape of this primitive cell are arbitrary (see Figure 1.6).

For all the lattices, there is only one unique and general way of delimiting a convex region of space that is primitive in the crystallographic sense. This construction is the Wigner–Seitz cell. It collects all the points of space closest to a given origin. Its limits are fixed by the first set of bisector planes to the segments linking that origin to its translational equivalents. Its construction is shown in Figure 1.6. The Wigner–Seitz cell is not extensively used in crystallography because its shape is hard to represent for non-Cartesian lattices.

There are 14 such possible combinations of translations, forming the Bravais lattices, also listed in Table 1.1. The 14 Bravais lattices combined with the 32 point groups generate all the possible 230 space groups that form the crystallography of the three-dimensional space. This limitation of the number of space groups comes from the major restriction that in crystallography, the space has to be completely filled by regular polyhedra

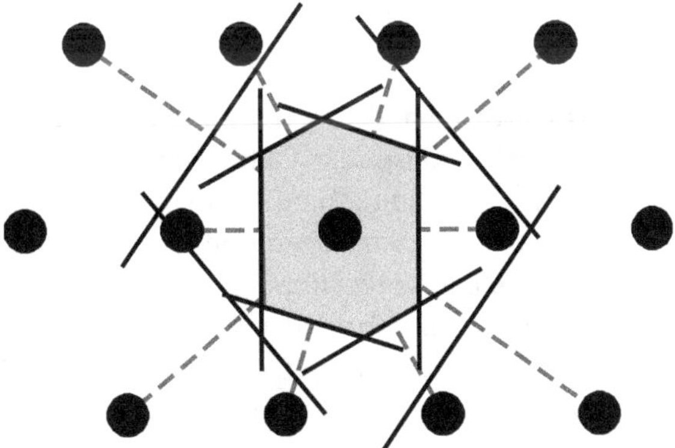

Fig. 1.6 The Wigner–Seitz cell represents the real primitive cell. It is built around a point of the direct lattice using the Voronoi construction. The segments linking that point to its equivalents by translation are bisected by perpendicular planes. The limits of the Wigner–Seitz cell are the closest bisectors to the reference lattice point.

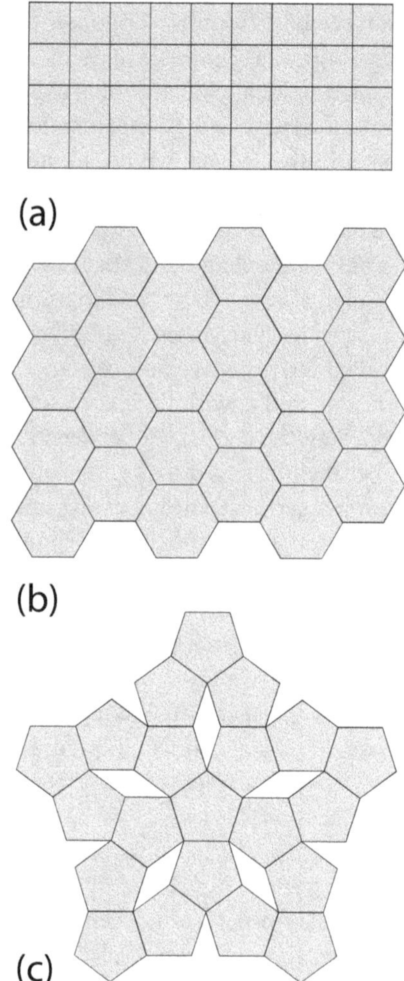

(a)

(b)

(c)

Complete tiling of the plane/space can be obtained only using polyhedra with 2-, 3-, 4-, and 6- fold symmetry (a or b). The use of regular pentagons (c) cannot cover the 3D space completely, that is, without leaving empty spaces.

Fig. 1.7

(Figures 1.6 and 1.7). This means that only certain symmetry operations fulfil this constraint: Only rotations of 2-fold, 3-fold, 4-fold, and 6-fold orders are allowed in 3D space. This is represented in Figure 1.7, where a 5-fold symmetry operation is applied to a space, generating pentagons. But the planar repetition of these pentagons will always leave holes in the space, which is forbidden.

The use of space groups in describing the symmetry of a crystal has profound implications for understanding the relationship between physical properties, not only within the same crystal but also during the phase transitions. Physical properties are the response of a crystal to external driving forces. They are represented using tensor, that is, matrices. The relations between the elements of these matrices are related by symmetry. For example, in the tetragonal system, because of the vertical fourfold symmetry axes, we expect the response of the structure to a driving force acting along the Cartesian x axis to be the same as the response to the force acting along the y axes. The same would be for the hexagonal and trigonal axis, but not for the orthorhombic or the triclinic classes. Then second-order phase transition can be traced, explained, and understood in terms of symmetry group-subgroup relations.

A very useful way of seeing the crystal is as a vectorial space (Box 1.1). The three unit cell parameters $\vec{a}$, $\vec{b}$ and $\vec{c}$ define the basis of the vectorial space. They are the only connection between the crystallographic vectorial space and the Cartesian space. Once the basis of the crystallographic vectorial space is defined, then the atoms populate this basis. Their positions are expressed with respect to the three vectors of the basis, with correspondingly three fractional coordinates in the $[0, 1)$ interval. The crystal is obtained by the periodic repetition of the basis along the three unit vectors. This vectorial space is left invariant by the crystallographic space group. For this reason, the matrix representation of the symmetry operations can also be defined with respect to the basis.

However, all these considerations are not limited to three dimensions. The same conditions for filling the space and forming symmetry groups

---

Box 1.1   **What is a crystal structure?**

In purely descriptive terms, a crystal structure is the periodic arrangement of atoms in space.

Mathematically, the crystal structure is a vector space populated with atoms. For this, consider a set of three non-collinear and non-planar vectors $\vec{a}$, $\vec{b}$, and $\vec{c}$, whose components with respect to the x, y, and z Cartesian space are defined as:

$$
\begin{matrix}
a_x & a_y & a_z \\
b_x & b_y & b_z \\
c_x & c_y & c_z
\end{matrix}
$$

The three vectors are considered unitary, and they form the basis for a vectorial space. Each atom α is positioned inside and relative to this basis with fractional coordinates x_α, y_α, and z_α, each larger or equal to 0 and strictly smaller than 1. The crystal lattice is the vectorial space obtained by the ensemble of atoms populating this space. It is obtained by the integer translation of the basis along the three vectors of the basis. The atom α from the unit cell i_x, j_y, and k_z lies at coordinates:

With respect to the cartesian space this atom lies at:

$$
R_x = (i_x + x_\alpha)\, a_x + (j_y + y_\alpha)\, b_x + (k_z + z_\alpha)\, c_x
$$
$$
R_y = (i_x + x_\alpha)\, a_y + (j_y + y_\alpha)\, b_y + (k_z + z_\alpha)\, c_y
$$
$$
R_z = (i_x + x_\alpha)\, a_z + (j_y + y_\alpha)\, b_z + (k_z + z_\alpha)\, c_z
$$

The advantage of using the vectorial description of the crystallographic space is that any transformation acts on the basis preserving its metric, and then the transformation propagates directly to all components of the entire space. This procedure is used, for example, in the implementation of the elastic constants tensors from linear response (see Section 15.2).

apply to crystallographic spaces with more than three dimensions. The descriptions of crystals as vectorial spaces and of crystallography as applied to these vectorial spaces allow us to easily work with spaces of any dimensionality. It all comes down to the number of unit cell parameters, which represent the rank of the vectorial basis. The group theory of the crystallographic space is the same, regardless of the dimension of the space.

However, as you increase the dimension of your space, more and more symmetry operations become valid. For example, the hypercube of a four-dimensional space allows for 5-fold rotations. While at first sight, this may sound like science fiction, there are actually crystals, even natural crystals, that exhibit such symmetry. They are called quasi-crystals and were first obtained by rapid cooling of specific Al-based metallic alloys [223], with thermal gradients on the order of 10^4K/s. There is a large literature on the synthesis and the crystallography of these materials, fuelled in part by their industrial applications (hard, resistant coatings, resistance to oxidation, etc.). Three decades after their synthesis, a natural quasicrystal was discovered in the remaining shocked pieces of a meteoritic impact [32]. The diffraction pattern of quasicrystals is characterised by sharp and well-distinct Bragg reflections, which witness an ordered atomic lattice. However, they exhibit unusual crystallographic symmetry, like a 5-fold symmetry axis.

Another special case of crystals described in modern multi-dimensional crystallography is the incommensurately modulated crystals. These are found oftentimes also in the mineral world. They are obtained from the superposition of two regular lattices that have at least one periodicity, which is incommensurate with each other. An excellent example is the result of the freezing-in of an unstable phonon mode in an irrational point of the reciprocal space. The underlying average lattice has basis vectors $\vec{a}$, $\vec{b}$ and $\vec{c}$, and the phonon periodicity is for example $q = \epsilon \vec{c}$, where $\epsilon \neq m/n$, for example, $\epsilon = \sqrt{1/2}$. In this way, the phonon periodicity is irrational with respect to the average lattice, as we cannot define a supercell with basis vectors $\vec{a}$, $\vec{b}$ and $\vec{c}$. The solution is to define a new vectorial space whose basis is $\vec{a}$, $\vec{b}$, $\vec{c}$, and $\overrightarrow{1/q}$. The crystallography of such an object is described in a four-dimensional space.

There are different physical mechanisms that lead to the formation of incommensurately modulated crystals [76, 52, 226, 209]. The balance between large-range and short-range interactions induces displacive transitions that appear in minerals like åkermanite, elpasolite, or feldspathoids. The strong electron-phonon coupling leads to electronic instabilities in the Fermi surface nesting and charge-density-wave transitions in minerals like the transitional metal chalcogenides. The inter-penetrations of two lattices, that is, cations inside tunnels formed of anionic groups, as for some zeolites, feldspathoids, or Mn-oxides, lead to mismatch and aperiodicity. Magnetic correlation and exchange lead to the occurrence of magnetic spin waves in selected magnetic materials.

The diffraction pattern of modulated crystals exhibits the strong main Bragg spots of the average lattice and satellite peaks, which are found close to these main peaks. If the position of the satellite peaks is irrational with respect to the underlying principal lattice, the structure is incommensurate; otherwise, it is commensurate. This latter case represents a superlattice.

1.4 The Reciprocal Space

The discovery of the atomic structure of crystals using X-rays in the beginning of the last century revigorated the field of crystallography. The diffraction experiment identified the presence of parallel and regular sets of planes. As the interatomic distance in crystals is compatible with the wavelength of the X-rays, then X-ray diffraction can occur. A schematic view of the phenomenon of diffraction is shown in Figure 1.8. X-rays sample the distance between the parallel sets of planes, as the diffracted waves parallel atomic planes show constructive interference. Consequently, it is useful to define another crystallographic space using the normal distances to the families of planes rather than the distance between atoms. The normals to the atomic planes form the reciprocal space, which is of fundamental importance in solid-state physics.

Using the elementary translations of the direct space, which directly reflect the distance between atoms, we can identify the set of distances between the atomic planes. The elementary translations (a_i, a_j, a_k) form the unit cell in the direct real space. The relations to the unit cell of the reciprocal space (a_i*, a_j*, a_k*) are:

$$\mathbf{a}_i^* = 2\pi \frac{\mathbf{a_j} \times \mathbf{a_k}}{\mathbf{a_i} \cdot (\mathbf{a_j} \times \mathbf{a_k})}, \text{ where } \mathbf{i}, \mathbf{j}, \mathbf{k} \text{ are the three indexes of the space} \quad (1.11)$$

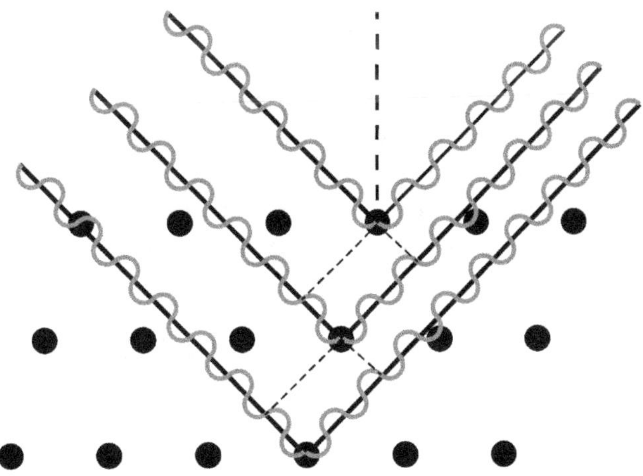

Fig. 1.8   Diffraction is the reconstruction of a wave packet as individual waves are reflected on a grating.

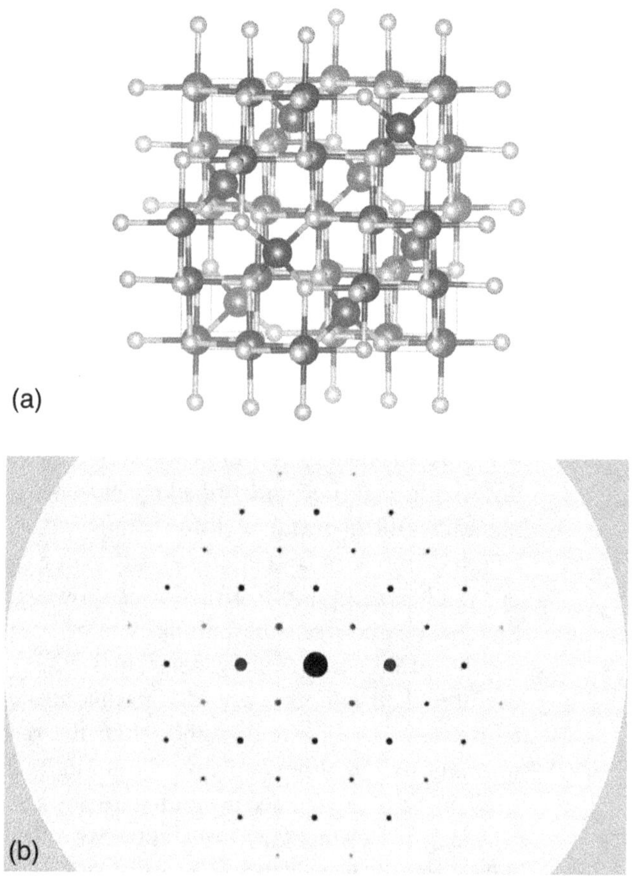

(a)

(b)

Fig. 1.9 Real-space (a) and reciprocal space (b) image of the crystal structure of magnetite, Fe_3O_4. The real space shows the two crystallographic sites, typical for spinels, where Fe is 4-fold and 6-fold coordinated.

The three vectors resulting from the above equation are called the reciprocal lattice vectors, denoted as $\mathbf{a}_1^*, \mathbf{a}_2^*, \mathbf{a}_3^*$ or sometimes as $\mathbf{b}_1, \mathbf{b}_2, \mathbf{b}_3$, or even $\mathbf{a}, \mathbf{b}, \mathbf{c}$.

By construction, the reciprocal space is populated with the normal distances to the interatomic planes of the direct space. Then the passage between the direct space and the reciprocal space can be done via the Fourier transform. This feature makes this transformation hugely important in solid-state physics, where we need to go back and forth between the direct and the reciprocal spaces. Figure 1.9 shows the real space and the reciprocal space representations of the magnetite, Fe_3O_4.

1.5 Tensors

In a broad sense, the physical properties of a crystal structure are related either to the different components of the energy or to changes of the

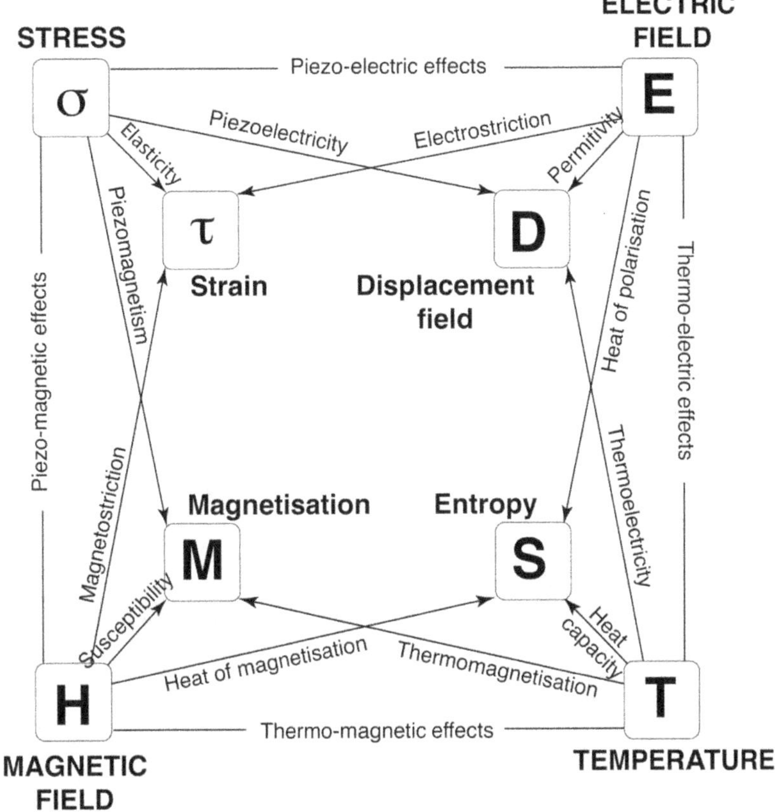

Fig. 1.10 Tensorial description of the physical properties as lattice responses to various driving forces. Note that every driving force has multiple effects and the resulting physical properties are all interconnected.

energy under the action of external driving forces (Figure 1.10). Mathematically, these relations between driving forces and lattice responses are described by tensors, that is, $n \times m$ dimensional matrices, where both n and m are integers not necessarily equal to each other. As with matrices, the number of dimensions of a tensor represents its rank. Throughout this book, we will look at a series of physical properties in the tensorial framework.

The simplest tensors are the 0th rank ones. They describe physical properties or physical states that are independent of direction. Scalars are also invariant with respect to a change of coordinates. Examples of such properties are the specific volume, the density, the heat capacity, or the electric charge.

Physical properties that are directional are described by first-order tensors, that is, vectors, such as the interatomic forces, which are derivatives of the energy as a function of atomic displacements, or the electric polarisation, which are the derivatives of the energy under the action of an electric field.

Second-order tensors describe directional properties that are dependent on a directionally dependent driving force. By construction, they relate two vectorial quantities. For example, the refractive index is the polarisation vector induced by the electromagnetic force. They are both vectors.

Higher-order tensors are matrix-type responses of the crystalline structure to the action of matrix-type driving forces. A typical example in mineralogy is the piezoelectric tensor, which is a third-order tensor that relates a polarisation vector (vector, first-order tensor) to the deformation (3×3 matrix, second-order tensor). Another important example in geophysics is the elastic constants tensor, which is a fourth-order tensor relating the deformation (3×3 matrix, second-order tensor) to the stress (3×3 matrix, second-order tensor).

The tensorial properties of crystals are translationally invariant. The application of the symmetry operations can be written as:

$$T' = RT \tag{1.12}$$

where R can be a rotation axis, a reflection, or an inversion. Each element of the T' tensor is obtained after applying all the R symmetry operations, in order, as:

$$T'_{ijk..} = \sum_a \sum_b \sum_c \sum_{..} R_{ia} R_{jb} R_{kc} .... T_{abc..} \tag{1.13}$$

This results in certain rules. For example, vectorial properties, like polarisation or ferromagnetism, cannot occur in a centrosymmetrical crystal, as the action of the centre of symmetry makes the sum of total local vectors vanish. On the other hand, all second-order tensors are symmetrical, as the reversal of the driving force induces a reversal of the response, and the only possible solution is for the tensor property to be invariant with respect to inversion. In other words, $T_{ij} = T_{ji}$.

The tensors can be diagonalised, which means the initial form of a tensor is reduced to a diagonal or near-diagonal form. For this, the non-diagonal tensor T_{nd} is transformed by multiplication with a matrix C to obtain the diagonal tensor, T_d:

$$T_d = C^{-1} T_{nd} C \tag{1.14}$$

This is mostly done and used for second-rank tensors, though (more complicated) formalism exists for higher-order tensors. The resulting diagonal components are the eigenvalues. As the diagonalisation also corresponds to a change in the orientation of the reference system, the basis vectors of this new system are the eigenvectors. Both the eigenvalues and the eigenvectors are highly sought-after quantities, as they define the most concise description of the tensorial physical properties and can be most easily compared to actual measurements.

1.6 A Few Unix/Linux Commands

The large majority of ab initio packages run at the command line, that is, in a Unix-like terminal and not in a dedicated graphical interface. The codes are actually notorious for lacking almost any graphical interface. The lack comes from the huge computing and memory resources such codes require. Adding a graphical interface to such a code would only complicate things for developers, while it would not bring any improvement to the final user. It is then extremely necessary to familiarise yourself with the terminal, where you will spend so much time building input files, checking and running simulations, and extracting results for further analysis. In the era of 3D computer graphics, smartphones, and tablets, it might seem weird and outdated to work in such a boring and dull environment. However, once you become at ease using the terminal, you will see what a powerful tool this can be.

Before we go and discuss a few Linux commands, you should know that all computers are run by a software program, which supervises the entire activity taking place inside. This software is called a DOS = disk operating system. The DOS is the one that allows communications between the main parts of the computer, between the peripherals, that starts and closes different applications, and so on. You should also know that all the stuff you have on your computer, folders, files, documents, pictures, and so on, is organised like a tree. The root of the main stem is the name of the *drive*, the physical place where your files are located. Because, back in the days of floppy disks and no graphical interface, the computers were running the DOS from an external floppy disk, this took the name of **a:**. This was quickly adopted by the MSDOS system, where MS comes from Microsoft. The second floppy disk was called **b:**, and when the hard disks, which are the non-volatile memories present inside the computers, became available, they naturally took the name of **c:**. Today the Windows computers, which evolved over several decades from MSDOS, conserved the **c:** notation. In all flavours of Unix, which include Linux and the terminal of Apple, the root of the main stem has simply the name 'backslash'.

In linux/unix, the names of the commands that you can use in the terminal are straightforward; their names are only abbreviations of the English words corresponding to the action you want to perform. For example, if you want to navigate in the tree that represents your hard disk, you actually change the directory or folder, so you make use of the command **cd**. The command **cd/home/your_user_name** brings you to your home directory. If you want to copy a file called FileName1 to a file called FileName2, it is enough to type **cp FileName1 FileName2**. And so on. Table 1.2 lists a few basic commands you are most likely to use while in the terminal.

Then there are several shortcuts in the terminal, which are not proper commands but are there to make your life easier. Typing fast twice the

Table 1.2	A few basic unix commands.
Command	Short explanation
pwd	Print Working Directory
cd	Change Directory
ls	LiSt content of a directory
cp	CoPy a file or a directory (-r)
mv	CoPy a file or a directory
rm	ReMoves a file
mkdir	MaKes a new DIRectory
rmdir	ReMoves empty DIRectory
cat	conCATenates the content of a file
grep	extracts from a file the lines containing a given pattern
head	lists the HEADing lines of a file
tail	lists the last lines of a file
tar	creates an archive by concatenation
gzip	compresses a file
top	lists the TOP processes running on a machine
kill	kills a given process on the machine
man	shows the MANual of a command
ssh	Secure SHell, opens a terminal on a different machine
scp	Secure CoPy of files between machines

TAB key autofills what you are typing; in case of several choices, it will list these choices. If you accidentally freeze the terminal, typing **Ctrl + S**, you can unblock it with **Ctrl + Q**. If you want/need to stop the command that is currently running at the terminal, type **Ctrl + C**. If you only want to pause it, type **Ctrl + Z**. Finally, **Ctrl + A** brings the cursor to the beginning of the line, while **Ctrl + E** brings it to the end of the line. At the terminal's command line, the processes are executed in the order in which they were typed. A sequence of **pwd; ls** will first print the name of the working directory (pwd) and then (;) will list its content (ls). The 'lower than' sign, <, feeds the content of the file that follows the sign to the process to the left of the sign. The 'greater than' sign, >, redirects the output into the file to the right of the sign. As our examples involve running the abinit package, you will type:

$$abinit < test.files > test.log \tag{1.15}$$

will start the abinit process and will feed the content of the test.files file to abinit (<), and will redirect the output of abinit to the test.log file (>). One 'greater than' sign initialises the file to the right, in the former case the test.log file, and replaces its content. But two 'greater than' signs append the output to the file to the right, the test.log file in this example.

The **&** sign at the end of the line runs the process in the background: **abinit < test.files > test.log**.

Finally, the vertical bar, |, feeds the output from the process at its left as input to the process at its right.

All the Linux/unix commands come with many options and in many flavours. The solution to navigate and pick what you need is to regularly search in the manual, which can be easily brought on the screen with the **man** command. This will not only list all the options but also give you a few common examples of how to use a given command. While in general, you cannot destroy much; only be careful with the **rm** and **rmdir** commands, as once you remove/delete something, it is done forever.

Finally, apart from the terminal, you will need some text editor. If you are running your simulations remotely on some supercomputer in a (super)computing centre, you will not have access to anything graphical on that machine. You connect there using the **ssh** command. Once on that machine, there is no graphical interface or working mouse. You navigate between folders with the **cd** command, list the content of those folders with the **ls** command, copy files with the **cp** command, and so on. A typical sequence of commands looks like this:

$$ssh\ rcaracas@super.computer.edu$$
$$pwd$$
$$cd\ /work/rcaracas/$$
$$ls\ -l$$
$$cd\ PROJECTS/Raman \tag{1.16}$$
$$ls\ -l$$
$$cd\ SiO2/Quartz/P10Gpa$$
$$cp\ ../P00GPa/quartz.P00GPa.files\ quartz.P10GPa.files$$
$$vi\ quartz.P10GPa.files$$

To visualise the content of a file and modify it, you will need a basic simple editor. The most powerful editor, probably ever invented, is the **vi**, or in its *modern* form, **vim**. This is the editor I was invoking in the example earlier. However, **vi** might seem cumbersome and outdated due to its lack of any interface with the user. A very strong and powerful alternative is **emacs**. You can find more about supercomputers in Chapter 18 of this book.

1.7 Some Special Notations in Quantum Mechanics

Because of the number of equations involved in developing almost any part of the theory in quantum mechanics, theorists invented a few specific notations. They look strange and can be intimidating. However, once you get used to them, you can learn to love them for their elegant simplicity.

We will see later on in this book that the fundamental tools of quantum mechanics are vectors and operators. The vectors represent *states* of one or an ensemble of quantum particles. The operators act on these vectors. Usually, the result of these operations is obtained by performing integrals.

The simplified notation covers both the vectors and their operations. Inspired by the English word bracket, the notation has two parts, a *bra* part, $<|$, and a *ket* part, $|>$. The operators are found in between the two vertical bars. When a vector, f, is integrated in the *ket* part, this is interpreted as:

$$|f(r)> = \int_r^\infty f(r)d(r) \tag{1.17}$$

When the same vector is in the *bra* part, the meaning is its complex conjugate. A multiplication of two vectors, g and f, is represented as:

$$<g(r)|f(r)> = \int_r^\infty g^*(r)f(r)d(r) \tag{1.18}$$

The operations are marked by a vertical line. The operator always acts to the right, on the *ket* part.

The action of an operator H that acts on a vector f over the entire space is represented as:

$$|H|f> = \int_r^\infty H(f(r))dr \tag{1.19}$$

The simplicity of this notation allows for an easy representation of equations and formulas and new operations. For example, the previous equation can be multiplied to the left by a new vector g, like:

$$<g|H|f> = \int_r^\infty g^*(r)H(f(r))dr \tag{1.20}$$

You can also imagine the *bra* and the *ket* parts of an equation as a vector arranged respectively as a row or column. With this, the product of two vectors, like $<g|f>$ represents:

$$<g|f> = \int_r^\infty g^*(r)f(r)dr = \sum_i^n g_i^*(r)f_i(r) \tag{1.21}$$

Of course, the dimensions of the two vectors f and g need to match. If the two vectors, g and f are different, the $<f|g>$ product represents the projection of the f vector onto the g vector. If the two vectors are identical, the product $<f|f>$ gives the norm of the vector. If f represents the physical states of a particle or ensemble of particles, then:

$$<f|f> = \int_r^\infty f^*(r)f(r)dr = 1 \tag{1.22}$$

We will try as much as possible to avoid this notation because it is so unfamiliar to geoscientists. However, there will be a few places in the book where its simplicity and conciseness are favoured over more classical but cumbersome full notations.

Atomistic Calculations Using Interatomic Potentials

2.1 General Considerations about Potentials

The energy of an arrangement of atoms is its most fundamental property. All the physical properties and the chemical behaviour of that material stem from its energy and the derivatives of the energy. As such, obtaining this energy theoretically was the centre of effort for generations of computational scientists. For a long time, such simulations have extensively used empirical interatomic potentials to describe the bonding in crystals. It is only in the last couple of decades that quantum mechanical calculations have become widespread.

The underlying idea of the classical simulations is to express the energy of the lattice as the sum of various terms, U, counting for single-body, two-body, three-body, four-body, and so on terms. For a system with N atoms, such a general formula or the interatomic potential is:

$$E(\vec{r_1}, \vec{r_2}, \vec{r_3}, ..., \vec{r_N}) = \sum_i U_i(\vec{r_i}) + \sum_{i,j>i} U_2(\vec{r_i}, \vec{r_j}) + \sum_{i,j>i,k>j} U_3(\vec{r_i}, \vec{r_j}, \vec{r_k}) +$$
$$\sum_{i,j>i,k>j,l>k} U_4(\vec{r_i}, \vec{r_j}, \vec{r_k}, \vec{r_l}) + ...$$

$$(2.1)$$

where the vectors $\vec{r_1}$, $\vec{r_2}$, $\vec{r_3}$, ..., are the positions of the atoms i, j, and k. Each function U_n has a number of parameters, which must be fit on available data, representing measured or computed physical properties, like interatomic bond lengths and angles, vibrational frequencies, elastic constants, and so on. The highest simplification of this equation sees truncation of all the many-body terms, with only pair-wise interactions considered. In complex simulations, various many-body terms are specifically included, sometimes in sophisticated formulations.

When calculating the actual value of the energy, the terms U_n must be evaluated for all the atoms i, j, k, and so on in the system, regardless of their distance. Considering only pair-wise interactions, the number of evaluations increases with the square of the number of atoms, and thus it can easily become prohibitive. For a small nanocrystal of quartz containing $10 \times 10 \times 10$ unit cells along the three lattice vectors, there are 9,000 atoms, corresponding to 81 million evaluations at each step. This theoretical nanocrystal along each axis is one order of magnitude smaller than the smallest silica nanocrystal available and used in the industry today. Real-size crystals, even the ones in the petrographic microscope, have orders of magnitude more atoms than a nanocrystal. Clearly, the evaluation of all interatomic interactions quickly becomes intractable.

There are several numerical solutions for the problem of the number of interactions, and they all involve various approximations. One of the most straightforward is to adopt a cut-off distance for the potential. The potential has the normal analytical form that depends on the interatomic distance but only up to a certain cut-off radius and is zero beyond that. As any function ends abruptly, the potential exhibits a discontinuity at the cut-off, which can create other supplementary problems related to the convergence and conservation of the energy. This problem goes away if the cut-off distance is large enough for the potential to decay asymptotically to zero, though at the expense of a larger amount of calculations. An alternative elegant way to deal with this is to keep a small cut-off radius and to replace the abrupt discontinuity with a continuous and hence derivable function around the cut-off distance. This replacement ensures that the potential goes smoothly to zero for larger distances.

Then the interactions between translationally equivalent atoms need to be computed only once and then just multiplied by the number of equivalent subsets of atoms. Further numerical techniques, like the Ewald summation, allow estimating the effect of long-range interactions in a crystal with a reduced computational effort.

2.2　Pair-Wise Interatomic Potentials

The pairwise, or two-body, interatomic potentials are the simplest and most used potentials describing chemical bonds. They usually have two parts: an attractive component, acting at a large distance, which brings the atoms together, and a repulsive component, acting at a short distance that prevents the atoms from merging. The variation of the energy with inter-atomic distance describes a potential well. With the usual energy reference as the sum of the individual isolated atomic energies, the two components cancel each other at the energy minimum. The corresponding distance at the energy minimum is the interatomic bond distance at equilibrium. The force acting on the atoms, that is, the energy derivative with respect to atomic displacements, is zero at the equilibrium, and the atoms lie at the bottom of the potential well. Figure 2.1 shows the variation of the energy as a function of interatomic distance for several diatomic molecules; the energy curves are computed from standard static density functional theory (DFT) simulations described in detail in Chapter 4.

The Lennard–Jones pair-wise potential [150], which is one of the most common potentials, can be used to fit these potential curves. This potential

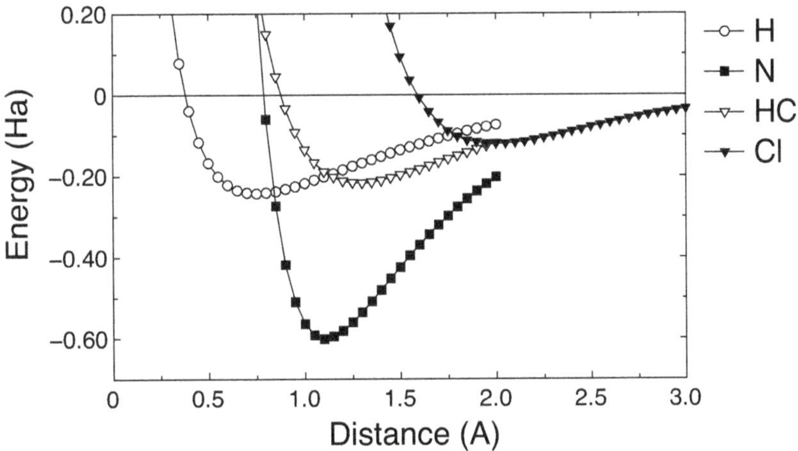

Fig. 2.1 Variation of energy as a function of interatomic distance for a few selected diatomic molecules. The reference 0 energy corresponds to the dissociated molecule, with the atoms at infinite distance. The minimum energy corresponds to the equilibrium distance at which the interatomic forces, that is, the slope to the energy–distance curve) are zero.

is also known as the 6–12 potential and is particularly useful for treating non-polar molecular systems, characterised by weakly interacting van der Waals bonds. Because of its simplicity, it is frequently used as a concept model when studying various physical properties or when developing code. Its form has two power terms:

$$E(r_{ij}) = 4\epsilon[\frac{\sigma}{r_{ij}^{12}} - \frac{\sigma}{r_{ij}^{6}}] \tag{2.2}$$

where $r_{ij} = |\vec{r_j} - \vec{r_j}|$ is the distance between the atoms i and j. ϵ and σ are two fittable parameters.

For highly asymmetric bonds, with an interaction stemming from a real chemical bond, for example, for OH in hydrous minerals [216] or for certain metallic systems, **the Morse potential** was very successful:

$$E(r_{ij}) = \epsilon[e^{-2\alpha(r_{ij}-r_0)} - 2e^{-\alpha(r_{ij}-r_0)}] \tag{2.3}$$

where r_0 is the distance at equilibrium, r_{ij} is the distance between atoms i and j, and ϵ and α are two fittable parameters.

For ionic systems and many minerals in the class of oxides, silicates, sulphates, carbonates, and so on, **the Buckingham potential** is widely used:

$$E(r_{ij}) = Ae^{-(r_{ij}/r_0)} - Br_{ij}^{-6} \tag{2.4}$$

The first term accounts for short-range interaction, and the second for the long-range one, just like in the Lennard–Jones potential. A and B are parameters that need to be fitted.

In many simulations on oxides, an improvement was brought by adding a term accounting for the electronic polarisability around the ions. This was realised by decoupling the electronic shell from its corresponding nucleus and relating them by a harmonic force:

$$E(core - shell) = \frac{1}{2}K_{cs}r^2 \tag{2.5}$$

If such polarisable shells are present, then extra pair-wise interactions are added between the cores and the shells of different atoms. Still, the entire description of the energy remains based on two-body interactions only. The use of the shell model also helps avoid the unnatural divergence of the second term in r^{-6} at very short interatomic distances, which cannot be otherwise counterbalanced by the first term in $exp(-r_{ij}/r_0)$. Eventually, this led to the development of a series of pair-wise potentials for simple oxides, where the atoms are not necessarily spherical – the aspherical ion model description [1], which specifically takes into account the instantaneous state of deformation of the ions and thus offers a more accurate description of the electronic clouds during vibrations and variations of the chemical bonding, for example, during compression.

Buckingham designed an extension of the Lennard–Jones potentials, with an application for noble gases equation of states [44]. After several iterations of the initial formalism, new modified forms containing Coulombian terms, core-shell interactions, and the addition of a supplementary exponential term to improve the description of the covalent bonds are widely used to describe major oxide and silicate mineralogical systems:

$$E(r_{ij}) = \frac{Q_i Q_j}{r_{ij}} + f_0(b_1 + b_j)exp(\frac{a_i + a_j - r_{ij}}{b_i + b_j}) - \frac{c_i c_j}{r_{ij}} +$$
$$[D'_{ij}exp(-\beta'_{ij}r_{ij}) + D''_{ij}exp(-\beta''_{ij}r_{ij})] \tag{2.6}$$

where Q_i are fractional charges depending on composition, r_{ij} are the interatomic distances, and all the rest, that is, a, b, c, and D, are parameters that need to be fit.

In general, once the cut-off radius is implemented, because of the simplicity of the formalism, pair-wise potentials can be easily used to study systems with several million atoms. This makes a series of mesoscopic properties available in atomistic simulations. For example, crack formation and propagation can be studied in grain-scale simulations to understand the rheology of materials. In other simulations, the number of atoms is kept to a minimum. Still, the time of the simulation is extended to be able to obtain enough statistics to capture even rare phenomena.

2.3 Multi-Body Potentials

But with increasing structural and bonding complexity, the addition of new terms is quickly necessary. The pair-wise potentials are usually good for reproducing and exploring properties within a very limited range of coordination. This limitation becomes obvious for covalent crystals. For example, a potential that is suitable for sulphur in fourfold coordination by oxygen, as in sulphates, would fail when applied to sulphur in sulphides, where the bonding environment is very different. Similarly, a potential fit on quartz, where Si is fourfold coordinated by O, might fail on stishovite, where Si is sixfold coordinated by O. Such limitations can be overcome by adding many-body terms to the potentials.

The many-body terms cover the angular dependence of the bonds, like three-body interactions to describe interatomic angles, four-body interactions to cover torsion angles, and so on. Another series of many-body terms are dielectric-type and account for the asymmetry of the charge distribution [99]; they are polarisable and deformable. Each of these terms comes with certain parameters, which need to be determined. They are fitted to best reproduce given geometries together with their energies and forces acting on the atoms. Depending on the type of potential, supplementary constraints can be added related to various physical properties, like elastic constant tensors.

A derivation of the asymmetric ion model briefly discussed earlier is **the Jahn and Madden potential**, which was proposed for minerals in the Ca-Mg-Al-Si-O system [149], thus describing the large majority of rock-forming silicates and oxides. The potentials contain pair-wise-additive Coulomb terms, dispersion interactions, a polarisable part, and short-range repulsion terms. The dispersion interactions take into account dipole-dipole and dipole-quadrupole contributions, in r_{ij}^{-6} and r_{ij}^{-8}, respectively. The polarisation and repulsion terms contain a series of damping functions as corrections for the asymptotic dispersion terms at short-range, dipole, and quadrupole interactions and the polarisability of the electronic shell. Despite missing specific multi-body interactions, they were proposed from the beginning for a wide range of thermodynamic conditions covering the entire Earth's mantle.

Because of the obvious technological importance, the diamond-like lattices (for C, Si, Ge, GaAs, etc.) have been studied for a long time. The potentials developed for this monoatomic and regular lattice became, with time, extremely elaborate and allowed the exploration of a wide variety of physical properties. They all went beyond the pair-wise terms. For example, **the Stillinger–Weber potentials** have a first complex pair-wise part [233]:

$$f_2(r_{ij}) = A(B\frac{1}{r_{ij}^p} - \frac{1}{r_{ij}^q})e^{\frac{c}{r_{ij}-r_c}} \tag{2.7}$$

where $A, B, p, q,$ and c are all parameters that need to be fit, and r_c is the cut-off distance. Additionally to this part, they contain a supplementary triple-wise part that accounts specifically for three-body interactions:

$$f_3(r_j, r_k, r_l) = h(r_{ij}, r_{ik}, \theta(jik)) + h(r_{ji}, r_{jk}, \theta(ijk)) + h(r_{ki}, r_{kj}, \theta(ikj)) \tag{2.8}$$

where the h functions are dependent on the interatomic distances r_{ij}, and on the interatomic angles θ via a $cos(\theta + 1/3)^2$ term.

The Tersoff potential [242, 243] has several complex terms, which ensure a very good accuracy of the calculations. It still has two components, an attractive and a repulsive part, but they each depend further on the chemical environment. Each lattice site i contributes to the total energy with:

$$E_i = \sum_{j \neq i} f_C(r_{ij})[a_{ij} f_R(r_{ij}) + b_{ij} f_A(r_{ij})] \tag{2.9}$$

The repulsive and attractive functions, f_R and f_A, have respectively exponential forms like $A exp(-\lambda_1 r)$ and $-B exp(-\lambda_2 r)$. The $f_C(r)$ function ensures a smooth vanishing of the potential at the cutoff radius. It requires two parameters, D and R, which are obtained from fits and have units of length. Then the $f_C(r)$ function decreases from a value of 1 at interatomic distances less than R-D to a value of 0 at distances larger than R+D, according to:

$$f_C(r_{ij}) = \frac{1}{2} - \frac{1}{2} sin\left(\frac{\pi}{2} \frac{r_{ij} - R}{D}\right) \tag{2.10}$$

The a_{ij} and b_{ij} prefactors have similar forms to each other, both using power laws. The a_{ij} is:

$$a_{ij} = (1 + \alpha^n \eta_{ij}^n)^{-1/2n} \tag{2.11}$$

with η_{ij} depending on the chemical environment as:

$$\eta_{ij} = \sum_{k \neq i,j} f_c r_{ik} exp(\lambda^3 (r_{ij} - r_{ik})^3) \tag{2.12}$$

and the b_{ij} is:

$$b_{ij} = (1 + \alpha^n \chi_{ij}^n)^{-1/2n} \tag{2.13}$$

with χ introducing explicitly a three-body angular-dependent term as:

$$\chi_{ij} = \sum_{k \neq i,j} f_c r_{ik} g_c(\theta_{ijk}) exp(\lambda^3 (r_{ij} - r_{ik})^3) \tag{2.14}$$

where finally:

$$g_c(\theta_{ijk}) = 1 + \frac{c^2}{d^2} + \frac{c^2}{d^2 + |h + cos(\theta_{ijk}|^2)} \tag{2.15}$$

The Tersoff potential is highly accurate. But fitting all the parameters is extremely time-consuming, and eventually, this potential saw limited use, mainly materials with high technological importance, which would justify the expensive fitting, like the tetrahedrally bonded semiconductors.

Hence, various solids whose structures derive from that of diamond, with strong tetrahedral bonding, can be modelled using Tersoff-like potentials. Every pair of atoms will have its own set of fitted parameters. However, the results are highly dependent on the system and on the availability of experimental or theoretical data for the fits. The quality of the resulting potentials is then highly variable. Various other possible structural types, like graphene or graphite, could be studied with modified versions of the Tersoff form.

However, most of the potentials described earlier have limitations related to the chemical environments for which they were designed and where they were fitted. They have difficulties adapting to new environments and coordination, and as a result, would limit chemical reactions and artificially extend hysteresis loops and metastability regions.

2.4 Interatomic Potentials for Metals

For metals, the pair-wise potentials, even with certain multi-body corrections, are still not good enough. The problem comes from the delocalised character of the electrons. Consequently, in most cases, a different approach is adopted, inspired by the tight-binding model and the DFT, discussed in the Chapter 3. This is the embedded atom method (EAM), whose central idea is to take into account (1) the interaction between a given atom i and the total electronic density at that site, ρ_i and (2) the pair-wise interactions with the atoms j. Then the total energy is:

$$E = \sum_i F_i(\rho_i) + \sum_{i,j>i} \phi_{ij}(r_{ij}) \tag{2.16}$$

where ρ_i is the electronic density at site i due to the other atoms j, lying at distances r_{ij}: $\rho_i = \sum_j \rho_j(r_{ij})$. The pair-wise interactions ϕ typically have several formulations depending on the distance.

A generalisation of these potentials implies the explicit use of n-body terms [Moriarty, JPCM, 2002], with n larger than 2, of the type:

$$\frac{1}{n} \sum_{i,j,..,n} v_n(i,j,..,n;\Omega) \tag{2.17}$$

where these terms depend on the relative position of the n atoms, and the unit cell volume Ω. Various sets of EAM potentials have been developed over the last two decades for various groups of metals, as well as for particular types of materials or structural settings, such as surfaces, bulk face-centred cubic structures, alloys, and so on. Accordingly, various codes are available to fit such EAM-type potentials on first-principles data [88].

Tremendous effort has been put into simulating various physical properties of metals, including rheology, fracture, alloying, elasticity, and so on. These come from the obvious importance of day-to-day application. Along these lines, one of the most remarkable and early such simulations was the study of the texture of a tantalum aggregate obtained by freezing the corresponding melt (Figure 2.2). The simulations used interaction potentials based on customised many-body, angular-dependent models, coming from the generalised theory [237]. A proper convergence test of the system size revealed that only simulation boxes containing one million atoms or more could yield the adequate texture observed in experiments. Indeed, the use of the periodic boundary conditions generated a series of artificial small patterns that repeated in the solid with the periodicity of the simulation cell.

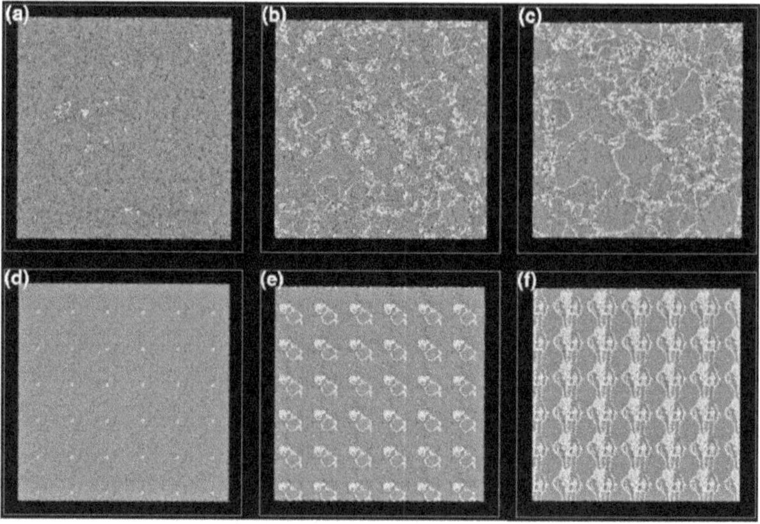

Fig. 2.2 Cross sections of the 16 million atoms (a–c) and 64,000 atoms (d–f) simulation taken at different stages during the solidification [Reproduced from [237]].

2.5 Force Fields

The developments of EAM potentials for metals inspired other complex attempts for potentials for ionic and covalent materials. This is how the reactive force fields were developed. They were initially developed for hydrocarbons because of their obvious industrial and technological importance. Such force fields contain a plethora of terms, which follow closely the development in Equation 2.1. First, they try to correctly describe the observed coordination; then, they correct for multi-body angular components and for long-range interactions. **The ReaxFF force field** has the form [90]:

$$E = E_{Coulomb} + E_{bond} + E_{over} + E_{under} + E_{val} + E_{pen} +$$
$$E_{tors} + E_{conj} + E_{vdWaals} \tag{2.18}$$

where the first term is the Coulombian pair-wise interaction. The next four terms deal directly with the coordination number. The second term defines a *bond order* parameter, which depends on the chemical system and has an exponential form. The coordination number is adjusted for *over*- and *under*-coordination; the effect of a possible electronic lone-pair and its correction on the coordination number is done in the E_{val} term, which also covers three-body interactions. A further E_{hb} can be added for systems where hydrogen bonds are essential for structural construction. The four-body interactions are captured in the torsion term, E_{tors}, which contains a specific dependence on the torsion angle between pairs of planes defined each by three atoms.

Other multi-body effects come first with the E_{conj} term, which corrects for long carbon-carbon chains, as commonly found in the various organic polymers. The electron delocalisation along these chains cannot be properly captured only from the previous valence-order terms.

The long-range weak bonds are taken into account in the $E_{vdWaals}$ term, which contains several factors decaying exponentially with distance.

As the force field is developed to reproduce the coordination number, a bond order parameter appears specifically in the formulation of all of the above terms except the van der Waals and the Coulombian terms. The bond order function is defined as a sum of three exponential terms describing respectively the sigma, the first π, and the second π bonds. Each of these terms has the form

$$BO_{ij} = exp(p'_{bo}) \left(\frac{r_{ij}}{r_0^{p''_{bo}}} \right) \qquad (2.19)$$

where p', p'', and r_0 are parameters. The resulting ReaxFF force field has more than 40 parameters to be fitted.

Further developments adapted the force fields to mineral and inorganic systems, like oxides and hydroxides, silicon and silicon oxides [90], iron [11], titanium [157], calcium [175], lithium [191], and so on. For example, hydrous silicate phases can be well described up to shallow mantle conditions by **the ClayFF force field** [77]. Of course, they are adapted to the particularities of hydrous silicates. The E_{bond}, E_{over}, and E_{under} terms are replaced with bond stretching and bending terms, which both have harmonic descriptions. The partial charges for the Coulombian terms are derived from the Mulliken analysis from ab initio molecular data.

In recent years a completely new category of interatomic potentials started to emerge. They are still based on general phenomenological aspects, but this time they are described using complex abstract functions, which a machine *learns* how to derive and construct. These are the so-called machine learning potentials.

The ChIMES model [174], which stands for Chebyshev Interaction Model for Efficient Simulation, involves two-body and three-body interactions. They are highly versatile and can be easily transferred and adapted from one structure to another. They can describe various extreme coordination environments and chemical reactions. The CHIMES model was fit on the forces and energies from first-principles molecular dynamics simulations and then applied to describe various materials such as molten carbon, carbon monoxide, or liquid water. The CIMES potentials contain a sum of polynomials:

$$E_{CiMES} = \sum_i \sum_{j>i}^{N} E_{ij} + \sum_i \sum_{j>i} \sum_{k>j}^{N} E_{ijk} \qquad (2.20)$$

Each term is described using a series of Chebyshev polynomials of order n that depend on the interatomic distance r_{ij}. The pair interactions are described using a single sum over polynomials:

$$E_{ij} = f_P(r_{ij}) + f_S(r_{ij}) \sum_n c_n T_n(s_{ij}) \tag{2.21}$$

where $T_n(s_{ij})$ is a Chebyshev polynomial of order n, which depends on a transformed pair distance s_{ij}. The f_P function is a penalty function to prevent interatomic distances below a minimum distance sampled in the training trajectory. The f_S function ensures the potential goes smoothly to zero at a given cutoff radius. The three-body terms require a sum over products of three polynomials, each depending on the same type of transformed pair interatomic distances as above:

$$E_{ijk} = f_S(r_{ij}) f_S(r_{jk}) f_S(r_{ki}) \sum_m \sum_n \sum_o c_{mno} T_m(s_{ij}) T_n(s_{jk}) T_o(s_{ki}) \tag{2.22}$$

The approach used in fitting Chebyshev polynomials can be extended even further. Nowadays, with the use of neural networks and machine learning techniques for fitting large-scale data, we witness the development of new classes of potentials. **Gaussian** potential [25] assumes the energy of the crystal to be described as a sum of weighted basis functions, ϕ_h:

$$\epsilon_i = \sum_h w_h \phi_h(d_i) \tag{2.23}$$

where the weights w_h are Gaussian, and the basis functions depend on descriptor functions d_i. The descriptors are abstract functions describing the local atomic coordination.

The abstractness can be pushed even further to replace the atoms with symmetry functions that describe the local coordination [100]. Once these functions are defined, the machine learning techniques allow them to be fitted over a wide range of structural models obtained from first-principles calculations results and/or experimental data. They are highly adaptive and transferable. They were successfully employed to reproduce the phase diagram of the Lennard–Jones system, of solid molecular water ice [101], and of various binary oxides.

Nowadays, machine learning potentials is a field with exponential growth and tremendous research potential. A large literature starts to develop with many models for potentials. We are still in the phase of exploring various schemes for their generation. The first review papers have just been published, and more and more models are being proposed, tested, and applied to various atomistic simulations. These are just a few references, far from being exhaustive [253, 110, 20, 184, 194, 156, 28, 71, 25]. But they can offer a good starting point for discovering this world, which deserves a collection of books by itself.

2.6 Requirements for Good Interatomic Potentials

The quality of the potentials reflects how well they can satisfy a series of criteria listed later [236].

The potential should be accurate. As they are fitted on a series of experimental or calculated values, they should be able to reproduce as well as possible all these data.

The potentials need to be stable and not generate numerical instabilities. These can typically happen when changing the thermodynamic conditions from the ones where the potentials were initially fit. Stability issues would fail to reproduce observed phase transitions, promote metastable phases, create new phase transitions, or simply not converge.

The potentials should show transferability. This is a very difficult task, but also extremely important. They need to cover as much as possible of the phase space with one set of potentials. This allows for comparing multiple structures and establishing stability and metastability relations. The transferability should be satisfied concerning thermodynamic conditions, symmetry (for example, they should describe equally well both clino- and ortho-pyroxenes, or quartz and stishovite), different redox states (for example, ferric and ferrous iron, or sulphur in sulphates and sulphur in sulphides), and of course, the chemical environment. In practice, all these transferability criteria cannot be equally satisfied, so different sets of potentials are optimal for specific groups of minerals. For example, the Buckingham-type potentials work well for oxidic environments, like salts (sulphates, silicates, oxides, etc.) at ambient conditions; the Jahn–Madden potential works best for dry silicates at mantle conditions; and ClayFF is optimal for hydrous phases, like brucite, micas, or clays.

PART II

STATIC PROPERTIES

The solution of our quest from the Chapter 2 to find the exact energy of a system of atoms lies in the *ab initio* methods. *Ab initio* means *from the beginning*. The term *first-principles* is the English translation from the Latin *ab initio*. Both first-principles and *ab initio* terms are employed alternatively in the literature, sometimes even in the same article or presentation. The *ab initio* methods do not involve any experimental input (except for the values of the fundamental constants) and aim to solve/describe the behaviour of an ensemble of quantum particles, constituting an atomic system. These methods can be applied to both the electronic ground state and excited states. Here the ground state is defined as the lowest energy configuration of the electrons at 0 K. The excited states lie at higher energy than the ground states. Solving the first-principles equation for the excited state is more complicated than for the ground state and usually requires expressing the functions dependent on time – the time that a given excitation is alive.

The Schrödinger equation is the fundamental equation of all first-principles methods that apply to an atomic ensemble of electrons and nuclei. This is a wave equation describing the nuclei and the electrons as interacting particles of that atomic system. The equation states that the total energy of the system is the sum of the kinetic energy of its particles and the potential energy.

The time-independent formulation of the Schrödinger equation, which describes the electronic ground state of a system of atoms, is written as:

$$H|\Psi> = E|\Psi> \tag{3.1}$$

which is equivalent to:

$$H\Psi(\mathbf{R})dr = E\Psi(\mathbf{R}) \tag{3.2}$$

where H is the Hamiltonian operator, which has the form:

$$H = T + V = \frac{i\hbar^2}{2M}\nabla^2 + V \tag{3.3}$$

With the explicit form of the Hamiltonian, the Schrödinger equation becomes:

$$\left[\frac{i\hbar^2}{2M}\nabla^2 + V\right]\Psi(\mathbf{r}, \mathbf{R}) = E\Psi(\mathbf{r}, \mathbf{R}) \tag{3.4}$$

The Schrödinger equation is written as a time-independent equation and, as such, gives the state of the system in its ground state: All the orbitals are filled with up to two electrons, one in each spin state, $+1/2$ and

$-1/2$, in increasing order of energy, till the last available electron. Each orbital, i, is described by a wavefunction, $\psi_i r$, R.

The wavefunctions are wave-like functions that describe the probability of the distribution of particles in space. They depend on the position of all particles, electrons ($r_N e$) and nuclei ($R_N n$), in the system:

$$\psi(R) = \psi(r_1, r_2, r_3, \dots, r_{Ne}, R_1, R_2, R_3, \dots, R_{Nn}) \qquad (3.5)$$

The first term of Equation 3.4 yields the kinetic energy, and the second term is the potential energy. $\hbar$ is the Planck's constant, h, divided by 2π. The Planck constant is a fundamental constant whose value is $h = 6.62607004 \times 10^{-34}$ J/s. It defines the quantum of electromagnetic energy. Its division by 2π is useful for unit transformation and simplification for writing further equations. $\hbar$ defines the Hartree as the atomic unit of energy: $\hbar = 1$ Ha $= 27.211396641308$ eV (the latter represents the energy gained by an electron when raising the potential by 1 volt).

The potential, V, is the energy resulting from the nuclear and electronic configuration obtained as Coulomb interactions and from the contribution of external fields. The vectors r, R describe the positions of all the particles in the system, both electrons and nuclei. For simplification, we will only employ a general R notation to designate the position of all particles in the system.

Solving the Schrödinger equation equates to determining the values and energies of these wavefunctions. This makes these methods transferrable to various conditions and constraints but also extremely complicated to resolve mathematically.

In purely mathematical terms, the Schrödinger equation is a system of differential equations. The eigenvalues of this system are the energy levels of the wavefunctions, where electrons can exist. The sum of these eigenvalues is the internal energy of the system. The eigenvectors of the Schrödinger equation are the many-body wavefunctions, describing the mutually dependent positions and velocities of all electrons and nuclei. The wavefunctions are the most important pieces that must be determined when solving the atomic puzzle. They characterise the physical state of the particles in the system. They give the probability spread of particles in space, their mutual relations, and the occupancies of the energy levels.

For the practical applications of the Schrödinger equation, solving it, unfortunately, is easier said than done, as there is no analytical solution to it, except for the hydrogenoid atom, that is, an atom with only one electron. This is not of much help when dealing with realistic systems. Consequently, for a long time after its initial formulation, the Schrödinger equation remained in the realm of theoretical physics. It did not make a real impact in day-to-day numerical scientific research. One major part of this problem comes from the many-body character of the wavefunctions, which even today has no analytical solution, as explained in Section 3.2. A second part of the problem is the limited computational power available today.

Even numerically, solving the Schrödinger equation comes down to estimating an equivalent of $n!$ evaluations for an n-particle system. This might not seem too much, but it represents 720 evaluations for a C atom with 6 electrons, 40,320 evaluations for an O atom with 8 atoms, goes up to 328,800 evaluations for a Ne atom with 10 electrons, and quickly goes beyond anything pronounceable even for a simple molecule: $8.68E^{36}$ evaluations for an SiO_2 molecule with 30 electrons and 3 nuclei.

The practical alternative to the Schrödinger equation arrived in the mid-sixties, with the seminal papers of Kohn and Sham [160] and Hohenberg and Kohn [142]. Kohn and co-workers could re-write the Schrödinger equation in terms of the electronic density using several approximations detailed later. This reduced the problem to working with electronic densities, which are defined on a finite number of points in space. This solution opened the path to decades of intensive and successful applications in physics and chemistry, and more recently mineralogy, at the atomic level. Its success is such that today we can talk about computational numerical experiments as the third branch of physics, at the intersection of its two traditional branches: experimental and theoretical. The theory that Kohn and Sham founded is known as the density functional theory (DFT). To acknowledge the importance of this work, Walter Kohn received the Nobel Prize for Chemistry in 1998.

3.1 Kohn–Sham Equations

The resulting DFT makes use of several approximations. First, the large contrast between the mass of the atomic nuclei and the mass of electrons allows us to separate the movement of the nuclei from the movement of the electrons. The Born–Oppenheimer approximation states that the electrons instantaneously follow the nuclei; they instantaneously adapt to the field generated by the nuclei. This allows for transforming the general wavefunction from the Schrödinger equation into a purely electronic wavefunction, defined for a specific fixed configuration of nuclei.

In a second major approximation, the electrons are treated as independent particles, which lie in an effective potential bath generated by all the other electrons and nuclei of the system. This allows for the simplification of the total electronic wavefunction, which becomes a product of non-interacting electronic wavefunctions:

$$\psi(\mathbf{r}) = \psi(\mathbf{r}_1)\psi(\mathbf{r}_2)\psi(\mathbf{r}_3)...\psi(\mathbf{r}_{Ne}) \tag{3.6}$$

where $\mathbf{r}$ denotes all the electrons, and $\mathbf{r_e}$ are the individual electrons, whose total number is Ne. This simplification brought by the relative 'independence' of the electronic wavefunctions allowed both the theory and the practice to move towards solving more realistic systems.

The electron density in a point $\mathbf{r}$ of space is defined in terms of double-occupied wavefunctions, $\psi_i(\mathbf{r})$, as:

$$n(\mathbf{r}) = 2 \sum_{i=1}^{N_{el}/2} \int \psi_i^*(\mathbf{r})\psi_i(\mathbf{r})d\mathbf{r} = 2 \sum_i |\psi_i(\mathbf{r})^2| \tag{3.7}$$

The sum runs over all the electronic orbitals, i.

In Density-Functional Theory, commonly abbreviated as DFT, the energy is obtained by applying the Hamiltonian on the electronic density rather than on the individual electrons. A function (Hamiltonian) that is applied to another function (electronic density) is called a functional, hence the 'F' from the DFT. The electronic density, the 'D' in DFT, is the function that describes the distribution of electrons in space and the probability of finding them in a given part of the space.

The electronic Hamiltonian in the DFT formalism becomes the Kohn–Sham functional. The expression for the Kohn–Sham energy for the ground state of an electronic system is:

$$E[n(\mathbf{r})] = H[n(\mathbf{r})] \tag{3.8}$$

The Hamiltonian, H, represents a series of functions detailing various components of the energy, all expressed in terms of the electron density. This yields:

$$E[n(\mathbf{r})] = T[n(\mathbf{r})] + E_{e-e}[n(\mathbf{r})] + E_{e-ion}[n(\mathbf{r})] + E_{ion-ion}[\{R_I\}] + E_{xc}[n(\mathbf{r})] \tag{3.9}$$

Here, $T[n(\mathbf{r})]$ is the kinetic energy:

$$xT[n(\mathbf{r})] = 2 \sum_{i=1}^{N_{el}/2} \int \psi_i^*(\mathbf{r})(-\frac{1}{2}\nabla^2(\psi_i(\mathbf{r})))d\mathbf{r} \tag{3.10}$$

The following three terms are Coulombian in nature. They contain the interactions between the point-like charged particles: nuclei–nuclei ($E_{ion-ion}$), electrons–nuclei (E_{e-ion}), and electrons–electrons (E_{e-e}). The terms involving the effect of nuclei, together with energy terms induced by any external fields, E_{exfi}, are grouped in a specific 'external' term, $E_{ext} = E_{ion-ion} + E_{e-ion} + E_{exfi}$.

The E_{e-e} term is also known as the Hartree energy, E_H, and is written as:

$$E_H[n(\mathbf{r})] = \frac{e^2}{2} \int \frac{n(\mathbf{r})n(\mathbf{r}')}{|\mathbf{r} - \mathbf{r}'|}d\mathbf{r} \tag{3.11}$$

Another aspect that made the DFT such a successful practical theory is that in the Hartree expression for non-interacting wavefunctions, the charged particles exert forces onto each other as in a vacuum, without screening. Due to the background electron density, for which an analytical expression lacks today, the screening is effectively hidden in the many-body character of the interactions. The approximation of non-interacting electrons removed the constraint of the dielectric permittivity of the electronic cloud. In the Kohn–Sham equations, this correction is moved to the last term, the exchange-correlation term, discussed in Section 3.2.

Coming back to the basic DFT equation (Eq. 3.9), its actual physical meaning is to find the electronic density that minimises the total internal energy of the system. This is done in a minimisation loop, which is an iterative procedure where each step should bring us closer to the minimum energy. At each step, the energy of the system is computed as a function of the electron density, and compared to the previous instances. Then the distribution of the electronic density is modified so that the value of the energy is smaller than the one in the previous steps. The modification of the density is improved using algorithms that take into account its distribution at the previous steps and that can bring the system to the electronic configuration with the lowest energy. The procedure is stopped when a minimum is achieved.

The numerical techniques employed to solve self-consistently the Kohn–Sham equations come from linear algebra. This is why most of the DFT codes require libraries of algebra. Different codes offer different numerical techniques to reach convergence, like Davidson iagonalisation [80], Anderson minimisation [7], conjugate-gradient, and so on.

3.2 Exchange–Correlation

The exchange-correlation term contains two parts: the electronic exchange and the electronic correlation. One way of visualising these two terms is to imagine the interplay between bringing together on the same band two electrons of opposite spin versus two electrons of equal spin. This corresponds respectively to the exchange part and the correlation part; the first term lowers the energy, and the second one raises it [202]. In practice, the exchange-correlation term is solved numerically by adopting various approximations.

The most commonly used is the local density approximation (LDA). Here the $E_{xc}(n(\mathbf{r}))$ energy depends only on the value of the electronic density $n(\mathbf{r})$ in that specific point, $\mathbf{r}$. Therefore, its 'Local' character. LDA was proposed in the original papers of the DFT [142] and it quickly became widespread. Over the years, LDA has been and is still extensively used for a variety of systems, where it performs surprisingly well. Part of the reason for its success is due to the conservation of the electronic norm, that is, the number of electrons is preserved during the electronic minimisation. To numerically calculate $E_{xc}(n(\mathbf{r}))$ in LDA, one needs to consider its two components, the exchange and the correlation. The exchange is estimated using the Dirac formula:

$$E_x(n(\mathbf{r})) = -\frac{3}{4}\left(\frac{3}{\pi}\right)^{\sqrt{1/3}} \int n(\mathbf{r})^{\sqrt{4/3}} d\mathbf{r} \tag{3.12}$$

The correlation part is equal to that of a uniform electronic gas with the same density $n(\mathbf{r})$. The uniform electron gas existing in a uniformly

---

Box 3.1   What is a low-dimensional solid?

Some solids are so anisotropic that physical properties are strikingly different along only one or two of the three possible directions. When you look at the habitus of a crystal, this immediately indicates the topological anisotropy of the structure. For example, the pyriboles, like pyroxenes, pyroxenoids, and amphiboles, have elongated crystals along one direction due to the formation of the chains of silica tetrahedra. Micas, such as biotite, hence the 'bio' part of the biopyribole series, are dominated by the sheets of octahedra and tetrahedra. Similarly, the minerals in the stibnite series have one pronounced dimension corresponding to the chains of polymerised sulphur and antimony, while molybdenite has a two-dimensional character corresponding to the layers of S-Mo-S. In all these materials, the electronic density exhibits the same anisotropy as the external habitus. The anisotropy of the electronic structure can sometimes be so pronounced that there are materials that are insulators along one direction and conductors along another. A famous example is graphite, which is metallic parallel to the basal hexagonal plane and insulating along the perpendicular c direction. In general, the low-dimensional solids are better described using GGA than LDA functionals (see also Box 3.2).

distributed positive charge background (representing the nuclei) is known as 'jellium', and its correlation energy can be parametrised by performing a Monte Carlo calculation [165, 70, 202]. The $E_{xc}(n(\mathbf{r}))$ in LDA is the exchange-correlation energy of the homogeneous electronic gas with the same density $n(\mathbf{r})$.

Unfortunately, LDA may fail for systems with quickly varying electronic density, like molecular crystals (ices, clathrates, zeolites, etc.), low-dimensional solids (Brucite, micas, clays, etc. – see Box 3.1), or transition metal compounds (metallic oxides and sulphides). The failure is seen in largely overestimated interatomic distances or non-realistic phase stability relations. A considerable improvement over LDA is the consideration of the gradient of the electronic density in the form of the generalized gradient approximation (GGA). There are many flavours of GGA, as proposed by different authors. Usually, these formulations are obtained from fitting various expressions of the exchange-correlation energy to match structural and physical experimental parameters of simple solids and binary compounds.

For example, one of the most widely used GGA forms was proposed by Perdew, Burke, and Ernzerhof in 1996 [204] and is typically abbreviated as PBE96. In this formalism, the exchange-correlation term is expressed as an expansion of the electronic density that includes the logarithm of a polynomial expansion in terms of the density gradient. The resultant exchange-correlation is a smooth function that is a constant close to the nuclei where the gradient is null and vanishes at large distances. Most of its parameters result from fundamental constants. This choice of using fundamental constants conferred it a strong physical background, which eventually made it such a successful formalism. An improved PBE96 was

> ### Box 3.2   How to choose the exchange–correlation formalism?
>
> This is a fundamental question in DFT calculations that still does not have a unique and universal answer. Instead, this depends on the particular mineral system, the availability of implementations, and the computing resources. In general, the rule of thumb is to carefully test several functionals and compare the theoretical results with as much experimental data as possible. Start with LDA as much as possible, particularly for relatively homogeneous systems dominated by atoms from the s and p groups. Compare your theoretical geometry with the experimental one, and be sure to attain about 2% accuracy in interatomic distances, about 1 degree in interatomic angles, and less than 10% in formation energy and phonon frequencies. If LDA fails you, that is, the accuracy of your simulation is not good, then switch to GGA. There are many forms of GGA, and you will need to choose from what you have available. There are particular minerals, with small electronic gaps and containing transitional metals, for which standard GGA fails as well [178]. Magnesiowüstite, $(Mg,Fe)O$, is a famous example [145] that comes out metallic in DFT simulations, while it is an insulator in reality. For this type of materials, you need to go beyond standard GGA and DFT and use DFT+U or even other techniques mentioned further in this book.

proposed by Wu and Cohen [267], and then by Perdew and collaborators [206]. In these versions, fits on more diverse physical properties, like the heat of formation and the interatomic bond distances and angles, of more and more unary and binary compounds, provided better numerical values for the exchange-correlation parameters.

GGA brought a net improvement over LDA for many systems. It has been observed that, as a rule of thumb, LDA tends to underestimate the bond distances and the heat of formation compared to experiments, while GGA tends to overestimate them. Differences in bond distances and lattice parameters on the order of a couple of per cent, negative for LDA and positive for GGA, are considered acceptable. Deviations of bond angles and unit cell angles on the order of one degree and in the heat of formation and phonon frequencies on the order of 10% are all deemed acceptable. While there is no general consensus about which exchange-correlation functional to use, that is, LDA versus GGA, or between various GGAs, a few suggestions are given in Box 3.2.

An improvement over LDA and GGA formalisms was brought by the use of the meta-GGA functionals [241, 205, 164]. Here, an additional second derivative of the density is also included, the Laplacian. Even higher-order terms could be considered, but the computation of all these terms requires further operations with the energy, and as such, the calculation quickly becomes intractable, without necessarily improving the final result.

In the last years, a new class of exchange-correlation functionals was developed, which combined traditional functionals, like LDA or GGA, with an amount of exact Fock exchange. Because they combine DFT and Hartree–Fock, they are called hybrid functionals. For many chemical systems, like solids in general, and also for many molecular solids, in particular hydrides, and so on, there were considerable improvements in the estimation of their physical properties [207, 106]. There are a few widely used such functionals, like HSE or B3LYP. However, many benchmarks show that hybrid functionals can be up to 50 times more expensive computationally than standard LDA or GGA [39].

In HSE, named after the three authors, Heyd, Scuseria, and Ernzerhof, the short-range part of the exchange is corrected with a screened Coulomb potential [138].

In B3LYP, the exchange-correlation is a linear combination of LDA exchange (E_x) and correlation (E_c), GGA in the formulation proposed by Becke[27], and the Hartree–Fock exchange:

$$E_{xc} = E_x^{LDA} + a_0(E_x^{HF} - E_x^{LDA}) + a_x(E_x^{GGA} - E_x^{LDA}) + a_c(E_c^{GGA} - E_c^{LDA})$$

$$(3.13)$$

Here the parameters a_0, a_x, and a_c are fitted to various atomic and molecular data provided by experiments, like the energy of formation, interatomic bond distances, interatomic angles, and so on.

In general, nowadays, more and more authors test hybrid functionals, or reviewers ask authors to perform such tests. The truth is somehow mitigated [271], as none of the present-day functionals gives exact answers. The degree of approximation varies from one form to another, and different functionals give better or worse results on various minerals and materials.

Probably the latest trend that will also develop considerably in the following years is to employ machine learning and neural network techniques to offer numerical solutions to the exchange-correlation problem. A recent model employing two hidden neural levels generated numerical functionals that accurately reproduced properties not only for three-electron idealised systems but also for Si and Ge crystals [218]. However, as always when using fitting, just adding pure numerical terms that lack a physical significance might not necessarily be a good idea [179, 129], and some of these new functionals fail for materials where they were not tested from the beginning, that is, they lack transferability. It is most likely that improved analytical solutions will eliminate, sometimes in the future, some of the inherent approximations of DFT related to the use of numerical approaches of the exchange-correlation term.

3.3   Planewaves and k-Points

There are two fundamentally different ways to represent the wavefunctions that describe the electron density: using localised basis sets mimicking the atomic orbitals or using periodic planewaves.

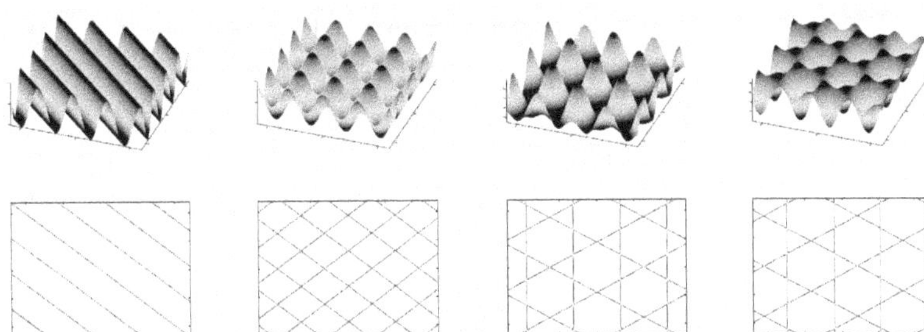

Fig. 3.1 The smooth distribution of the electron density can be reconstructed from a sum of many enough planewaves, each having its wavevector and amplitude. The upper panels show such a sum of planewaves, whose periodicity is shown in the lower panels. More and more terms are added from the left to the right panels. (The Matlab code that served to build this image is courtesy of Ricardo Carretero at https://carretero.sdsu.edu/)

The first method is based on the direct representation of the atomic orbitals, like H1s, Ne2s, Ne2p, hybridised Csp^3, and so on, as specific basis functions, or a sum of these. This collection of functions forms a set, which is called a basis set. A common example is the angular-dependent Gaussian basis set. The electron density is obtained as a linear combination of these individual components, whose coefficients are calculated during the calculation.

In general, the basis sets are highly localised, and they are mainly used today in molecular chemistry codes. They are particularly efficient for describing non-periodic systems, like isolated molecules and clusters. The size of the basis sets needed is in general small or modest, which allows for simulating large systems without requiring huge computing resources. However, the downside is the bad scalability with the system size, as these simulations are hard to parallelise.

The second method employs periodic planewaves. In a way, this is very similar to obtaining the electron density from the diffraction. In diffraction, the pattern is obtained as the Fourier transform of the electron density. It is obtained in the reciprocal space as a collection of maxima and minima of intensity. Each of the maxima, that is, the 'peaks', corresponds to a family of parallel lattice-periodic planes of alternating high and low electron density. Similarly, one employs periodic planewaves to represent the electron density in parallel planes of alternating high and low amplitude, as shown in Figure 3.1. The advantage of planewaves is that they preserve the information of the lattice periodicity.

The planewaves ϕ_i are defined by their wavevectors in the reciprocal space. These are anchored on the nodes of the reciprocal Bravais lattice or the Fourier transform of the lattice, denoted as the set of **G** points.

The grid of **G**-points is discrete, while the distribution of charge in the direct space is continuous. As such, the information that the grid contains is limited. The reciprocal space can be populated with further points on a regular and symmetric set to overcome this limitation. This set is called the **k** points grid [183]. The **k** points lie in-between the **G** points. There are alternative options to the Monkhorst–Pack grid, like the equivalent Baldereschi point [19] or the adaptive mesh of Chadi and Cohen [72].

Regardless of the choice of **k** points, the planewaves are eventually defined on the **k + G** grid of points. Thus, they preserve the lattice periodicity via the **G** grid and carry phase information via the **k** grid (Box 3.3). The wavefunctions are then expressed as a weighted sum of planewaves, $\phi_i^{(k+G)}$:

$$\psi_i = \sum_k \sum_G c_i^{(\mathbf{k+G})} \phi_i^{(\mathbf{k+G})} = \sum_k \sum_G c_i^{(\mathbf{k+G})} e^{i2\pi \mathbf{k}} e^{i2\pi \mathbf{G}} \tag{3.14}$$

where c_i is the weight of each planewave. With the expansion of the wavefunctions in planewaves, one obtains an expression that depends entirely on these coefficients c_i. Then the Kohn–Sham functional is applied to the sum of planewaves. The energy is obtained as:

$$E = \langle \psi | H | \psi \rangle = \sum_{i=1}^{N_{el}/2} \int \sum_k \sum_G c_i^{*(\mathbf{k+G})} \phi_i^{*(\mathbf{k+G})} H c_i^{(\mathbf{k+G})} \phi_i^{(\mathbf{k+G})} d\mathbf{r} \tag{3.15}$$

Each term of the Kohn–Sham Hamiltonian is applied to the sum of planewaves. For example, the kinetic energy term becomes:

$$T = \langle \Psi | \nabla^2 | \Psi \rangle = - \sum_{i=1}^{N_{el}/2} \int \sum_k \sum_G c_i^{*(\mathbf{k+G})} \phi_i^{*(\mathbf{k+G})} \nabla^2 c_i^{(\mathbf{k+G})} \phi_i^{(\mathbf{k+G})} d\mathbf{r} \tag{3.16}$$

Equation 3.15 corresponds to a square system of equations whose size is the number of planewaves. Solving the Kohn–Sham equations is a minimisation problem: finding the set of $c_i^{\mathbf{k+G}}$ coefficients that yield the minimum value of the energy, that is, the minimum value of the Hamiltonian (Box 3.4).

The planewaves corresponding to large **k + G** vectors vary rapidly in real space, as their wavelengths are very short. Thus, they do not contribute much to the electron density; instead, their contribution mimics background noise. Consequently, they could be disregarded. The threshold of including them or not in the development in Equation 3.14 is given by their contribution to the kinetic energy as a measure for their rapidly oscillating character. This is called the kinetic-energy cutoff and is expressed in units of energy. It is one of the most important parameters of a calculation; as such, it is always, or at least should always, be reported in the methodology section of a publication. Behind the scenes, the machinery of the DFT code transforms this energy into an equivalent radius, $\mathbf{r_c}$, in the reciprocal space (Figure 3.2). Only the planewaves with **k + G** vectors smaller than $\mathbf{r_c}$ are considered in the development of the planewaves.

The total energy is variational, meaning that some minimisation algorithms can reach its lowermost value. In practice, this is realised by converging the value of the kinetic energy cutoff for the planewaves and, thus, by converging the number of planewaves. In a series of calculations, the value of the kinetic energy cut-off is increased and the total energy

Box 3.3	How to choose the grid of k points?

The grid of **k** points in the reciprocal space can be determined similarly to the kinetic-energy cutoff, described in Section 3.3, by consecutively increasing its density until the convergence criterion of the energy is reached. The obvious choice of the grid of **k** points is such as to use at maximum the advantage of the crystal symmetry, applied on the **k** phase of the wavevector. This is achieved using regular grids along the three directions of the reciprocal space [183]. For example, for an orthorhombic two-dimensional (2D) crystal, the 2×2 grid of *k* points can be chosen in several ways:

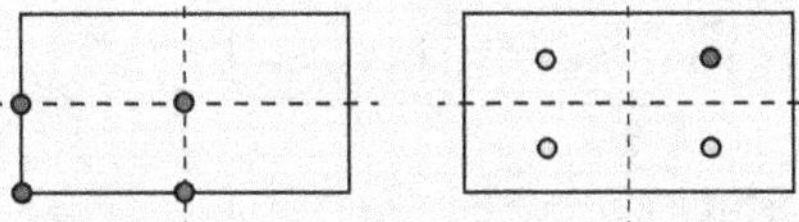

The left choice does not make use of symmetry, as every black point on the grid needs to be calculated. The choice on the right fully makes use of the symmetry: The coefficients are evaluated only in the black point, and the grey points are related by the simple symmetry operations (the mirror planes represented with dashed lines, the A_2 axis, and the symmetry centre, always present in the reciprocal space as time-reversal symmetry). By convention, some codes, like abinit, consider the phase shift of the grid to be 0 0 and 0.5 0.5 in the left and right panels, respectively. Other codes might do the opposite. Check the convention before doing the calculation!

It is also good to know that before going from a grid of 2×2 to a denser grid of 4×4, you can use additional shifts of the same grid:

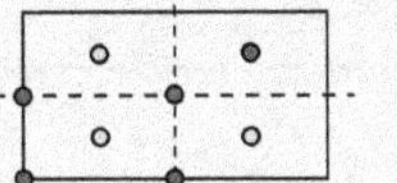 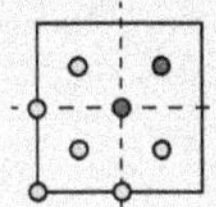

The 2×2 grid shifted twice (earlier) requires five points to be calculated in the rectangular case on the left or only three in the case of the square case on the right. It all comes down to symmetry: The more we can apply it, the better it is, and the fewer actual calculations are needed. The following table lists a few such optimal choices for various Bravais lattices of the crystal:

Bravais type	Grid construction		
P	nshiftk	1	# number of grids
	shiftk	0.5 0.5 0.5	# phase shift
Hexagonal P	nshiftk	1	# number of grids
	shiftk	0 0 0.5	# phase shift
I	nshiftk	2	# number of grids
	shiftk	0.25 0.25 0.25	# phase shift
		−0.25 −0.25 −0.25	
F	nshiftk	4	# number of grids
	shiftk	0.5 0.5 0.5	# phase shift
		0.5 0.0 0.0	
		0.0 0.5 0.0	
		0.0 0.0 0.5	

The DFT procedure aims to find the electronic density that minimises the energy. In practice, the Kohn–Sham equations are written in terms of the electron density obtained from the independent Kohn–Sham wavefunction. In the planewave formalism, each wavefunction is written as a sum of planewaves. By their construction, the pseudopotentials are used to initialise the electron density and thus the coefficients of the planewave expansions of the wavefunctions. The DFT code optimises the values of these complex coefficients to find the minimum energy. The procedure closely follows the workflow from Box Figure 1:

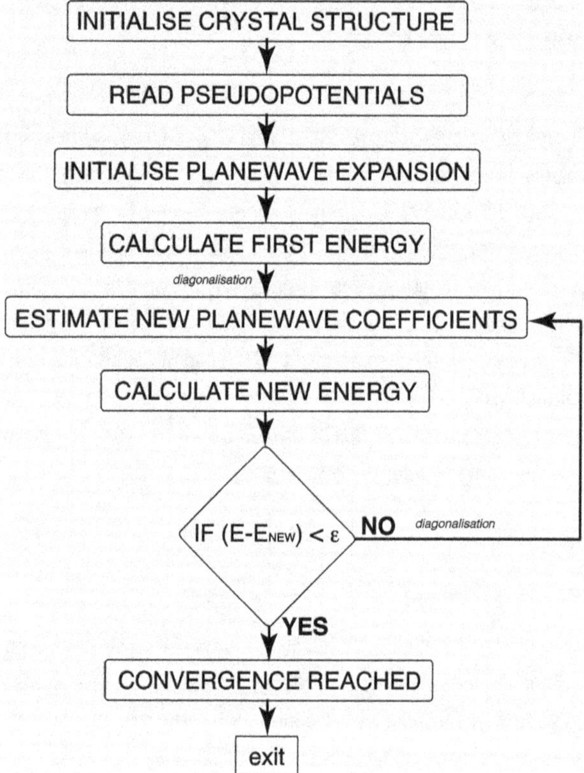

Box Figure 1. Self-consistent loop for computing energy in DFT.

This minimisation loop is called the *self-consistent (SCF) loop*. In practice, every time the Kohn–Sham equations are solved, the matrix of coefficients needs to be diagonalised. This step is what takes most of the time for the calculation. The coefficients are the eigenvalues of this matrix. Consequently, the DFT codes are heavy consumers of linear algebra techniques in the quest for the fastest and most efficient minimisation path.

value is monitored (Box 3.5). Figure 3.3 shows the variation of the total energy as a function of the kinetic energy cutoff obtained for quartz. The total energy follows a variational principle, according to which an increase in the number of planewaves yields a decrease in energy value.

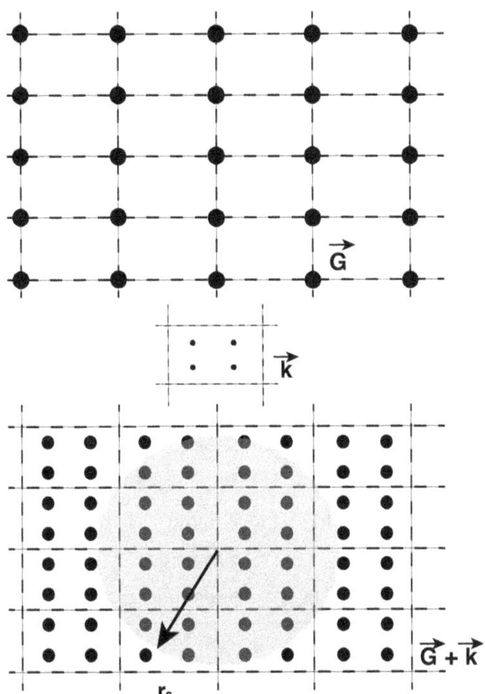

Fig. 3.2 Construction of the **G** (large black dots) + **k** (small grey dots) grids of wavevectors used in the planewave expansion of the electron density. **G** are the nodes of the reciprocal lattice, the equivalent of the Bragg diffraction points (upper panel). **k** are the fine grid regularly spaced inside each of the cells of the reciprocal lattice (middle panel). The planewave expansion is then realised on the ensemble of **G** and **k** points (lower panel), which are summed as vectors. Only the planewaves whose kinetic energies are lower than a certain cutoff (large black arrow) are taken into account for the expansion. This kinetic energy cutoff may be represented graphically as the radius of a sphere. The **k** + **G** points outside this radius are discarded.

In the same way, one needs to obtain the density of the grid of **k** points (Box 3.6). But this does not follow a variational principle, meaning that the total energy does not necessarily decrease monotonously while increasing the density of the grid. However, convergence can be achieved when the density of the grid is sufficient. Only numerical tests can prove that.

Convergence of the self-consistent electronic energy better than 10^{-6} Ha/atom can be considered safe to continue with the calculation of the other physical properties. Further convergence tests can be done for different physical properties, such as the pressure (Figure 3.4). Some specific simulations require a better convergence; for others, a less strict convergence criterion might be enough. Specifically, the physical properties obtained as energy derivatives, like the phonons, the Raman, or the elastic tensors, require tighter convergence criteria and very dense grids of **k**-points.

Sometimes, the minimisation loop can diverge, and as a result, the minimum energy is not found, or the optimisation procedure can get stuck in a local minimum. These difficult situations typically reflect either peculiar

The solution to the Kohn–Sham equations is variational in terms of planewaves, meaning it can be asymptotically achieved by employing a large enough number of planewaves. Finding the correct number for each is obtained with a convergence study. You need to iteratively increase the number of planewaves while monitoring the total energy of the system. Usually, well-converged energy (within at least the 6–7 significant digits) yields correct values for other physical properties, like pressure, elasticity, phonon frequencies, and so on. However, it is sometimes advisable to monitor the value of those properties in separate convergence studies. The convergence with respect to the number of planewaves involves the determination of the kinetic energy cut-off:

```
ndtset   11       # number of iterative calculations
cut:     10       # starting value (in Hartree)
ecut+    2        # increment
toldfe   1.0d-05  # convergence of the energy
```

You noticed that the energy convergence is set at only E-05. This is because we are usually looking for Six representative digits, which typically is on the order of the miliHartree per atom.

atomic configurations or electronic structures. Systems with large vacuum volumes, vacancies, or electronic spins are only a few examples of such cases. Different minimisation schemes should be tried in these cases, as most codes come with various options. Some of these, while slower, are better at converging the energy in different sensitive situations. In other cases, starting with a different magnetic configuration or with different amplitudes of the magnetic spins might help the system escape from a local potential well by, for example, breaking the symmetry of the wavefunctions.

3.4 Pseudopotentials

From the very beginning of the DFT, an important observation was made and fully employed. Not all electrons participate in the chemical bonds. The inner electrons, also called *core states*, which are the closest to the atomic nuclei, influence the charge distribution and the total energy. But they do not participate directly in the chemical bonding. They are quasi-inert with respect to the changes in valence electron density, their energy does not change significantly, and neither does their density distribution. Knowing only their influence on the valence electrons is enough, and they

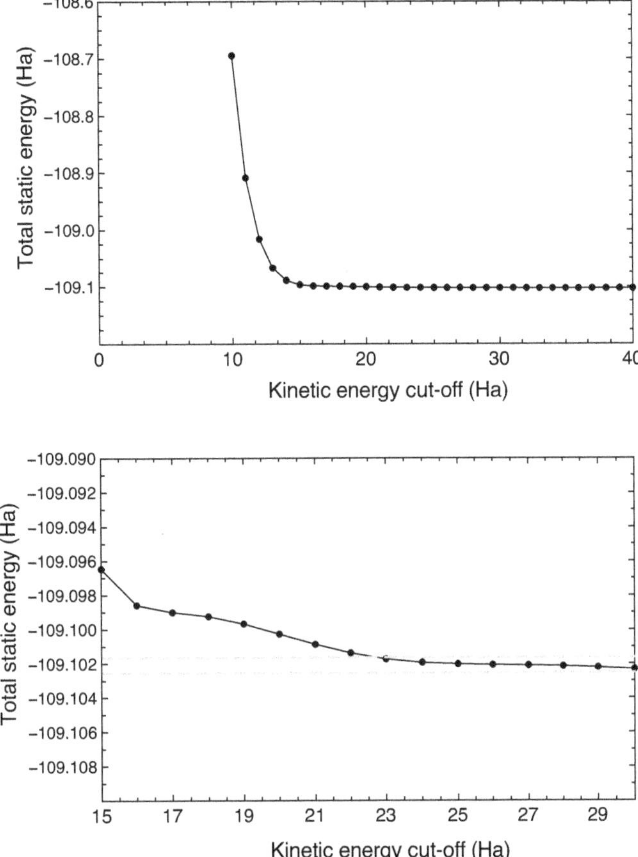

Fig. 3.3 Convergence study of the kinetic energy cut-off for SiO_2 quartz. GGA calculations were performed using the PAW formalism. At a cut-off of 16–17 Ha, the convergence is on the order of 1 mHa/unit cell, which should be enough for studying most physical properties of the crystal.

do not need to be explicitly calculated every time we perform a new simulation. Then we may approximate the core states as being frozen, that is, equal to their state for an isolated atom. In DFT, this approximation is realised using pseudopotentials.

The kinetic energy contribution of the core electrons to the total energy is hard to converge using planewaves because these states are highly localised, and the planewaves are, by construction, fully delocalised. Because the pseudopotentials replace the potential exerted by the core electrons on the valence electrons, their use tremendously simplifies and accelerates the DFT calculation.

The region of space where the core electrons lie can be delimited by a sphere defined by the radius and centred on the atomic nuclei. The pseudo-wavefunctions inside this region have a much smoother variation than the actual wavefunctions and do not reproduce the actual atomic orbitals. They reproduce only the part of the valence electrons physically located inside the spheres. Outside of this region, beyond that radius, they need to exactly reproduce the all-electron wavefunctions. The same constraints are

Box 3.6 — How to check convergence with respect to k points?

The Kohn–Sham equations are not variational with respect to the **k**-points sampling of the reciprocal space. However, if you increase the density of the grid of **k** points, you can reach a numerical convergence of the total energy of the system. This stems from the coverage of the reciprocal space.

You can chain several calculations on increasingly denser grids in abinit. Remember that the grid can be shifted such as to make maximum use of the symmetry. For a face-centred cubic lattice, like the one in the structure of halite, NaCl, the ideal coverage is obtained using four shifted grids, like in the following example:

acell	3*5.6404	angstroms # Halite unit cell parameter
rprim	0.0 0.5 0.5	
	0.5 0.0 0.5	
	0.5 0.5 0.0	
natom	2	# define the atoms
ntypat	2	
typat	1 2	
znucl	11 17	# Na Cl
kptopt	1	# make use of the full symmetry
nshiftk	4	# consider the ideal shifts based on symmetry
shiftk	0.5 0.5 0.5	
	0.5 0.0 0.0	
	0.0 0.5 0.0	
	0.0 0.0 0.5	
ndtset	10	# number of iterative calculations
kpt:	2 2 2	# starting grid along the three directions
kpt+	1 1 1	# increment along the three directions
toldfe	1.0d-05	# convergence of the energy

You noticed that the energy convergence is set again at only E-05 for the same reason as in Box 3.5.

imposed on the potential: it reproduces exactly the all-electron potential outside of its radius.

When running a DFT calculation, special care has to be paid to the geometry of the system, such as not allowing, or at least ensuring, a minimum overlap of the spheres. In case overlap exists, this usually translates into a breaking of the orthogonality condition between the valence and the core states, that is, artificial mixing of electrons of a different character, and hence wrong electronic band occupancies, energies, geometries, and so on. In practice, however, there might be cases where the atoms are too close to each other.

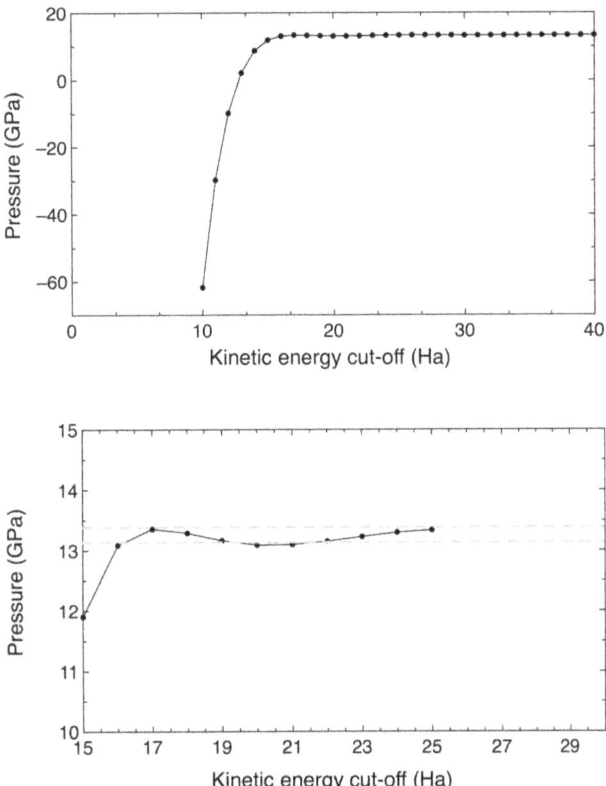

Fig. 3.4   Convergence study of the kinetic energy cut-off for SiO_2 quartz. GGA calculations were performed using the PAW formalism. At the convergence cut-off of the energy, which is 16–17 Ha (Figure 3.3), the pressure is converged within 2–3 kbars. It is always safe to check the convergence of the physical property that is studied, not only of the energy.

When constructing the pseudopotentials, the aim is to consider as many electrons as possible within the core states, simplifying the DFT calculation. However, this has to be carefully done, as sometimes electrons from the last core shell can also participate in the chemical bonds, depending on the chemical context. These states that sometimes participate in the chemical bonds and other times do not are known as semi-core states. Alkaline and calc-alkaline metals have such semi-core states, as well as some elements of the p bloc. These are the states on the second-last electronic shell, for example, the $3p$ and sometimes even $3s$ electrons of Ca. The semi-core states become increasingly important for the quasi-totality of the periodic system when simulating structures at high pressures (Figure 3.5).

There is a world of itself that developed around pseudopotentials. There are several software packages that build pseudopotentials [256, 213, 248, 97, 221]. And most of the *ab initio* codes come with libraries of pseudopotentials. The most widely used pseudopotentials today are built using numerical schemes. They are provided as long tables of numbers, covering the description of each electronic state on a radial grid and the potential exerted by the core electrons outside the cutoff radius. Other pseudopotentials are generated using analytical expressions of the orbitals/states [111].

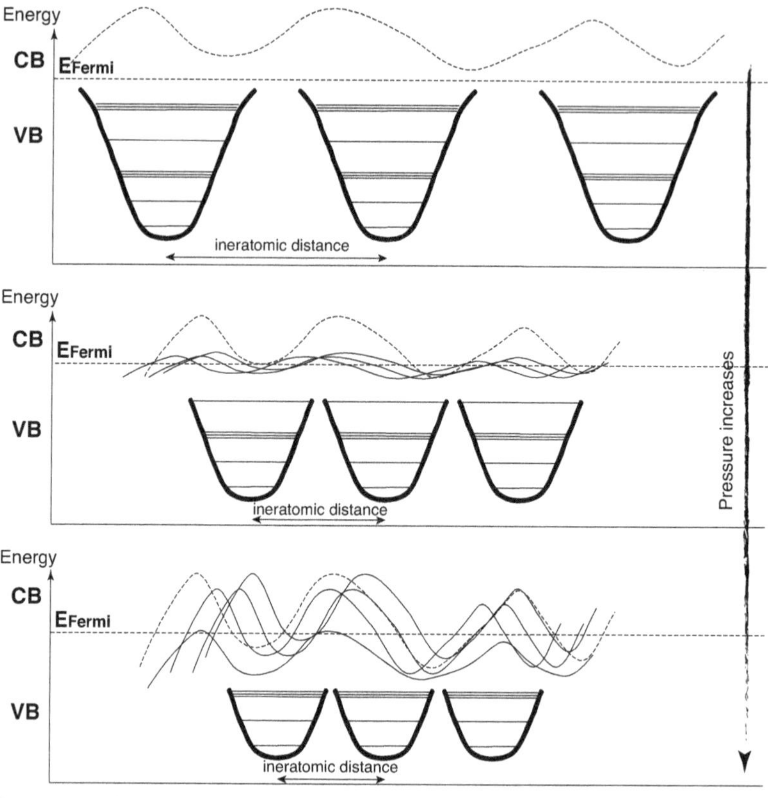

Fig. 3.5 At low pressure, the inner electrons are localised deep inside the potential well around the atoms. As the pressure increases and the atoms lie closer and closer to each other, the potential wells overlap, and more and more deep states start to delocalise. Hence, these states need to be explicitly calculated and cannot be frozen anymore inside the pseudopotentials. Under high enough pressure, the electronic gap can eventually close, and some of these bands cross the Fermi level.

An ideal should be transferable, meaning that it should work in a variety of chemical environments, be smooth, such as to ensure the continuity of orbitals and their derivatives at the cutoff radius, not have *ghost* states, meaning only physical orbitals should be present with no numerical artefacts, and be efficient, such as to require a minimal set of planewaves. It should reproduce exactly all the all-electron wavefunctions and potential.

The norm-conserving pseudopotentials are historically the oldest type. Their name comes from the constraint that they preserve the number of electrons, that is, the norm (which represents the total negative charge of the unit cell). They have a spherically symmetric part, which is local, and one part depending on each angular moment, which is non-local. They work for most flavours of exchange-correlation and, by construction, are particularly suitable to compute advanced properties, including high-order energy and wavefunctions derivatives. Most DFT codes with planewaves and pseudopotentials today employ such norm-conserving pseudopotentials. The ab initio norm-conserving pseudopotentials were proposed initially in 1979 [127]. Troullier and Martins later built a scheme to optimise the pseudopotentials [248], which was widely used during the following decades.

However, when using norm-conserving pseudopotentials, oftentimes it is hard to converge the wavefunction associated with those valence states that have a large part geometrically situated inside the core region. This is solved by using numerous planewaves to determine the electron density, especially close to the atomic nuclei, which renders the simulations heavy and slow. This is true, particularly for elements with highly localised orbitals, like the d and the f block. Only one such atom in the simulation is enough to enforce this requirement and thus considerably up the computational cost of the entire calculation.

Thus the ultrasoft pseudopotentials were introduced to relax the norm conservation requirement [256]. By construction, they act on the strong constraint of conserving the norm. They *'loose'* some of the electronic population inside the pseudo-spheres. This softens the variation of the pseudo-wavefunctions, which can be described with a smaller set of planewaves. This, in turn, helps achieve convergence faster. This approximation, of course, comes at a cost. As the norm is not conserved anymore, to preserve the charge neutrality and find the correct number of electrons in the entire system, the space needs to be split into two parts: the inside and the outside of the spheres around the atoms. The part of the valence wavefunctions that falls spatially inside the core region has to be 'artificially' augmented to retrieve the total number of electrons.

In contrast, the outer part is treated in the same standard way as in the case of norm-conserving pseudopotentials. But with this operation, the Hamiltonian has to be applied to two different regions and adjusted for the augmentation. Not every code can do that. The ultrasoft pseudopotentials were introduced in 1990 [256], and, assuming available implementations, they proved extremely efficient and fast [185]. Furthermore, they helped the development of a new flavour of DFT: the planar augmented wavefunction method, which we will discuss in the Section 3.6.

3.5 Beyond Standard Pseudopotential-Based DFT

Using pseudopotentials implies an inevitable loss of accuracy, as the core and sometimes semi-core states are frozen. Moreover, a certain amount of hardship has to be committed to achieving convergence. Consequently, considerable effort was put into finding better and faster computational solutions.

The all-electron methods were the first attempt to solve this problem. They solve the DFT equations for all the electrons in the system, including the core states. As such, they are computationally heavy and very impractical. But they are accurate, and even if they have seen limited use, they are important benchmarks to test against when using or when computing particular physical properties, which explicitly require the core electrons (like the electronic spectra at microprobe analysis).

To reduce the computational burden of the simulations, a first numerical simplification consisted of separating spherical regions around the atomic nuclei, similar to the approach described earlier. The potential was considered spherically symmetrical in this approximation, called the atomic sphere approximation (ASA) [225]. In more recent implementations, this constraint may be removed. Outside the spheres, the potential is flat. Inside the spheres, the DFT equations are solved numerically using a basis set mimicking atomic orbitals. The results are the exact core states, which are not frozen anymore. Outside the spheres, because the electron density is smooth, the DFT equations are relatively easy to solve using various approximant functions, like planewaves, Gaussians, or spherical waves. The connection between the two regions of the space is realised at the sphere boundaries, where the planewaves are adjusted to linear combinations of the wavefunctions inside the spheres. The potential, its derivative, the wavefunctions, and the derivatives of the wavefunctions must all be continuous at the boundaries of the spheres.

The planewaves approach later developed into the linear augmented plane wave (LAPW) method [192]. Of the all-electron methods, LAPW is probably the most in use. The LAPW method is often used as an ultimate reference to test the validity of standard DFT calculations and pseudopotentials [34]. The Wien2k code is a successful implementation of the LAPW method.

In the linear muffin-tin orbital (LMTO) method [219], the orbitals inside the atomic spheres are modelled using Hankel functions, which are modified spherical Bessel functions. They have the advantage of smooth character when compared to planewaves. The LMTO method takes its name from the shape of the potential that looks like a muffin tin.

3.6 Projector Augmented-Wave (PAW) Method

But the real breakthrough and the promising path for the future were realised when the LAPW formalism was combined with ultrasoft pseudopotentials. The result was the projector augmented wave (PAW) method [35]. PAW preserves the separation of space in two regions. In the region close to the atomic nuclei, the wavefunctions vary rapidly and are usually hard to converge. In the area far from the atomic nuclei, the wavefunctions are smooth and easily representable with planewaves. The two regions are delimited by spheres centred on the atomic nuclei. In the frozen core approximation, the core states are assumed to remain inside the PAW spheres and be orthogonal to the valence states [163]. That

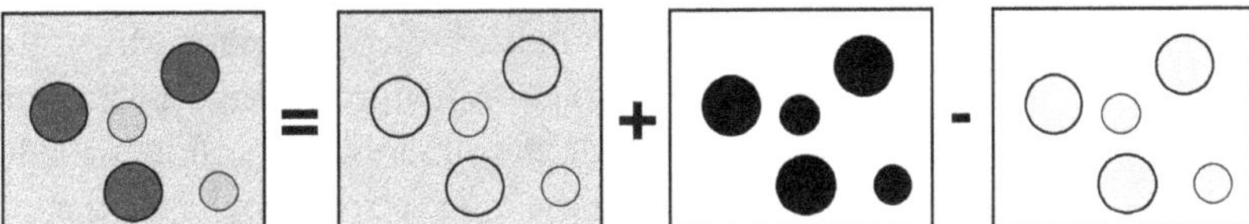

Fig. 3.6 Schematic representation of the PAW terms. From left to right, the total electron density is the sum of two terms: the planewave density computed in the entire space, the all-electron density inside the PAW spheres, from which the double counting of electrons inside the spheres has to be extracted.

means they need not be specifically calculated, and thus they are taken into account via pseudopotentials. For the rest, PAW deals only with valence electrons.

Inside the PAW spheres, the wavefunctions, and hence the density, are computed exactly using a linear combination of spherical harmonics [35, 163, 245], and not relying on the smooth planewave expansion. This is important because the electronic wavefunctions fluctuate highly when they are close to the nuclei. Reproducing these fluctuations correctly requires planewaves with higher frequency, meaning shorter wavelengths. This corresponds to larger $\mathbf{G}$ vectors in the reciprocal space. In extreme cases, there might not even be enough information provided by the planewaves expansion to reproduce the information stored in these fluctuations correctly. This implies that the smooth planewave expansion is not accurate enough to reproduce these fluctuations. Solving the wavefunctions with spherical harmonics close to the nuclei solves this problem. Moreover, because the solution is local, as the sum over harmonics is performed only inside the PAW spheres, it is readily obtainable during a calculation.

Outside the PAW spheres, the valence electron density is written and solved according to the standard planewave expansion already present in the case of norm-conserving pseudopotentials. By construction, this is the same as the all (valence) electron density outside the spheres. But because the fluctuations of the wavefunctions, and thus of the electronic density, are weaker far from the nuclei, they can be more easily modelled with a smaller number of planewaves.

So far, in the description of the wavefunctions, and thus of the electron density, we have two components: the all-electron smooth pseudo-wavefunction outside the PAW spheres and the all-valence-electron exact wavefunction inside the PAW spheres. But because the planewave expansion extends throughout the space, it also generates a smooth contribution of the pseudo-wavefunctions inside the PAW spheres. This contribution needs to be removed to conserve the norm, that is, the total number of electrons. Then this contribution is compensated by a third term of the expansion, which has the same value but the opposite sign as the smooth planewave contribution inside the PAW spheres. Hence the electronic wavefunctions, the electron density, the energy, and any other operator are described using these three components. Figure 3.6 shows a graphic representation of these three terms.

In practice, the connection between planewaves and the atomic wavefunctions is done using a transformation, a projection. This transformation is unitary outside the spheres. But inside the spheres, it projects the smooth density onto the all-electron one, hence the name *projector* of the method. This ensures the continuity of the wavefunctions and the electron density at the boundary of the PAW spheres. It is a constraint inherited from the previous LAPW method.

The all-electron valence states inside the PAW spheres are described on a radial grid that takes into account the angular dependence of the orbitals. When performing a PAW calculation, one needs to provide information about both the $\mathbf{k} + \mathbf{G}$ grid of planewaves, that is, the kinetic energy cutoff radius of the planewaves, and information about the size or density of the radial grid inside the PAW spheres.

There are several obvious advantages of using PAW over the other approaches. Because the pseudo-wavefunctions are smooth outside the augmentation sphere, PAW requires fewer planewaves, which is translated by a smaller kinetic energy cutoff radius than the norm-conserving pseudopotentials [163]. This makes the PAW method considerably faster. And because of the use of spherical harmonics inside the PAW spheres, the PAW method is just as precise as the all-electron LAPW method [143].

Because of its advantages, the PAW method was quickly embraced by the DFT community. After the year 2000, the developers of most DFT codes implemented the PAW method, not only for standard static and molecular-dynamics calculations [37, 105, 116] but also for dynamical properties, like phonons [13, 245, 78], electric charges, and even elastic tensors [177].

3.7   Quantum Monte Carlo

Complementary to DFT, alternative numerical techniques exist to find or explore solutions of the Schrödinger equation. One of the most used is the quantum Monte Carlo, which is based on the random sampling of the configurational space of the particles. The name of this method comes from the famous casino where randomness plays a central role [95, 69, 14]. Quantum Monte Carlo, and in general Monte Carlo methods, are based on performing enough trials until numerical convergence is achieved (Figure 3.8).

For example, let us say that you want to measure the surface of a circle with a radius of one without applying the analytical formula from geometry or that you want to determine the value of π. Then you can consider a square that is tangent to the circle. You randomly start to pick up points inside the square, and each time you measure the distance from the point to the centre of the circle and the centre of the square. If the distance is

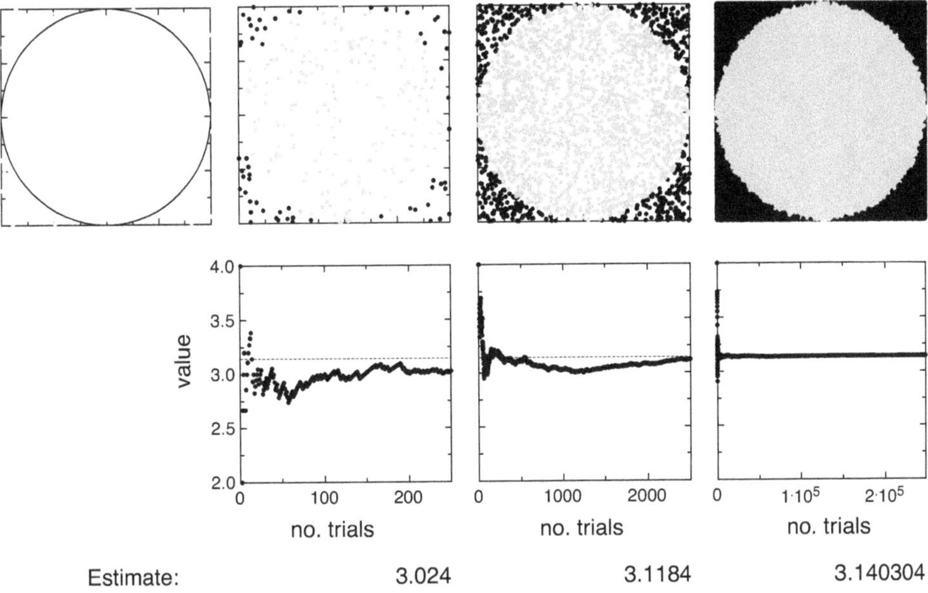

Fig. 3.7 Determination of the area of a circle with radius 1, or equivalently, approximating the value of π. Random points are chosen inside a square with a side of 2, tangent to the inner circle. Convergence is reached as more and more points are taken.

smaller than 1, the radius of the circle, then the point is counted towards the circle. If the distance is larger, then it is not counted for the circle. Figure 3.7 shows the sample of the circle and the improvement of the area estimate with the number of trials. Depending on the required precision, fewer or more trials are needed. Two-digit accuracy is obtained after 2.5 million trials. For 10 times more trials (not shown on the graph), the value is 3.1415136.

In QMC, every calculation yields the exact energy of one given configuration of particles. Then the energy of the entire system and the wavefunctions are obtained after many enough configurations are probed, as the weighted sum of all these trials:

$$E = \frac{\Psi(\mathbf{R})H\Psi(\mathbf{R})}{\Psi(\mathbf{R})\Psi(\mathbf{R})} = \frac{1}{N}\sum_{i=1}^{N} E_i \tag{3.17}$$

The number of needed configurations is typically in the order of millions, and just as in the case of the circle sampling, tests can ensure an asymptotic convergence.

When doing quantum Monte Carlo calculations, the key to success is the choice of sampling. In variational MC, the configurations are picked up randomly from the entire space of configurations. In diffusion Monte Carlo, one defines random walkers in the space of configurations that 'diffuse' from one configuration to the next. These are methods used for zero temperature calculations, and much work has been put so far into implementing and optimising the random walkers [95]. For simulations at high

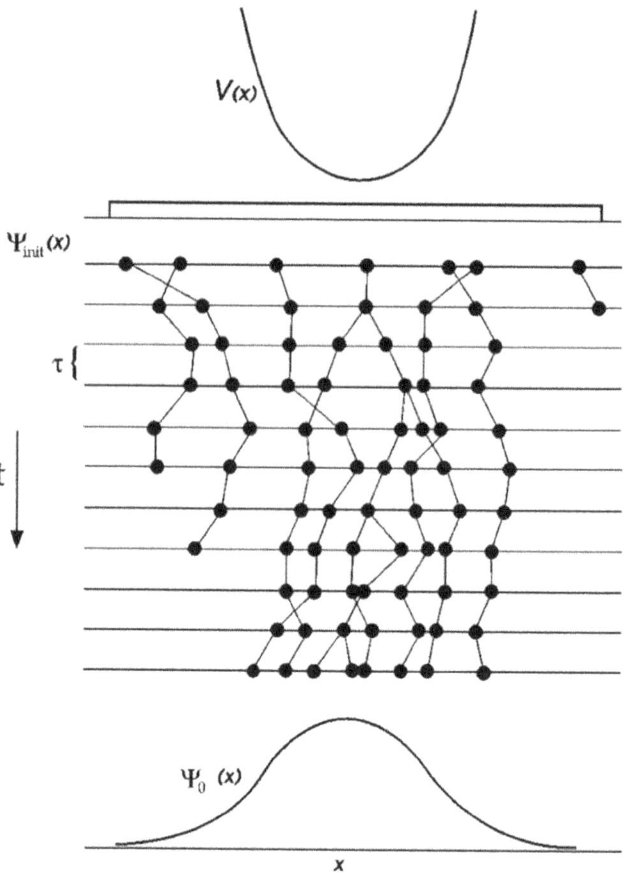

Fig. 3.8   The diffusion quantum Monte Carlo method is based on a series of walkers in the configurational space that starts with random probability from some initial state. As a function of their energy, quantum Monte Carlo makes the Metropolis algorithm propagate the steps these walkers take. Their energy is evaluated after each step; some walkers with lower energy continue their trajectory, while others with higher energy are stopped. After enough walkers and steps, the wavefunctions are properly sampled, and the energy is close to exact.

temperatures, several sets of connected states can be sampled simultaneously in what is called path-integral Monte Carlo [30]. This method is highly efficient for studying quantum effects, like describing H or He in hot planetary mixtures [84].

As the number of trials needs to be very large, quantum Monte Carlo is computationally very expensive. But because each trial is independent of all the others, this method is perfectly parallelisable. However, its cost remains so prohibitive that quantum Monte Carlo is seldom used in mineralogy or in any application on real materials. Usually, it is applied on relatively simple molecules or highly symmetric lattices with a small number of particles per unit cell. It often provides absolute reference numbers for the energy of such systems. Unfortunately, because of the discrete character of this random sampling, the energy derivatives do not have analytical solutions, or at least their development is at its beginning [16] and must be obtained by finite differences. This is a significant drawback as these methods can hardly be used to compute even relatively standard properties like interatomic forces or stresses.

Though a few recent developments try to address these problems, more and more applications will emerge in the following years [269]. And while the practical use of quantum Monte Carlo is still limited, it remains a powerful tool that can considerably improve the agreement between the standard DFT theoretical results and experiments [85].

Static Electronic Properties

As the main object of the DFT is the electronic density, its determination is provided by all the DFT codes. In this chapter, we analyse its components and see how we can extract and interpret a series of related physical properties.

4.1 Internal Electronic Energy

The Kohn–Sham equations consist of applying the Hamiltonian, H, to the set of electronic wavefunctions, ψ_i, in every point of the reciprocal space, $\mathbf{k}$:

$$\sum_j H\psi_j(k) = E_i(k)\psi_i(k) \tag{4.1}$$

Then the internal energy of the system is just the sum of all the energy eigenvalues of the wavefunctions in all the $\mathbf{k}$ points, weighted by their occupation $f_i(k)$ at each $\mathbf{k}$ point:

$$E = \sum_i \sum_k f_i(k) E_i(k) \tag{4.2}$$

The internal energy determined in DFT is the ground-state energy, computed at static conditions, that is, 0 K. According to its definition, it is characterised by the absence of any electronic excitation. All the electronic bands are fully occupied, with two electrons on each band, in increasing energy. Obtaining the value of the electronic energy is straightforward, as it results naturally from solving the Kohn–Sham equation. Box 4.1 shows the minimum required input by ABINIT to calculate the static internal energy for quartz at 0 K, using the experimental structure, in the projector augmented wave (PAW) flavour of DFT.

The DFT energy does *not* represent the absolute value of the energy. Because of the Kohn–Sham transformation of the Schrödinger equation, there is a constant but unknown energy shift. However, the relative values of the energy are correct. That means that the position of the electronic bands is correct relative to the Fermi level (see later) but not their absolute values. Similarly, the energies of the different structures do not have a physical meaning, but the differences between structures can be compared. This difference can be used to obtain the heat and the enthalpy of formation (Box 4.2).

Box 4.1	How to calculate the energy of quartz?

This is an example of the input cell used to obtain the energy of quartz in a static calculation at 0 K. The input parameters are kept to a minimum, though they ensure that a correct calculation is performed. Please note and pay special attention to the use of atomic units, like Bohr for distance, where 1 Bohr = 0.52917724 Å, and Hartree for energy, where 1 Ha = 27.2116 eV. The structure of the unit cell is described using the space group and the atoms listed in the reduced part of the unit cell.

		# parameters
		# k points options
kptopt	1	#use full symmetry
ngkpt	6 6 6	# size of the grid
nshiftk	1	# one (primitive) grid
shiftk	0.0 0.0 0.5	# most efficient shift of the grid
		# kinetic energy cut-offs , in Ha
ecut	17	# for planewaves
pawecutdg	40	# for wavefunctions in PAW spheres
toldfe	1.0d-09	# good convergence of the energy
		# crystal structure
angdeg	90.0 90.0 120.0	# hexagonal angles
xred		# atoms in reduced coordinates
	0.4727 0.0 1/6	# Si atom
	0.4151 0.2623 0.2904	# O atom
acell	9.2636 9.2636 10.2039	# unit cell parameters, in Bohr
natrd	2	# number of atoms in reduced cell
natom	9	# number of atoms in unit cell
ntypat	2	# number of types of atoms
typat	1 2	# atomic types
znucl	14 8	# atomic number of atoms, Si and O
spgroup	154	# space group number

4.2 Electronic Density

The calculation of the energy using the Kohn–Sham equations goes hand in hand with obtaining the electronic density, $n(r)$, which is the sum of the wavefunctions:

$$n(r) = <\psi_i(r)|\psi_i(r)> = \sum_i \psi_i(r)^* \psi_i(r) \tag{4.3}$$

Box 4.2 How to calculate the heat of formation from elements?

Because of the rewriting of the Schrödinger equations, the solution to the Kohn–Sham equations does not provide the absolute values of the internal energy, as the energies are shifted by an arbitrary amount. The energy differences between structures have full meaning. This gives us the possibility to compute the heat of formation as the difference between the internal energy of different structures and components. This allows us to estimate the direction of a reaction and the relative stability of various phases. As the components can be defined in various ways, this can also allow us to estimate the heat of formation from elements, which is illustrated below for quartz.

First, use Box 4.1 to obtain the internal energy of quartz. Then, using the same kinetic energy cut-offs, make a series of calculations in which you place a single atom of each type, first of Si, then of O, in the middle of a large cubic box of 15 Bohr on the side.

		# 1 Si atom
		# kinetic energy cut-offs , in Ha
ecut	17	# for planewaves
pawecutdg	40	# for wavefunctions inside PAW spheres
toldfe	1.0d-09	# good convergence of the energy
		# crystal structure
acell	3*15.0	# unit cell parameters, in Bohr
xred	1/2 1/2 1/2	# 1 atom in the middle of the cell
natom	1	# 1 single atom
ntypat	1	# 1 atomic types
typat	1	# 1 type of atoms
		# one calculation for each atomic type
ndtset	2	# two calculations in all
znucl1	14	# first, 1 Si atom
znucl2	8	# second, 1 O atom

The calculation in Box 4.1 gives a ground-state energy of −989.6 eV for one formula unit of SiO_2 in the quartz unit cell. The two calculations for the atomic ground-state energy give respectively −431.6 eV for one Si atom and −102.9 eV for one O atom, both for single atoms in large empty cubic boxes. The difference yields the energy of formation of SiO_2 of quartz from elements of −352.1 eV per formula unit. This energy, being calculated at null temperature, corresponds to the enthalpy of formation or the heat of formation relative to isolated atoms. Considering oxygen molecules instead of isolated atoms reduces the heat of formation to only −122 eV/formula unit.

The sum runs over all the electronic bands i. In Equation 4.3, the electronic density and the wavefunctions are written in the real space (r), while in the Kohn–Sham equation, they are written in the reciprocal space (k).

> **Box 4.3**   **How to calculate the electronic density?**
>
> The electronic density is intrinsically computed during the minimisation of the Kohn–Sham
> hamiltonian by all the DFT codes, though only sometimes explicitly written down in a separate
> file. The reason is to save I/O time and disk space. In abinit, if you want to save the electronic
> density and use it later, you need to activate the *prtden* keyword. Also, it might be a good idea
> to use a highly converged set of wavefunctions with a dense grid of **k** points, especially if you
> plan to reuse the density for other purposes. Beware that the density of the grid is material
> dependent, with considerably larger sets required for metals than insulators.
>
> ```
> toldfe      1.0d-09      # good convergence on energies in Ha
> prtden      1            # flag to print the charge density
> ```

The correspondence between the two spaces is quickly done using the
Fourier transformation. This construction allows for obtaining a well-
converged electronic density stemming from a finite and limited grid of
k-points in the reciprocal space, thus avoiding long, tedious calculations
in the real space. Box 4.3 explains how to compute the electronic density.

Equation 4.3 describes the total electron density. But the same procedure can be applied to a restricted subspace of the electronic bands. We
can perform the sum over a limited set of bands, whether they are well
separated from the others or not. We can also sum over the projections of
the total wavefunctions on specific states, like the s or p states. The result
is a partial electronic density corresponding to that set of electronic states.
This is particularly useful when trying to understand the participation of
various orbitals in the chemical bonding of minerals.

4.3 Static Atomic Charges

Once the electronic density is determined, one step further in the analysis
can yield the atomic charges. However, the actual problem resides in the
way of defining them. They can be defined as the difference between the
negative charge, that is, the number of electrons, and the positive charge,
that is, the number of protons, of a given atom. If the latter is straightforward to count, the former actually raises a fundamental problem. How
can the electrons be assigned to a given atom? Where is the limit of that
atom? How can we decide how much charge belongs to a certain atom?

Answering these apparently trivial questions is not obvious, and this
leads to various definitions of the atomic charge. These definitions are
based either on the analysis of the charge density or on the analysis of the
occupancy of atomic-like orbitals. These two categories of analysis yield

Box 4.4   **How to compute the Bader atomic charges?**

The main idea of the Bader approach to atomic charges and atomic volumes is that atoms are separated from each other by the minima of the charge density. These points correspond to the surface of zero charge flux. Finding the atomic volumes requires the determination of the derivatives of the charge density. Then one determines the saddle points and their intersections, the so-called critical points, which is a lengthy process. This is done on radial grids centred on the atomic centres. While this process is hard to parallelise, it has the advantage that all atomic volumes and charges can be determined at the end of the calculation. Finally, the core electrons (replaced in the calculation by the s) have to be specifically added to the final ground-state charge density using dedicated files. Once this is done, the actual Bader calculation with the *aim* tool of the abinit package requires the following variables with non-default values:

crit	2	# compute the critical points
surf	1	# construct the Bader surface
atom	1	# compute the Bader surfaces around 1 atom
irho	1	# flag for computing the Bader charge
ivol	1	# flag for computing the Bader volume
maxatd	7.0	# maximum distance to search for neighbouring atoms
maxcpd	5.0	# maximum distance to search for critical points
radstp	0.05	#radial step along the radius (=0.09 angstrom)
ratmin	1.89	# minimum atomic radius (=1 angstrom)
ntheta	32	# these two variables define the radial grid
nphi	48	# around the atoms to look for the critical points
		these values define a dense-enough grid for most cases

static charges, as they rely on the ground state of the atomic system at a given moment in time.

A very common approach to determining the atomic charges is to consider the limit of the atoms at the saddle points of the electronic density around every atom. This is a tedious way of delimiting atoms as it needs to compute the Laplacian of the density in all directions of the space on radial grids originating on each atom. But this method has the advantage of providing an obvious physical meaning of the resulting charges. The electronic density is highest closer to the atoms, becoming smaller and smaller away from the nuclei. In a certain place, it hits a minimum and then increases again while approaching the neighbouring atom. Hence the delimiting surface between the two atoms is defined by the minima of the electronic points, that is, the saddle points, along the interatomic bonds. These points do not need to lie in the middle of the interatomic bonds; generally, they lie closer to the atom with fewer electrons. At the intersection between three atoms, the saddle points are basin minima, corresponding to

Table 4.1 Bader atomic charges and volumes in stibnite, Sb_2S_3.

Atom	Sb1	Sb2	S1	S2	S3
Charge	1.12	1.11	−0.73	−0.69	−0.81
Volume ($\mathring{A}^3$)	22.78	21.63	26.06	26.37	25.09

Table 4.2 Bader atomic charges and volumes in jadeite, $NaAlSi2O_6$.

Atom	Na	Al	Si	O1	O2	O3
Charge	0.61	1.78	2.06	−1.08	−1.12	−1.07
Volume ($\mathring{A}^3$)	10.39	11.22	11.94	9.08	9.12	9.07

the points of zero charge flux. This method of delimiting atoms and counting their charges is called the Bader charges, after the chemist who invented it [15], sometimes also named the atoms-in-molecule approach (AIM).

Box 4.4 shows an example of how to obtain Bader atomic charges using the ABINIT extension program, *aim*. A few other stand-alone program packages perform the same analysis and can read the charge computed with other DFT packages. One example is the program provided by the Henkelman group (http://theory.cm.utexas.edu/henkelman/ code/bader/ [240]) that can read CHGCAR files with the charge density generated by VASP. Please note that the default for some codes, like ABINIT, is to have the electronic density written in a binary file. This has the clear advantage of rapidity, binary writing being always faster than ASCII writing, and of size, a binary file is smaller. But the binary files are machine-dependent and cannot be ported between the supercomputer and a personal computer. In this case, the binary file with the electronic density must be transformed into an ASCII file, which is independent of the operating system.

Regardless of the way the Bader charges are computed, the density of the core electrons needs to be added. In the computed valence electronic density, the inner electrons do not appear explicitly, and the atoms are actually represented as charged spheres devoid of electronic states – the *s*. ABINIT made available specific density files for all the atoms in the periodic table on the ABINIT website. The files are independent or only weakly dependent on the exchange-correlation functional.

Table 4.1 shows the Bader atomic charges and volumes of Sb and S atoms in stibnite, Sb_2S_3. The charges are fractional and deviate from the nominal values quite significantly. Sb values are around +1.1 compared to the +3 nominal value, and the S values are around −0.7 to −0.8 compared to the −2 nominal values. The deviation shows considerable covalent bonding, as values close to the nominal charges are specific to ionic systems. In all cases, integer values for the atomic charges are only found in chemical models of nominal values. There are, respectively, two and three symmetry-independent atomic sites for Sb and S, which explains why there are several values for each atomic type.

In quartz, SiO_2, the bonding between the Si and the O atoms has a covalent component. This is directly reflected in the Bader charges, which deviate from the nominal values. The Bader charge on the Si atom is about $+3.11$, and the Bader charge on the O atom is about -1.55. The Bader charges in jadeite, $NaAlSi_2O_6$ (Table 4.2), reflect the strong covalent bonding, which constitutes the entire backbone of the structure, the AlO_6 octahedra connected to the SiO_4 tetrahedra. They deviate from the nominal charges by almost two for Si.

Hirshfeld imagined a different way of computing atomic charges by approximating all atoms as spheres. The Hirshfeld charges [141] are obtained by subtracting the nuclear charge from the electronic charge inside these spheres. The radius of the spheres can be arbitrarily chosen and typically is larger than the cutoff radius. Choosing the ionic or covalent radius or the Wigner–Seitz radius are good choices. For example, the Hirshfeld charges in quartz are $+0.44$ on the Si atom and -0.22 on the O atom. There is a significant difference between the Bader and the Hirshfeld charges, a typical effect of the two analyses that stems from the different atomic limits.

In PAW, it is common practice to quickly estimate the charge based on the total electronic density inside spheres centred on atoms with the same radius as the PAW spheres. The values are typically listed as a table in the main output file resulting from the simulation. These values are also fractional. However, these values must be taken with extreme care, as they reflect only the amount of valence electrons inside the PAW spheres. For example, at ambient pressures, for quartz, where the Si and the O radii of the s, in the LDA formalism, are 0.77 Å and 0.59 Å, respectively, and the Si-O bond distance is on the order of 1.6 Å, the charges resulting from the integrated electronic densities inside the PAW spheres are underestimated. It is even better to name this 'the integrated amount of charge inside the sphere' rather than the actual 'charges'. For the Si atoms, the charge is 0.53 e^-, and for the O atoms, the charge is 3.33 e^-. Under pressure, as the Si-O bond decreases and the structure becomes more compact, the amount of integrated charge increases.

The static atomic charges, regardless of their different formalisms, are scalar values because they represent integrals of the electronic density over some regions of space. They are not directional, and thus they do not capture the changes in interatomic bonding along different bonds. The dynamical charges discussed later are tensorial properties, as they depend explicitly on the movement of the atoms and on their directional interaction with the neighbouring atoms.

Because the saddle points mark the limit of the atoms, it is then straightforward to calculate the volume of each atom. Volumes of the Sb and S atoms from stibnite are also listed in Table 4.1. The radii of an atomic sphere having the same volume as the Bader volume are about 1.76 and 1.73 Å for the two Sb atoms and 1.84, 1.85, and 1.82 Å for the three S atoms. These values can be used to approximate the Wigner–Seitz radii of the atoms. In the case of quartz, the Bader equivalent atomic radii are about 0.82 Å for the Si atoms and about 1.34 Å for the O atoms. But the atoms are not spherical, and the structure is neither a collection of hard spheres. This is clearly seen as the sum of the two Bader radii, at 2.15 Å, is larger than the Si-O bond distance at 1.61 Å, as measured geometrically in the quartz structure.

In the Bader analysis, the entire space is divided between the different atoms forming the crystal. The atoms are not spherical, but their shape is cut from the position of the attractors, that is, of the charge density. In this analysis, there is no overlap between the atoms. But in the case of the Hirshfeld analysis, the atoms can overlap because of the values of the radii of the Hirshfeld atoms, which are arbitrary.

The partitioning of the total molecular wave function onto atomic orbital-like functions may be used as an alternative to delimiting the spatial extent of atoms. The Mulliken charges stem from the overlap matrix between the total wavefunction and the atomic orbitals. This population analysis is widely used when the total electron density is obtained as a linear combination of molecular orbitals, like in many quantum chemistry codes. For plane-wave codes, the electron density must first be projected onto atomic-like functions before the Mulliken analysis can take place. This is a rather lengthy procedure, which is not common with mineral physicists.

4.4 Brillouin Zone

The electronic wavefunctions inherit the periodicity of the crystalline lattice. It is then enough to calculate the wavefunctions in one primitive unit cell and, from there, use translational symmetry to characterise the entire lattice. This requires the primitive unit cell. In the reciprocal space, the cell is centred on the Bragg positions and called the Brillouin zone. It is constructed using the same technique as the Wigner–Seitz cell of the direct space (Figure 4.1), described in Section 1.3. This cell was defined by Léon Brillouin in 1930 and bears his name. The Brillouin zone is the convex volume that contains the points in the reciprocal space closest to a Bragg position. It is delimited by the planes perpendicular to the middle of the segments joining the Bragg point at its origin with all the translationally equivalent Bragg points. Because of this construction, which relies heavily on translations, the shape of the Brillouin zone is determined not only by the size of the lattice but also by the type of Bravais lattice and the point group.

The concept of the Brillouin zone can be extended. Instead of considering the closest bisector planes to the segments between Bragg peaks, we can consider the second closest ones. This construction yields the

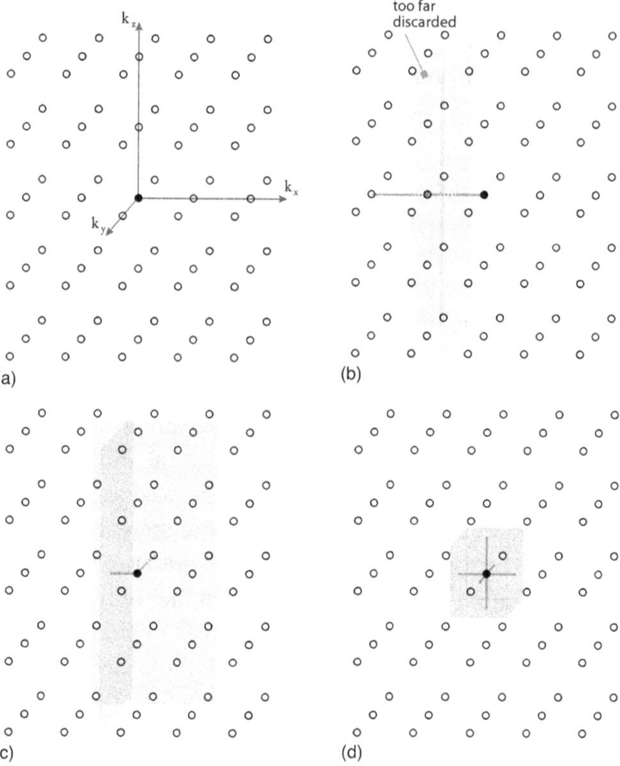

Fig. 4.1 The construction of the Brillouin zone in the reciprocal space is done in the same way as that of the Wigner–Seitz cell of the direct space. We start with one Bragg point of the reciprocal space, which becomes the origin of our construction (a). We draw line segments between this origin point and all of its neighbours. We construct the perpendicular planes passing at the middle distance on these segments (b,c). We consider the closest planes to the origin point, all the other planes being discarded. Thus, we delimit the smallest convex volume around the origin (d). This is the first Brillouin zone.

second Brillouin zone. We can also consider the third closest bisector planes, in which case we obtain the third Brillouin zone, and so on. These higher-order Brillouin zones can be useful when studying particular phase transitions involving electron-phonon coupling, for example, in alkali metals under extreme pressure [167]. In this case, the phase transitions usually involve multiplying the unit cell in the direct space due to collective distortions. The result is obtaining commensurate superstructures. In the following, we will refer only to the first Brillouin zone unless otherwise specified.

The centre of the Brillouin zone is called the Γ point, and its coordinates are [0 0 0]. Most of the high-symmetry points inside the Brillouin zone have specific labels, while the other points have a generic k notation. Figure 4.3 exemplifies the Brillouin zones for a few of the 14 Bravais lattices. The description of all the Brillouin zones is detailed in a few reference textbooks (e.g., [42], and, nowadays, online, for example, at the Bilbao Crystallographic Server, using the KVEC software (at www.cryst.ehu.es/cryst/get_kvec.html)). This package constructs on-the-fly the Brillouin zone for all the space groups. In all these options, all the high-symmetry points and lines of the Brillouin zone are labelled.

The symmetry of the centre of the Brillouin zone, that is, the Bragg peak, always preserves the symmetry of the point group. But the reciprocal space also contains the inversion symmetry. This supplementary symmetry stems from the time-reversal symmetry of the wavefunctions. The wavefunctions expressed in Γ are always real numbers, as their phase is 0. The other points of the Brillouin zone respect only part of the symmetry operations of the point group. This behaviour is similar to the different crystallographic sites of the direct space. Each of the symmetry operations of the respective point of the Brillouin zone acts on the wavefunctions and transforms them. How the wavefunctions respond to these transformations determines their character. All the possible responses of the wavefunctions can then be classified, and the resulting signatures form the ensemble of irreducible representations.

As the symmetry operations are applied to the entire volume of the Brillouin zone, different points, edges, faces, and volume fractions can be related to each other. This is a relief as fewer calculations are needed to cover the entire Brillouin zone. The smallest volume, which is eventually left invariant by all the symmetry operations, is called the irreducible part. The electronic properties are calculated, described, and often represented graphically only inside the irreducible part, as all the other space is obtained by symmetry from this part. Figure 4.2 exemplifies the symmetry relations inside the Brillouin zone of a primitive orthorhombic lattice.

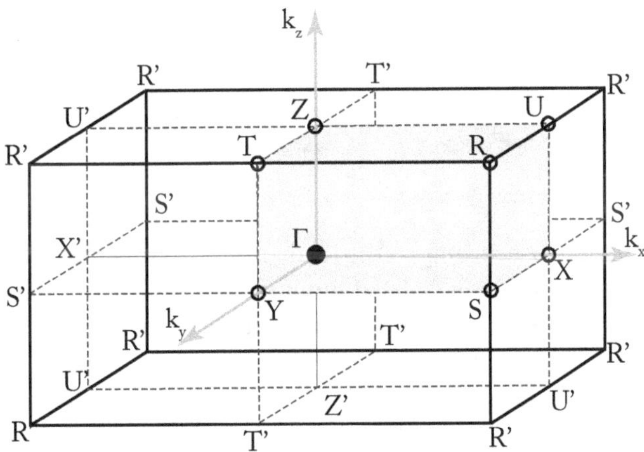

Fig. 4.2 The symmetry relations inside the first Brillouin zone of the primitive mmm orthorhombic lattice. The three symmetry planes, represented with dashed lines combined with the time-reversal symmetry characteristic of the reciprocal space, ensure that the entire Brillouin zone can be obtained from only 1/8th of its volume. This grey-shadowed volume fraction is called the irreducible part of the Brillouin zone. Different high-symmetry points are labelled. The points in the irreducible part are labelled with capital letters, and their equivalents by symmetry bear an extra ' to outline the relations.

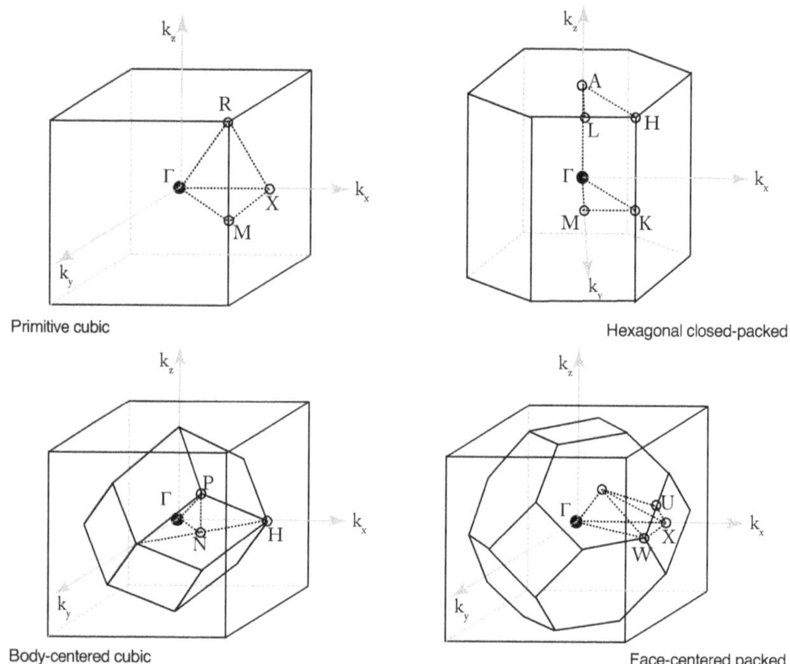

Fig. 4.3 Several examples of the most common high-symmetry Brillouin zones. The Brillouin zone of a primitive cubic lattice is a cube, and of a hexagonal lattice is a hexagon. But the first Brillouin zone of a face-centred cubic crystal (fcc) in the reciprocal cell is the same as the Wigner–Seitz cell of a body-centred cubic (bcc) in the direct space, and vice versa. The high-symmetry points of the asymmetric part of the Brillouin zones are labelled with capital letters. For simple cubic: $X = 1/2\,0\,0$; $M = 1/2\,1/2\,0$; $R = 1/2\,1/2\,1/2$. BCC: $H = 1/2\,0\,0$; $N = 1/4\,1/4\,0$; $P = 1/4\,1/4\,1/4$. FCC: $X = 1/2\,0\,0$; $W = 1/2\,1/4\,0$; $K = 3/8\,3/8\,0$; $L = 1/2\,1/2\,1/2$; $U = 1/2\,1/8\,1/8$. HCP: $M = 1/2\,0\,0$; $K = 2/3\,1/3\,0$; $A = 0\,0\,1/2$; $L = 1/2\,0\,1/2$; $H = 2/3\,1/3\,1/2$.

4.5 Electronic Band Structure

According to the Rutherford–Bohr model, postulated already in 1913, for a single isolated atom, its electrons sit in pairs with opposite spins on a series of atomic orbitals at discrete, energetic levels. The exclusion principle, elaborated in 1925, states that there cannot be two electrons with the same combination of quantum numbers in an atomic system, either an isolated atom or a collection of atoms. This was formulated back in 1925 by Wolfgang Pauli as a postulate or a principle, not as a theorem, meaning it is not proven but rather observed systematically.

The electronic energy of the static atom is the sum of these levels multiplied by their occupancy (Eq. 4.2). The low-lying orbitals, fully occupied and closer to the core, participate most in this energy. The external non-occupied orbitals bring no contribution to the ground-state energy because of their null occupancy. But they are important when and if electrons are excited, for example, at high temperatures or under the action of strong external electromagnetic fields. In these situations, the outer electrons gain energy and need to occupy a higher energy place in the system.

For a cluster of two atoms, the orbitals become doubly *degenerate*, meaning that their energies, initially at the same levels, now split into two values. The lowest energy level has a bonding character, as the electrons prefer this level. The highest energy level has an anti-bonding character, as its occupancy increases the total energy of the system. For small clusters and diatomic molecules, there is only a small energy separation between these levels.

Enlarging the size of the atomic cluster by bringing in another atom results in the orbitals splitting into three distinct but weakly separated orbitals. As this process continues with more and more atoms, the bands become more and more degenerate and fan out. In the limit of a large or infinite number of atoms, like is the case for a crystal lattice, this splitting and degeneracy of the orbitals are also pushed to an extreme. The individual atomic orbitals cannot be distinguished anymore. Instead, there are regions that represent the electronic bands. They lie at specific energy levels and are typically separated by energy gaps, where states are forbidden. They confer a discrete character to the electronic properties of a crystal (Figure 4.4).

The electronic wavefunctions can be described as a product between a function with the periodicity of the lattice and a planewave whose wavevector **k** lies inside the first Brillouin zone:

$$\psi_{j,k}(\mathbf{r}) = u_j(\mathbf{r})e^{i\mathbf{kr}} \tag{4.4}$$

This is known as the Bloch theorem. Because the $u_j(\mathbf{r})$ is periodic with periodicity **R** of the direct lattice:

$$u_j(\mathbf{r}) = u_j(\mathbf{r} + \mathbf{R}) \tag{4.5}$$

It can also be described as a Fourier series anchored on the **G** vectors of the corresponding reciprocal lattice, defined as $\mathbf{R} \cdot \mathbf{G} = 2\pi n$, with n as an integer:

$$u_j(\mathbf{r}) = \sum_{\mathbf{G}} c_{j,\mathbf{G}}e^{i\mathbf{G}r} \tag{4.6}$$

When all these pieces are put together, the wavefunction can be expressed as a weighted sum of planewaves, anchored on the **G** and **k** points:

$$\psi_{j,k}(\mathbf{r}) = \sum_{\mathbf{G}} c_{j,\mathbf{k}+\mathbf{G}}e^{i\mathbf{k}+\mathbf{G}r} \tag{4.7}$$

Equation 4.7 gives the theoretical basis for sampling the reciprocal space with grids of **k**-points. Extending this concept, the value of the **k**-point can vary independently of the symmetry. The possible variations are restricted to the inside of the boundaries of the Brillouin zone, as this represents the smallest volume of the reciprocal space that is translationally invariant. Then removing, by convention, the 2π factor, the **k**-space

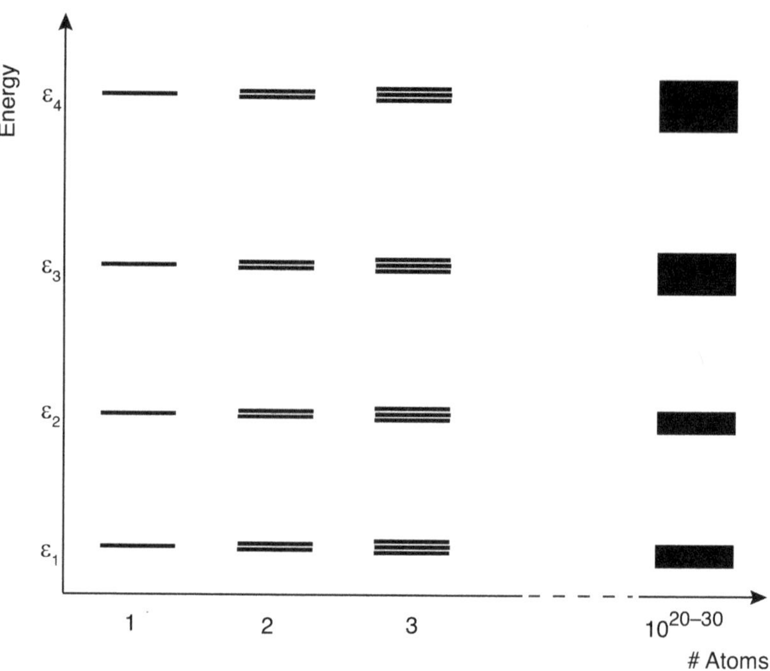

Fig. 4.4 The electronic levels (orbitals) are discrete in isolated atoms and small isolated clusters. But as the size of the cluster increases, the orbitals become degenerate. At macroscopic-sized crystals, the possible electronic levels fan out, and there are regions of energy, wide bands, where the former orbitals are placed.

spans the range $[-1/2, 1/2]$. As the k value varies inside this range, there is a dependence of the energy of the electronic bands with the planewave wavevector, $\mathbf{k}$. This dependence is at the origin of the dispersion of the electronic bands, j. The dispersion is represented in diagrams following certain, typically high-symmetry paths in the irreducible part of the first Brillouin zone. For example, for the simple cubic symmetry, such a path can be Γ - X - M - R - Γ. This path can be followed in Figure 4.3. For the face-centred cubic symmetry, such a path could be Γ - X - W - Γ - L - U. But there is no rule for where to start or end such a dispersion path. The last example can be rewritten as L - Γ - W - X - U - L - X - Γ, or reduced to the most interesting parts, L - Γ - W - X.

Once the high-symmetry path is chosen, the energy of the electronic bands can be represented at each discrete point along this path. This represents the dispersion diagram or the electronic band structure. When making the actual diagrams, it is common to represent all the valence bands as well as the first conduction bands. Because of the Pauli principle, from the number of valence electrons, it is straightforward to calculate the minimum number of bands needed. The dispersion diagram shows the variation of the energy with respect to the wavevector, regardless of the actual occupancy of the band. Figure 4.5 shows the electronic band structure of pyrope, the pure magnesium end-member term of the garnet series, $Mg_3Al_2Si_3O_{12}$. The path covers the high-symmetry points and the

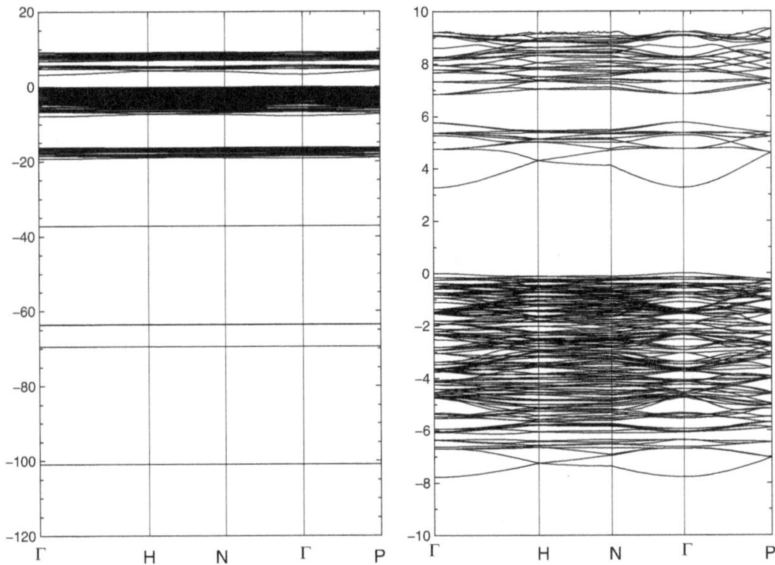

Fig. 4.5 Electronic band structure along a high-symmetry path through the Brillouin zone of pyrope garnet, $Mg_3Al_2Si_3O_{12}$. All the electronic bands arrive in Γ asymptotically horizontal. All the bands are continuous. The path through the Brillouin zone follows the high-symmetry points labelled in Figure 4.2.

high-symmetry lines of the body-cubic-centred symmetry. For pyrope, the chosen path was Γ - N - H - Γ - P. The relevant variables of the input file are shown in Box 4.5.

There are several topological rules concerning the dispersion diagrams [190]. For example, all bands are continuous and smooth; there cannot be any discontinuities. Because the reciprocal space contains the time-reversal symmetry, the wavefunctions are symmetric with respect to the zone centre, that is, $\psi_j(k) = \psi_j(-k)$ and their corresponding eigenvalues $\epsilon_{j,k}(\mathbf{r}) = \epsilon_{j,-k}(\mathbf{r})$. For the dispersion diagram, this rule implies that all electronic bands must arrive in Γ asymptotically horizontal. The same is true for the middle of all the segments relating to the centres of any two neighbouring Brillouin zones. For example, the bands approaching the X point (1/2 0 0), which lies at the Brillouin zone border, midway between the (0 0 0) and the (1 0 0) Bragg points, also need to arrive asymptotically horizontal because $\psi_j(k = 1/2 - \epsilon) = \psi_j(-k = -1/2 + \epsilon)$. The same takes place in the middle of the faces and of the edges [41].

Because of symmetry, certain bands can be degenerate in some high-symmetry points or along high-symmetry directions of the Brillouin zone, that is, they have the same energy. For example, the three p states in highly symmetric systems can be triply degenerate. The d orbitals of the transitional metals have all the same energy in a vacuum or in a uniform field. They split in the presence of a symmetric crystal field into two packets with twofold and threefold degeneracy, namely the 2 e_g and the 3 t_g electronic

 How to calculate the electronic band structure?

The dispersion of the electronic bands, which forms the electronic band structure, is usually calculated along specific directions of the reciprocal space. Unless for specific reasons, like charge-density-wave transitions, these directions are commonly chosen to be the high-symmetry directions, passing through the high-symmetry points. Because there are many points to plot along the different dispersion lines, and because they do not cover uniformly the Brillouin zone, the most efficient way to compute the electronic band structure is to do it in two steps: First, we compute the standard electronic density and the electronic wave functions on a dense regular grid of **k** points self-consistently (dataset 1 in the table below), and then we utilise these as a start to determine the wave functions and their eigenvalues non-self-consistently (dataset 2 in the table below). The example for the high-symmetry path through the Brillouin zone of a body-centred cubic lattice along the Γ - H - N - Γ - P path is shown in the table below. This input was used to obtain the electronic band structure of pyrope, shown in Figure 4.5.

ndtset	2	# two steps calculation
		# ground-state electronic density
toldfe1	1.0d-09	# good convergence on energies in Ha/Bohr
prtden1	1	# flag to print the charge density
kptopt1	1	# use the symmetry to reduce the number
		of **k** points actually calculated
		# remember that the grid of **k**-points is
		system-dependent and needs to be converged
iscf2	0	# non-self-consistent calculation
tolwf2	1.0d-14	# convergence criterion on the wave functions
prtden2	0	# do not print electronic density
kptopt2	-4	# do not use symmetry for the **k**-points
ndivk2	100 71 71 85	#
kptbounds2		# set the limits of the segments
	0.0 0.0 0.0	# of the path through the
	0.5 0.0 0.0	# Brillouin zone
	0.25 0.25 0.0	# these are high-symmetry points
	0.0 0.0 0.0	
	0.25 0.25 0.25	

states. In distorted environments, they can be further split down into individual bands and even individual polarised bands. These are detailed in Chapter 5.

Projecting the electronic band structure on the energy axis yields the population of electrons at a given energy or the electronic density of states (DOS). This represents the number of states as a function of energy. There are several numerical ways to perform this projection. First, a very dense

```
#
# ABINIT package : DOS file
#
# nsppol = 1, nkpt =  80, nband(1)= 122
# Smearing technique, occopt = 7, tsmear= 0.010 Hartree, tphysel= 0.000 Hartree
# For identification : eigen(1:3)= –2.924 –2.924 –2.924
#
# Fermi energy :    0.26690745
#
# The DOS (in electrons/Hartree/cell) and integrated DOS (in electrons/cell),
# as well as the DOS with tsmear halved and doubled, are computed,
# at 3601 energies (in Hartree) covering the interval
# between  –3.0000 and   0.6000 Hartree by steps of 0.00100 Hartree.
#
#    energy    DOS    Integr. DOS    DOS      DOS
#                        (tsmear/2)  (tsmear*2)
  –3.000   0.000000   0.000000    0.000    0.000
  –2.999   0.000000   0.000000    0.000    0.000
  –2.998   0.000000   0.000000    0.000    0.000
  –2.997   0.000000   0.000000    0.000    0.000


#
# ABINIT package : DOS file
#
# nsppol = 1, nkpt =  140, nband(1)= 144
# Tetrahedron method
# For identification : eigen(1:3)= –2.513 –2.513 –2.513
#
# Fermi energy :    0.13851273
#
# The DOS (in electrons/Hartree/cell) and integrated DOS (in electrons/cell) are computed,
# at 6801 energies (in Hartree) covering the interval
# between  -2.6000 and   0.8000 Hartree by steps of 0.00050 Hartree.
#
# energy(Ha)   DOS  integrated DOS
  –2.60000   0.0000   0.0000
  –2.59950   0.0000   0.0000
  –2.59900   0.0000   0.0000
  –2.59850   0.0000   0.0000


#
# ABINIT package : DOS file
#
# nsppol = 1, nkpt =  236, nband(1)= 144
# Tetrahedron method
# For identification : eigen(1:3)= –2.924 –2.924 –2.923
#
# Fermi energy :    0.26428784
#
# The local DOS (in electrons/Hartree for one atomic sphere)
# and integrated local DOS (in electrons for one atomic sphere) are computed.
# at 7401 energies (in Hartree) covering the interval
# between  –3.0000 and   0.7000 Hartree by steps of 0.00050 Hartree.
#
# Local DOS (columns 2–6) and integrated local DOS (columns 7–11),
# for atom number iat =  1  iatom =  1
# inside sphere of radius ratsph=   1.901579E+00 Bohr.
#
# energy(Ha) l=0   l=1   l=2   l=3   l=4  (integral=>) l=0  l=1  l=2  l=3  l=4
  –3.00000   0.0000   0.0000   0.0000   0.0000   0.0000      0.00  0.00  0.00  0.00  0.00
  –2.99950   0.0000   0.0000   0.0000   0.0000   0.0000      0.00  0.00  0.00  0.00  0.00
  –2.99900   0.0000   0.0000   0.0000   0.0000   0.0000      0.00  0.00  0.00  0.00  0.00
  –2.99850   0.0000   0.0000   0.0000   0.0000   0.0000      0.00  0.00  0.00  0.00  0.00
  –2.99800   0.0000   0.0000   0.0000   0.0000   0.0000      0.00  0.00  0.00  0.00  0.00
```

Fig. 4.6   Examples of the beginning of various DOS files that can be printed by ABINIT. While with other codes, the format can be different, they typically all follow the same pattern. Here the header contains the Fermi energy (note that in abinit, it is in Hartree!), followed by a table where the columns give the density of states and the integral.

grid of **k** points is needed. But even on a dense grid, a simple counting of the states might yield a spiky pattern, which is obviously unrealistic.

A Gaussian interpolation is a good first option, available in most codes, as it is straightforward to implement. The electronic states are computed on the grid of **k** points and summed up as a function of energy. A standard smoothing function with the width of the smearing parameter (*tsmear* in abinit) removes the unnatural spikes. The value of the smearing can be adjusted for the electronic DOS to best replicate the dispersion of the electronic bands. Figure 4.7 shows the electronic DOS for pyrope, with several default options of the smearing parameter. While the smearing reflects the temperature, the physical temperature (*tphysel* in abinit) can be used when adopting the Fermi distribution of electrons around the Fermi level.

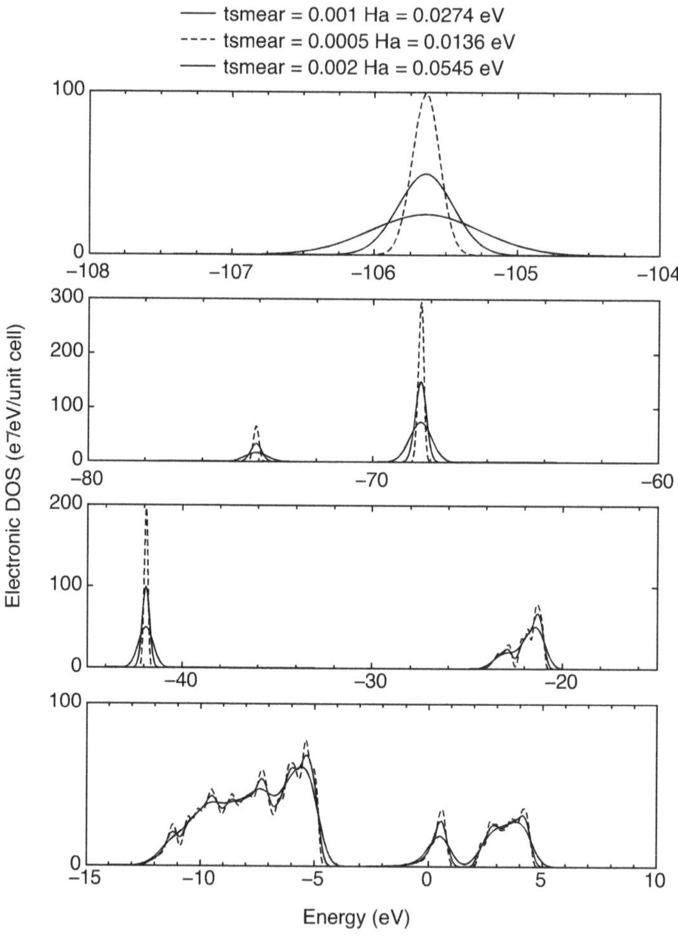

Fig. 4.7 The effect of the smearing parameter, *tsmear*, on the electronic density of states of pyrope. The default value of 0.001 Ha (=27.211 meV) gives a more natural image, the halved value yields a spiky pattern, while the doubled value removes many details and features of the spectrum.

Another more elegant and realistic way to compute the DOS is to use the tetrahedron method [36]. According to this method, the Brillouin zone is covered by a dense grid of **k**-points. The smallest convex volumes that can be constructed with such a grid are tetrahedra joining the closest four points to each other. The resulting tetrahedra cover the entire space and do not have any superposition between them. The number of states is estimated inside each tetrahedron by interpolation. The resulting sum yields directly the DOS without requiring smearing. Provided the grid of **k** points is dense enough, this sum can yield a smooth DOS. The tetrahedron method, devoid of smearing, offers a more exact image of the details of the DOS.

Figure 4.8 compares the electronic DOS of pentlandite, Fe_9S_8, which was calculated with the two methods, Gaussian interpolation and tetrahedron. Box 4.6 gives the computational details for obtaining the electronic DOS in the two cases. The smearing method provides a much smoother DOS, though it can miss some details of the electronic band structure. The tetrahedron method provides a more detailed DOS, whose spikiness can be exaggerated by the insufficient sampling with **k** points.

The electronic DOS needs to be converged with respect to the density of the grid of **k** points. The denser the grid, the better and more accurate the resulting DOS. This is even more stringent for the tetrahedron method. For minerals with large unit cells, like pentlandite, the Brillouin zone is small, so a reasonable grid might yield acceptable results. Figure 4.9 shows that for pentlandite, the electronic DOS is converged for grids of $8 \times 8 \times 8$ k points. Even at grids of $6 \times 6 \times 6$, there are some issues around the Fermi level. But for minerals with few atoms in the unit cell, that is, large Brillouin zone, the results are even more sensitive to the density of the grid. The necessity for dense grids is accentuated in the case of metallic minerals or small-gap semiconductors. Figure 4.10 exemplifies the electronic DOS in calcite, $CaCO_3$, computed with various grids of **k** points. The electronic DOS is already converged for $6 \times 6 \times 6$ grids, with the exception of the last set of peaks, at the highest energy, which, being non-occupied, usually requires more **k** points.

Oftentimes the electronic DOS is characterised by the existence of several peaks separated by energy gaps. The integral over each such peak gives the number of electrons populating that packet of electronic bands (Figure 4.9). Each time, this is, of course, double the number of bands in that packet. The exception is the last packet of bands below the Fermi level for metals, where the bands may have fractional occupation. Each packet of bands constitutes subspaces of the electronic bands corresponding to a certain symmetry and a certain combination of states. For example, s states of cations and/or anions, weakly hybridised d states of transition metals, p-p hybrids of elements from the p-group, and so on. Further examples of electronic band structures and their corresponding electronic DOS are shown in the various figures of this chapter. The examples show several

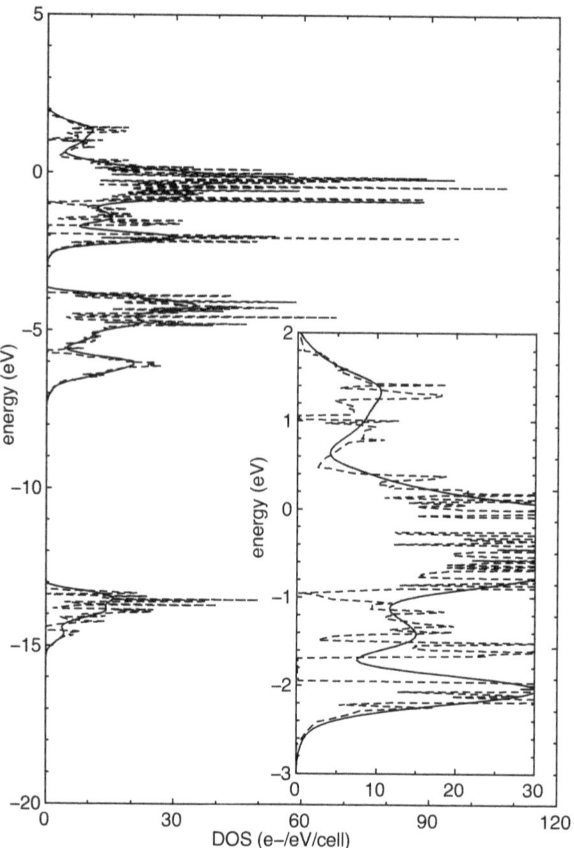

Fig. 4.8 The density of states of pentlandite, Fe_9S_8, computed using the smearing method (solid line), with prtdos 1, and with the tetrahedron method (dashed line), with prtdos 2. There are considerably more details seen with the tetrahedron method. Moreover, the absence of an arbitrary smearing parameter makes the tetrahedron better reflect the electronic band structure.

packets of bands, the position of the valence bands, and the electronic gap between the valence and the conduction bands. They were obtained similarly to the example in Box 4.5 and are all discussed in detail in the following pages.

It might be useful sometimes to perform a decomposition of the total electronic DOS in contributions of the different orbitals present in the system. This allows for mapping the hybridisation that contributes to the formation of the chemical bonds, as well as the ionisation of various orbitals. For norm-conserving pseudopotentials, this task is quite complicated because these pseudopotentials do not carry an easy representation of the orbitals. Total wavefunctions need to be projected on the different orbitals, but this is not straightforward. The widespread use of PAW allowed us to easily solve this problem. It is easy to project the total wavefunction on the different states inside the PAW spheres because of the construction of the PAW itself.

Box 4.6   **How to calculate the electronic density of states?**

In abinit, there are two major ways of computing the electronic density of states (DOS): Gaussian interpolation (prtdos 1) and the tetrahedron method (prtdos 2). In the first case, you may perform the first calculation on a regular grid of **k** points with the standard occupancy (occopt 1 for insulators and 3 or 4 for metals). In a second step, use the electronic density and wavefunctions and compute the DOS on a much finer grid of **k** points, using metallic smearing with occopt 7. If you use the tetrahedron method, no smearing is needed and the density of states is computed directly from the grid of **k** points; hence, you need to use a dense grid, and perhaps even be sure that your density of states is converged with respect to the grid (see also Figures 4.9 and 4.10). The following example illustrated the computation of the density of states of an insulator:

```
ndtset      3          #
toldfe      1.0d-09    # good convergence on energies in Ha/Bohr
kptopt      1          # use the symmetry to reduce the number
                         of k points actually calculated
ngkpt1      6 6 6      # regular dense grid of k points
prtden      1          # flag to print the charge density
occopt1     1          # standard insulator, like magnesia or forsterite
prtdos2     1          # print the density of states
                       # using a Gaussian smearing scheme
getden2     1          # helps with convergence
ngkpt2      12 12 12   # dense grid of k-points
occopt2     7          # use a metallic-like occupancy to ensure smearing
prtdos3     2          # print the density of states
ngkpt3      12 12 12   # dense grid of k-points
getden3     1          # helps with convergence
```

The result is printed in a _DOS file where the header contains the Fermi energy (note that in abinit it is in Hartree!), followed by a table where the columns are in order: the energy (in Hartree), the total density of states, the integral, the halved density of states, and the double of it. The last two columns just show two different possible smearing values. Before choosing the one to plot, the resulting smeared density of stats must correctly reflect the dispersion of the electronic bands. The two are frequently plotted close to each other, with the energy along the vertical axis. For the tetrahedron method, with prtdos 2, the output _DOS file contains only the total density of states without additional smearing.

Box 4.8 exemplifies the computation of the partial DOS on the Mg, Fe, Si, and O atoms of an intermediate member of the olivine series, $(Mg_{0.5}Fe_{0.5})_2SiO_4$. Figure 4.11 shows the total DOS and the projections on the individual orbitals of the four atomic types.

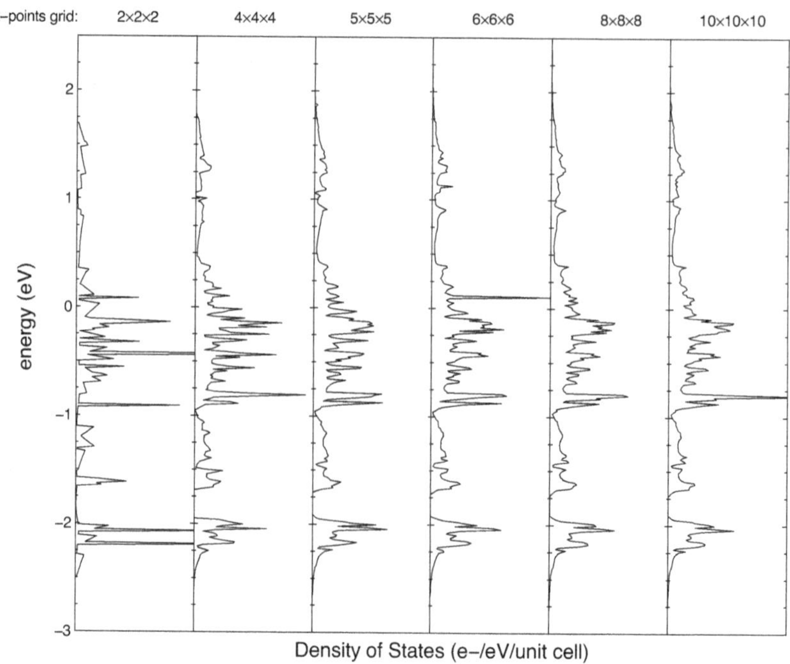

Fig. 4.9 Convergence of the density of states of pentlandite, Fe_9S_8, computed using the tetrahedron method (dashed line), with prtdos 2. Each panel shows a different grid of **k** points. While the general picture is revealed even for coarse grids, only starting with a grid of $8 \times 8 \times 8$ the details are well converged.

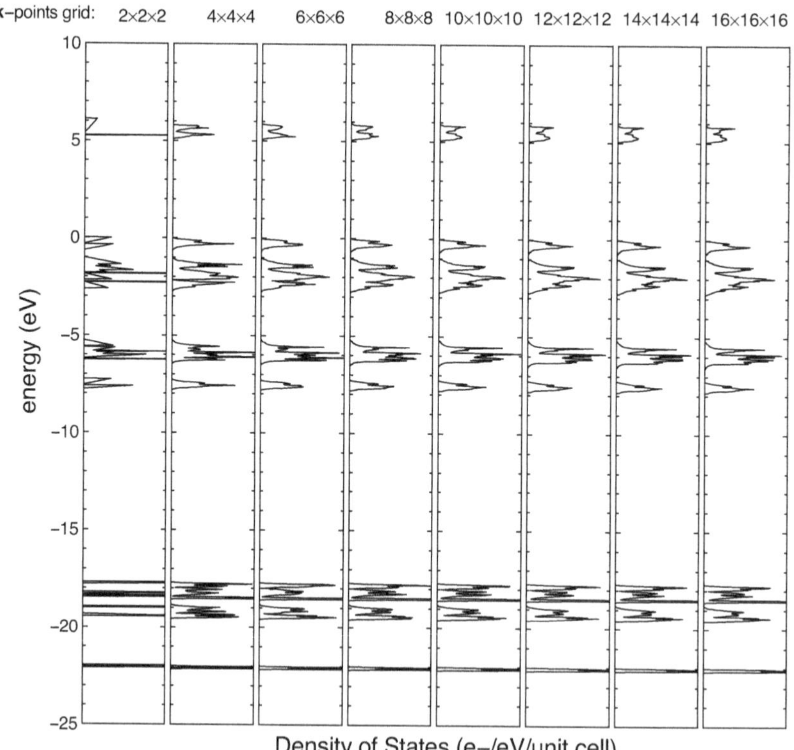

Fig. 4.10 The density of states of calcite, $CaCO_3$, computed with the tetrahedron method, reveals a wide gap insulator. Convergence is achieved at $6 \times 6 \times 6$ grids for the valence bands. The first peak of the conduction bands requires a denser grid.

Box 4.7   **How to automatise the electronic DOS calculation in solid solutions?**

The total electronic density of states can be represented as a function of Fe content in olivine. While in real samples, there is a large amount of order disorder on the two cation sites of olivine, the M1 and M2, here we assume ideal ordered structures. For a high-quality publication, the effect of Mg-Fe distribution in the structure needs to be specifically considered.

acell	4.749 10.1985 5.9792	angstroms
rprim	1 0 0 0 1 0 0 0 1	# can be omitted
natom	28	# Z=4 for olivine
ntypat	4	# keep four atomic types
znucl	12 26 14 8	# even for the end-members
ndtset	9	# in each dataset 1 more Mg
		# is replaced by 1 Fe atom
typat1	1 1 1 1 1 1 1 1 4*3 16*4	# $(Mg)^{M1}(Mg)^{M2}SiO_4$
typat2	2 1 1 1 1 1 1 1 4*3 16*4	# $(Mg_{0.75}Fe_{0.25})^{M1}(Mg)^{M2}SiO_4$
typat3	2 2 1 1 1 1 1 1 4*3 16*4	# $(Mg_{0.5}Fe_{0.5})^{M1}(Mg)^{M2}SiO_4$
typat4	2 2 2 1 1 1 1 1 4*3 16*4	# $(Mg_{0.25}Fe_{0.75})^{M1}(Mg)^{M2}SiO_4$
typat5	2 2 2 2 1 1 1 1 4*3 16*4	# $(Fe)^{M1}(Mg)^{M2}SiO_4$
typat6	2 2 2 2 2 1 1 1 4*3 16*4	# $(Fe)^{M1}(Mg_{0.75}Fe_{0.25})^{M2}SiO_4$
typat7	2 2 2 2 2 2 1 1 4*3 16*4	# $(Fe)^{M1}(Mg_{0.5}Fe_{0.5})^{M2}SiO_4$
typat8	2 2 2 2 2 2 2 1 4*3 16*4	# $(Fe)^{M1}(Mg_{0.25}Fe_{0.75})^{M2}SiO_4$
typat9	2 2 2 2 2 2 2 2 4*3 16*4	# $(Fe)^{M1}(Fe)^{M2}SiO_4$
xred	0.0 0.0 0.0	
	0.9913 0.2773 0.25	
	....	
	0.7774 0.3370 0.5329	
ecut	18	
pawecutdg	64	
ngkpt	13 7 9	# use a very dense grid
kptopt	1	
nshiftk	1	
shiftk	0 0 0	
toldfe	1.0d-07	# check convergence
prtdos	2	# tetrahedron method
occopt	4	# treat olivine as a metal because
		# the presence of Fe
nstep	100	

The result will be printed in another _DOS file (Figure 4.6).

How to calculate the partial electronic density of states?

The use of PAW simplifies considerably the computation of the different orbital contributions to the electronic wavefunctions.

ndtset	2	# use two steps to compute the density of states
toldfe	1.0d-09	# good convergence on energies in Ha/Bohr
kptopt	1	# use the symmetry to reduce the number
		of **k** points actually calculated
ngkpt1	6 6 6	# regular dense grid of **k** points
prtden	1	# flag to print the charge density
occopt1	1	# standard insulator, like magnesia or forsterite
prtdos2	1	# print the density of states
		# using a standard smearing scheme
ngkpt2	12 12 12	# remember that the grid of **k**-points is
		system-dependent and needs to be converged
occopt2	7	# use a metallic-like occupancy to ensure smearing

The result will be printed in a _DOS file where the header contains the Fermi energy (note that in abinit, it is in Hartree!), followed by a table where the columns are in order: the energy (in Hartree), the total density of states, the integral, the halved density of states, and the double of it. The last two columns just offer two different possible smearing values. The resulting smeared density of stats must correctly reflect the width of the dispersion of the electronic bands. They are usually plotted close to each other, with energy on the vertical axis, which is shared by the two.

For the tetrahedron method, with prtdos 2, the output _DOS file contains only the total density of states without additional smearing. Once again, be sure to converge the resulting density of state with respect to the grid of **k** points.

4.6 Metals versus Insulators

The electronic bands are occupied in increasing order of energy by all the available electrons. The occupied bands are called the valence bands, and the unoccupied ones are called the conduction bands. The highest energy at which the last electrons are placed is called the Fermi energy (Figure 4.12). Hence the valence bands lie below the Fermi energy, and the conduction bands lie above the Fermi level. Because of their relatively low energy, the electrons in the valence bands are localised, and they cannot travel freely through the crystal. They participate in chemical bonds, which

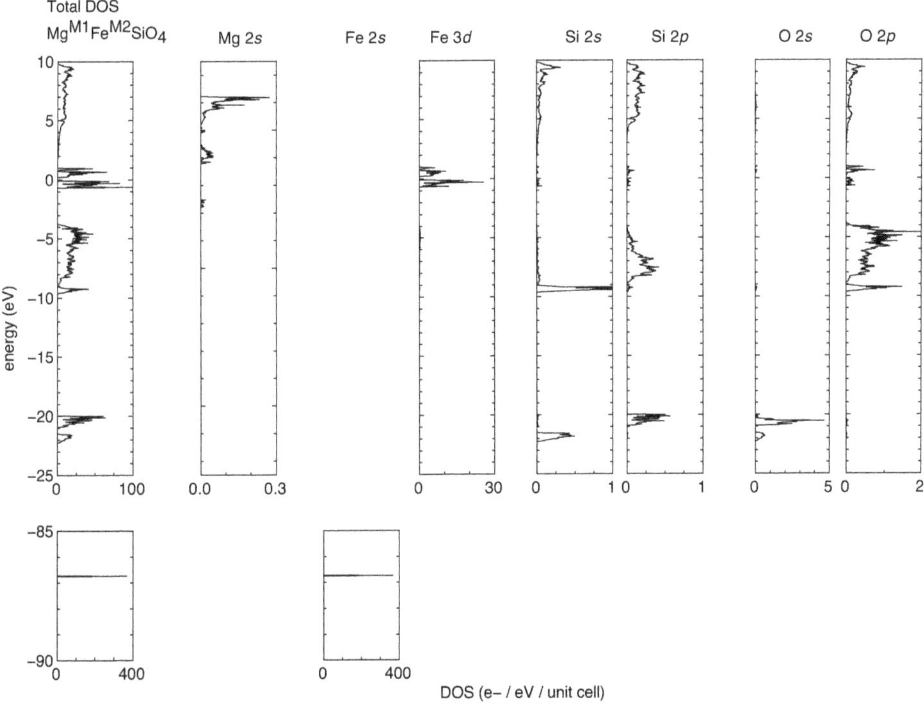

Fig. 4.11 The decomposition of the density of states on different orbital-like contributions for an intermediate term in the olivine solid solution with $(Mg_{0.5}Fe_{0.5})_2SiO_4$ composition. The leftmost image shows the total electronic density of states as a reference. The different orbitals are well limited to different energy levels. The orbitals that occupy similar levels would typically hybridise to form the chemical bonds. The calculation is done without correction for the strong correlation of the Fe d electrons, and without assuming a magnetic structure. The calculations in this figure were done using the model input file in Box 4.8.

are known in chemistry as either ionic bonds or covalent bonds. The electrons that occupy the conduction bands have enough energy to escape the local atomic potential wells. Consequently, they are delocalised and participate in metallic bonds. The electrical conduction in a crystal is realised by these delocalised electrons.

The energy difference between the lowest energy of the conduction bands and the highest energy of the valence bands defines the electronic gap. The Fermi level can be placed anywhere inside this electronic gap. In chemical molecular language, the equivalent of the top of the valence bands is the highest occupied molecular orbital or HOMO, and the equivalent of the bottom of the conduction bands is the lowest unoccupied molecular orbital or LUMO. Hence, the HOMO-LUMO difference in molecules (the terminology in quantum chemistry) is the electronic gap between the valence and the conduction bands in periodic crystals (the terminology in solid state physics). If the top of the valence bands has the same wavevector as the bottom of the conduction bands, the gap is direct; otherwise, it is indirect.

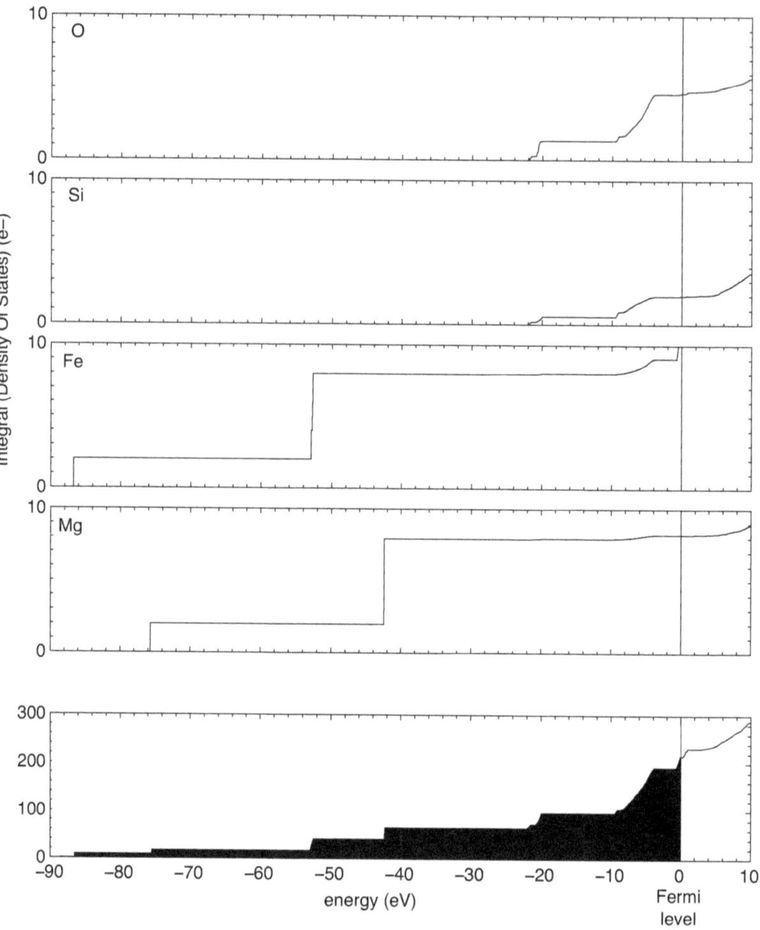

Fig. 4.12 The integral of the density of states as a function of energy gives the number of electrons that lie up to that value of the energy. Obviously, this is an increasing function, but that increases in steps. Integer values are added after each complete subset of electronic bands. The integral to the Fermi level gives the number of electrons in the unit cell.

If the right amount of energy is provided to the valence electrons, for example, by electromagnetic radiation, heating, electric fields, and so on, they can become excited and jump into the conduction bands. In the extreme case, if the amount of energy is sufficient, they can absorb it and break all bonds to the underlying lattice; the electrons leave the structure, leaving behind an ionised crystal.

Depending on the existence or not of the electronic gap, solids are classified into insulators and metals. If the electronic gap between the valence and the conduction bands is finite and positive, the material is an insulator. Furthermore, if the gap is wider than 3 eV, then it is a proper insulator; otherwise, it is a semiconductor [12]. The distinction at 3 eV is arbitrarily fixed. Using the Boltzmann constant, there is an easy way to transform the energy of the gap in electronic temperature: 1 eV = 11604.5250061657 K. This shows that almost 35,000 K in temperature is needed to percolate the

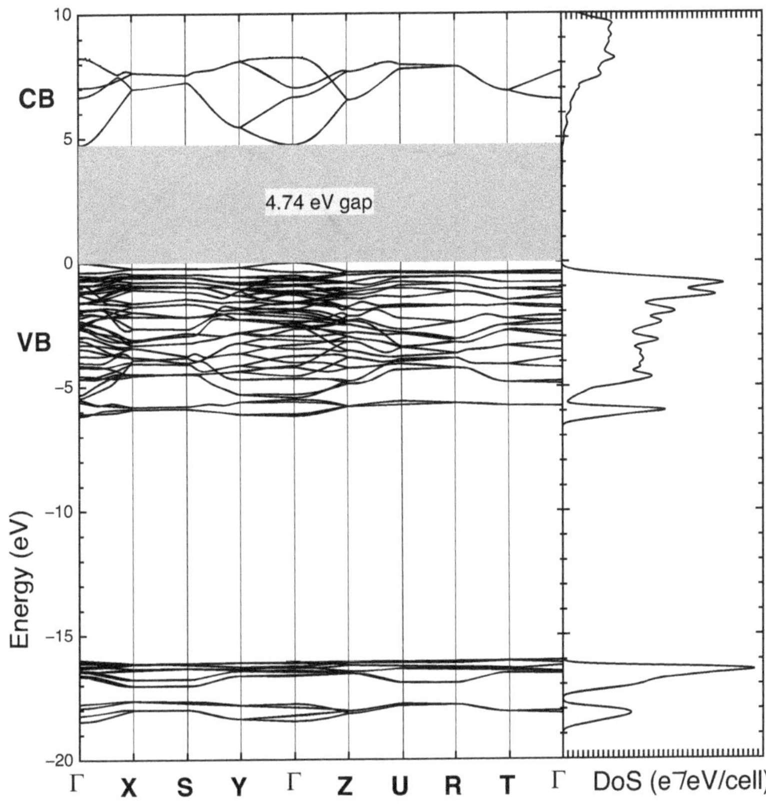

Fig. 4.13 Electronic band structure (a) and the corresponding electronic density of states (DOS) (b) of forsterite, Mg_2SiO_4. Several distinct sets of bands are clearly visible. The gap between the top of the valence bands (VB) and the bottom of the conduction bands (CB) is also shown. With a gap of 4.75 eV, forsterite is a wide-gap insulator. Remember that, as usual with DFT and LDA, the gap is underestimated. The gap is direct, in Γ. The rest of the points in the Brillouin zone are X = 1/2 0 0; S = 1/2 1/2 0; Y = 0 1/2 0; Z = 0 0 1/2; U = 1/2 0 1/2; R = 1/2 1/2 1/2; T = 0 1/2 1/2.

gap of an insulator, solely by heating. In the literature, the insulators are usually described as wide gap for gaps on the order of many eV or tens of eV. Also, the notation of small gap semiconductors can be found, but again, these two notations are arbitrary. As the distinction between proper insulators and semiconductors shows, it is much easier to induce conduction into a semiconductor, especially a small-gap semiconductor, than into a proper insulator. As an example of a good insulator, Figure 4.13 shows the electronic band structure and the corresponding DOS of forsterite, Mg_2SiO_4. The LDA calculation yields a gap of 4.74 eV, which is an underestimation of the actual experimental value, on the order of 8.4 eV [222].

An interesting category of materials is zero-band-gap semiconductors, which are semiconductors in the sense that the valence bands and the conduction bands are well delimited and do not cross at any point. However, the valence bands and the conduction bands touch each other at one or more points of the Brillouin zone. This happens because symmetry makes

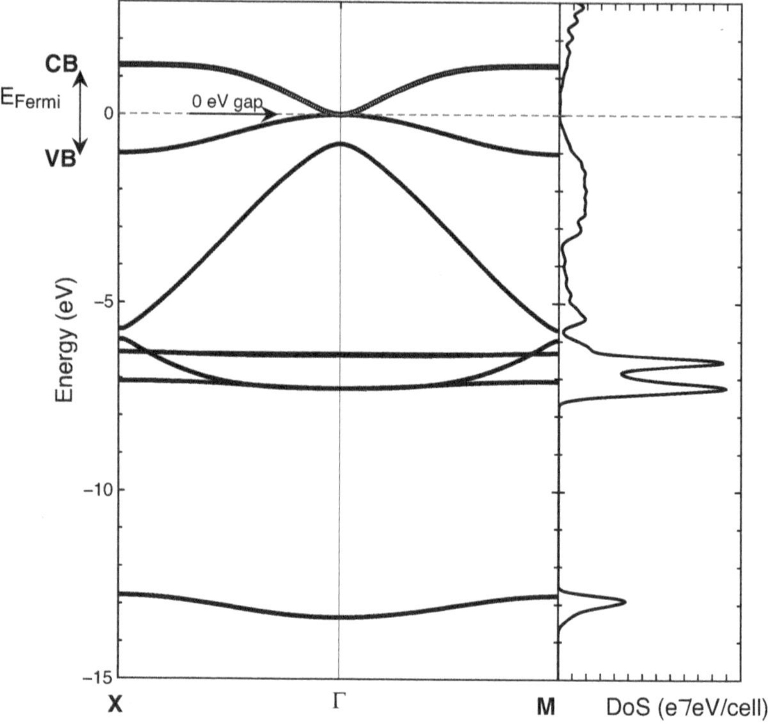

Fig. 4.14 Electronic band structure (a) and the corresponding electronic density of states (DOS) (b) of metacinnabar, HgS. The gap between the top of the valence bands (VB) and the bottom of the conduction bands (CB) is null. In Γ three bands (two VB and one CB) are triply degenerate, leading to an insulator with no gap. The occupancy of the Fermi level is null. $\Gamma = [0\,0\,0]$; X $= [1/2\,0\,0]$; L $= [1/4\,1/4\,1/4]$.

orbitals degenerate in a given point of the Brillouin zone, meaning they have the same energy, while there are not enough electrons to fill all these orbitals. HgS meta-cinnabar is such an example (Figure 4.14): The top two of the valence bands and the lowest conduction band are all triply degenerate in Γ because of the symmetry and hybridisation. There are only four electrons available to fill the three bands. This results in one unoccupied band with the same energy as the two other occupied bands. Away from the Γ point, the symmetry is reduced, and the bands split.

The non-existence of a gap separates metals from the rest of the solids. This can be done either by having bands crisscrossing in one or more parts of the Brillouin zone (Figure 4.15) or by having the top of the valence bands at higher energy than the bottom of the conduction bands, without the bands crossing each other. In the first case, this is a proper metal, and in the second case, it is a semimetal. There are considerably fewer semimetals than insulators and metals. Some of the metalloids, like some arsenic chalcogenides, might turn out to be semimetals. But with the DFT underestimating the band gap, the calculations provide a metal instead of a semi-metal for many of these materials.

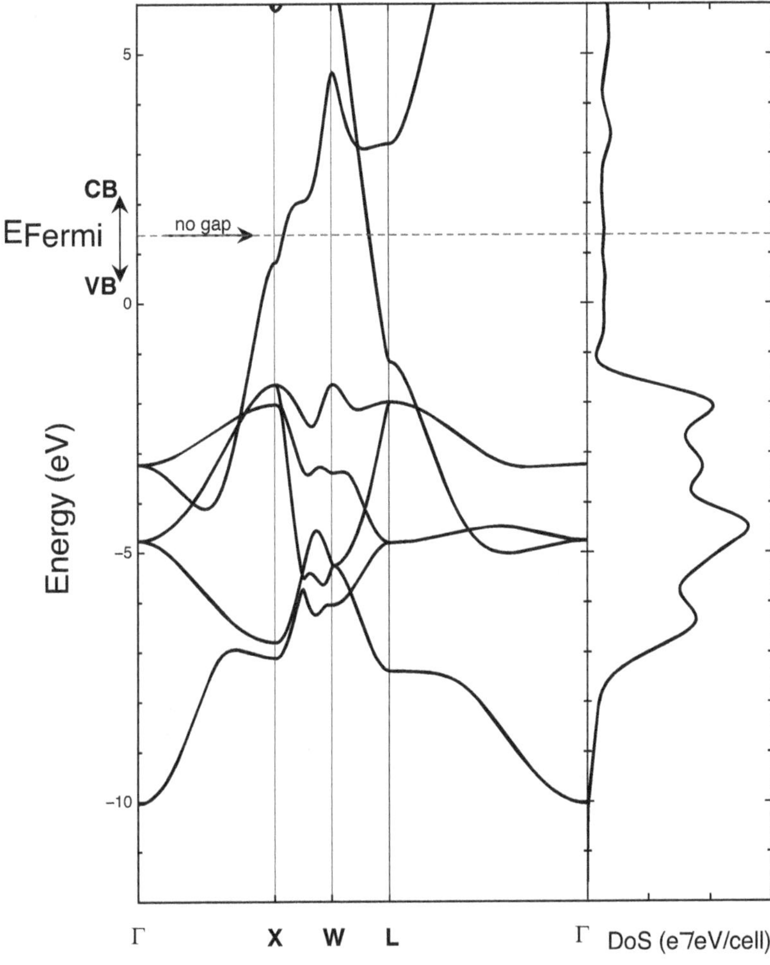

Fig. 4.15 Electronic band structure (a) and the corresponding electronic density of states (DOS) (b) of gold, Au. The five *d* bands form the top of the valence bands (VB). The bottom of the conduction bands (CB) is an *spd* hybrid. Two bands cross the Fermi level, whose occupancy is positive, making Au a metal. $\Gamma =$ [0 0 0]; X = [1/2 0 0]; W = [1/2 1/4 0]; L = [1/4 1/4 1/4].

As a result of the electronic band structure, for insulators, the occupancy at the Fermi level is strictly zero, but for proper metals, the occupancy of the electronic bands around the Fermi level is partial, finite, and positive. Because of the zero gap, the slightest thermal excitation can excite electrons, to populate the first conduction bands and thus to delocalise the electrons. The distribution of electrons as a function of temperature around the Fermi surface follows the Fermi–Dirac distribution:

$$g(\omega) = \frac{1}{1 + e^{\frac{E - E_F}{k_B T}}} \tag{4.8}$$

Because the electrons obey this Fermi–Dirac distribution law, they are called *fermions*. The Fermi distribution is a function dependent on temperature. As the temperature increases, there are more and more electrons in excited states. When performing a simulation, the electronic temperature needs to be taken explicitly into account. This is true especially at high temperatures, for example, during molecular-dynamics simulations, while for static 0 K simulations or even for ambient temperature ones, this requirement is less strict. The Fermi distribution is the correct physical way of describing the electrons in a metal. However, other distribution functions, like Gaussians or various polynomials, are more numerically efficient, so they are used in practice. They may ensure faster calculations, as they typically have a smaller number of partially occupied bands that need to be computed. However, this comes at a cost, as these alternative distribution laws might exhibit small numerical perturbations, which result in weird occupancies: The last valence bands fully occupied have an occupancy slightly larger than 2.0, and the first unoccupied bands have a small negative occupancy.

For metals, the intersection of the electronic bands with the Fermi level defines the Fermi surface. If only one band crosses the Fermi level, then this intersection defines one Fermi surface. The shape of the Fermi surface can be irregular or not, but it will preserve the symmetry of the point group. If several bands cross the Fermi level, the Fermi surface is composed of several such surfaces. Figure 4.16 shows the Fermi surface for copper, Cu, as a model for fcc metals. In general, the Fermi surface delimitates pockets of holes and pockets of electrons. The holes correspond to the **k** region where the bands are above the Fermi level and thus unoccupied – there are missing electrons in the total count for those particular bands. The pockets correspond to the **k** region where the bands are below the Fermi level and thus occupied – there are electrons populating the bottom parts of these bands that intersect the Fermi level.

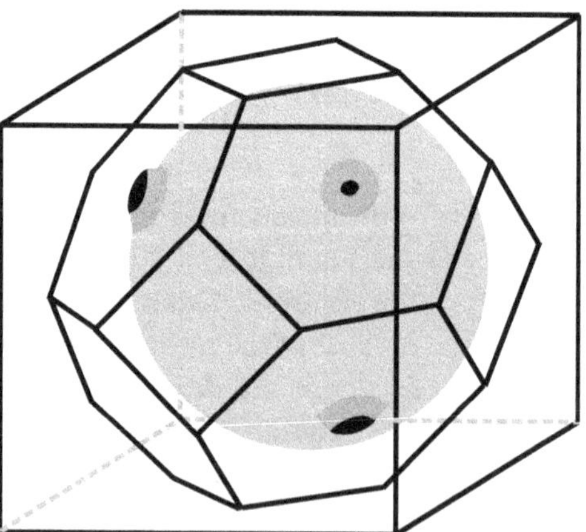

Fig. 4.16 Schematic representation of the Fermi surface of fcc metals, such as Cu, Ag, and Au. The Fermi surface is a sphere with protuberances called *necks* that touch the middle of the hexagonal faces of the Brillouin zone. These necks ensure the connectivity of the Fermi surface from one Brillouin zone to the next. The electrical conductivity in these metals is realised by the electrons, which flow along these necks and are thus delocalised.

The shape of the Fermi surface is highly variable and depends on symmetry and chemistry. In some cases, like the chalcogenides of transitional metals or alkali metals at high pressure, regions of the Fermi surface(s) may be parallel to each other, even if not necessarily flat. This is called Fermi surface *nesting*. In this case, electronic wavefunctions can change phase (i.e., the **k** vector) at equal energy (i.e., the Fermi energy) without energy loss. This phase change may result in a collective motion of electrons in the direct space, which entrains a collective displacement of the atoms via electron-phonon coupling. The mechanism is efficient if there is a phonon whose displacement pattern and wavevector correspond to this collective change. The final result is a distortion of the underlying crystalline lattice and thus a phase transition. This phase transition is triggered by a charge-density wave resulting from the coupling between the electrons and a phonon – an electron-phonon coupling. For this to happen, the electronic population of the corresponding bands that form the nesting should be relatively high. Such transitions occur typically at low temperatures, as the decrease of the temperature increases the occupancy of the Fermi level, and thus the population potentially affected by the nesting. There is a large amount of literature on this topic. Many chalcogenides of transitional metals exhibit such instabilities and phase transitions. A wide variety of possible symmetry relations occur during the transition: commensurately or incommensurately modulated structures can easily form in this way. Other times the electron-phonon coupling is such that electrons are trapped, paired in Cooper pairs, and thus induce the appearance of superconductivity [98].

4.7 A Few More Examples and Interpretations of the Electronic Band Structure for Minerals

The in-depth analysis of the electronic band structure reveals a lot of information about the bonding state of the system. The groups of bands well delimited by energy gaps are relatively easily identifiable in terms of hybrids of atomic orbitals.

For example, in the case of cinnabar, HgS, the s electrons of Hg lie at the lowest energy levels, followed at higher energy by the s electrons of S and the Hg p and S p, which show a certain amount of hybridisation. For metallic Au, the five valence bands are the d states, while the last of the valence bands and the first conduction band is a hybrid of the $s, p,$ and d states.

In most of the ionic minerals, like oxides and salts, the electronic bands are well distinct, with the lower s states being highly localised and the p states being hybridised and participating in the bonds. The anionic groups, like silicates, sulphates, and so on, are good examples.

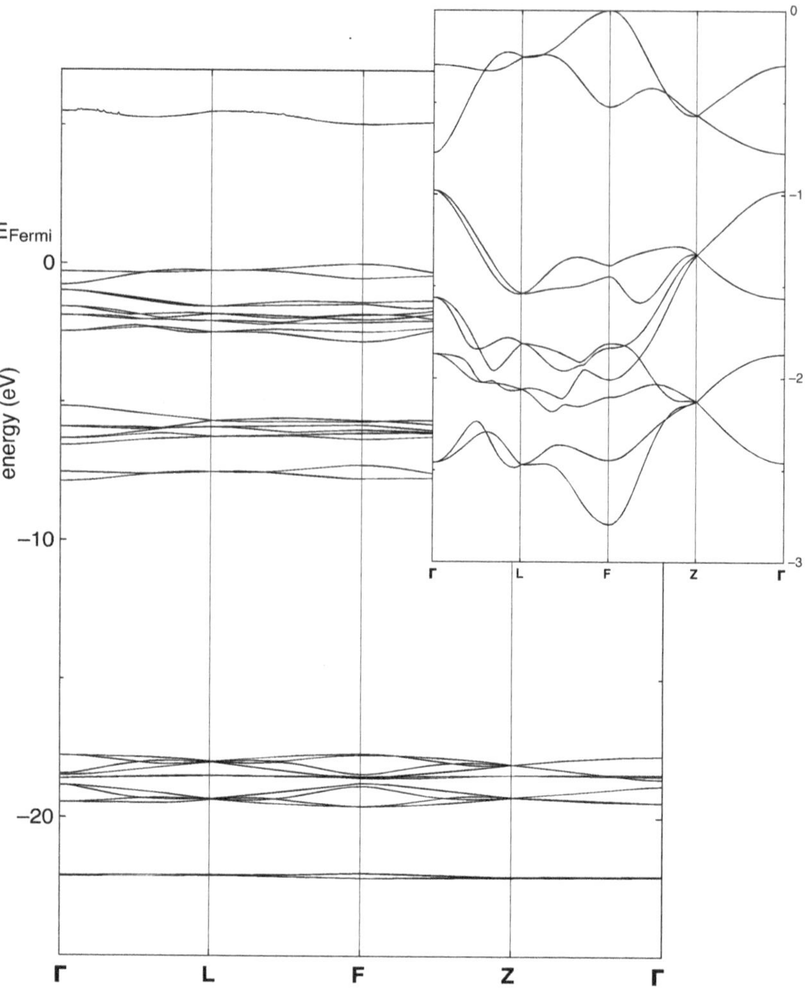

Fig. 4.17 The top of the valence electronic bands and the first conduction band of calcite, $CaCO_3$. There is no preferential direction in the structure. The bands lie in several groups corresponding to the different orbitals, like Ca 3s, O2s, and so on, separated by large gaps. $\Gamma = [0\,0\,0]$; $L = [1/2\,0\,0]$; $F = [1/2\,1/2\,0]$; $Z = [1/2\,1/2\,1/2]$.

The electronic band structure of calcite, $CaCO_3$, shows a series of groups of bands that are separated by wide gaps. Within each group, the bands are dispersive, the top of the valence bands spreading over 3 eV (Figure 4.17).

The valence bands in orthoenstatite, $MgSiO_3$, show two large groups of bands, corresponding roughly to s and p hybrids (Figure 4.18). The bands perpendicular to the reciprocal directions of the SiO_4 chains are dispersive. The bands parallel to the $[SiO_4]_n$ chains, characteristic of pyroxenes and amphiboles, are weakly dispersive due to electron localisation. These features highlight the quasi-one-dimensional character of the pyroxene structure (Box 3.1).

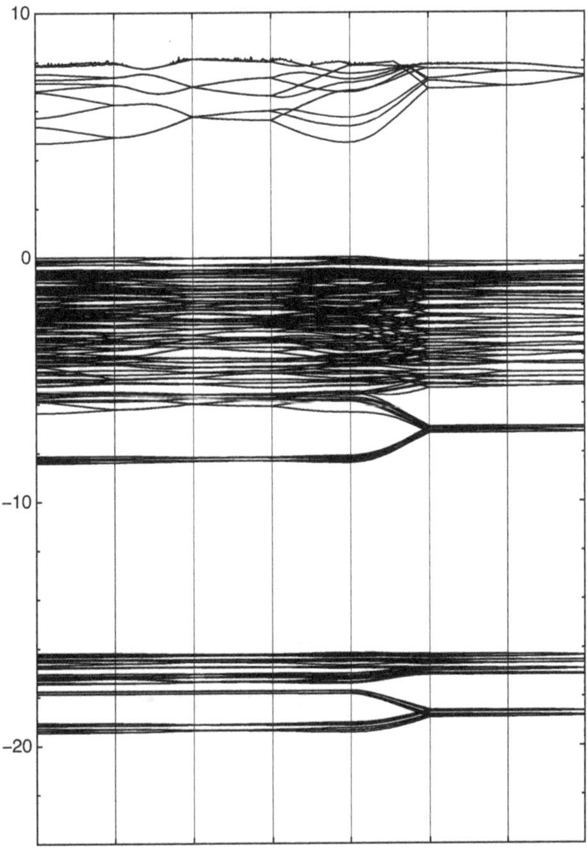

Fig. 4.18   Electronic band structure, MgSiO$_3$. In this figure, the last conduction band has an unphysical shape, which is very wiggly. This is an indication that the residuals of the last empty band were not well converged. It is not a problem, as neither the energy nor the other properties of the mineral change, as its occupancy is zero, but this band might be left out of a graph submitted for publication. This feature appeared in several other graphs from this chapter and was purposely left out. The path through the orthorhombic Brillouin zone follows the same high-symmetry points as for olivine, labelled in Figure 4.2.

The electronic band structure of åkermanite, CaMgSi$_2$O$_7$, shows a clear sequence of electronic bands, weakly dispersive in the reciprocal space (Figure 4.19), hence highly localised in the direct space. In increasing order of energy, we can distinguish Ca 3s, then Ca p hybridised with O orbitals, with the rest of the Si and O orbitals higher in energy. This mineral is an excellent example that shows the extensive hybridisation between the Sis, Sip, Os, and Op electrons to form one of the most ubiquitous anionic groups of the mineral world – the SiO$_4$ group. The role played by the semi-core states for the Ca s is also clearly seen by the high energy of the Ca p states: They hybridise with some O electrons to form the second DOS peak, and consequently, they need to be explicitly included in the valence states of the Ca [64].

In forsterite, shown in Figure 4.13, the bands are more dispersive than in åkermanite, because there is no preferred direction in the structure. The

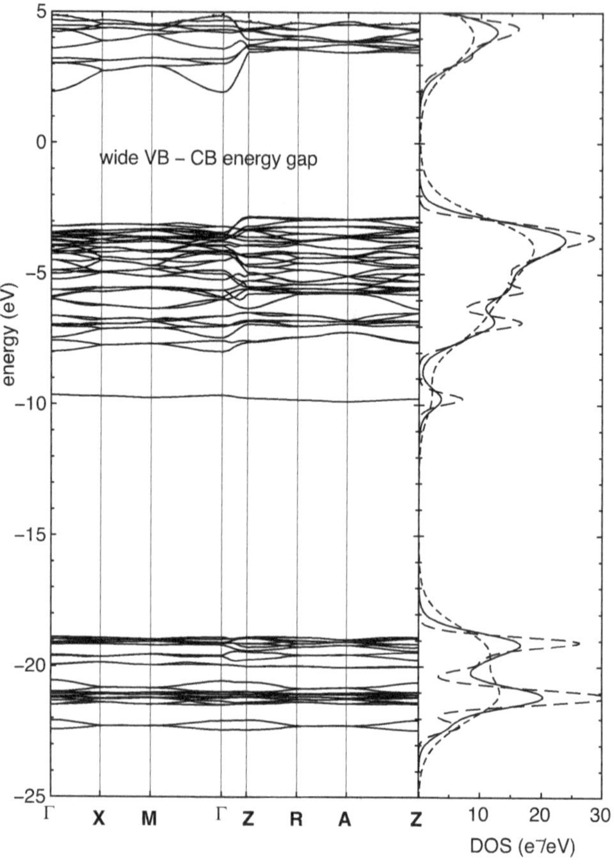

Fig. 4.19 Akermanite, $MgCa_2Si_2O_7$, is a refractory sorosilicate whose structure is formed by alternating layers of silica tetrahedra connected to alumina octahedra and large cations in octahedral coordination. The bands are dispersive perpendicular to the layering. Along the directions parallel to these planes, the bands are weakly dispersive. These features highlight the quasi-two-dimensional character of the structure. X = 1/2 0 0, M = 1/2 1/2 0, Z = 0 0 1/2, R = 1/2 0 1/2, A = 1/2 1/2 1/2.

electrons are not localised along certain crystallographic directions. This is also the reason why olivine does not have cleavage, for example. The electronic gap is wide, at about 6.5 eV, making it a good insulator. This is reflected in the low conductivity values of the upper mantle, which is dominated by olivine and its high-pressure polymorphs.

Like many sulphide and sulphosalt minerals, pyrite, FeS_2, while it has a clear metallic lustre, turns out to be a small gap semiconductor in its pure phase (Figure 4.20). For the experimental structure, the predicted LDA gap is about 0.4 eV, while the experimental gap is 0.95 eV. While the gap is still smaller than the experimental value, it is remarkable that the LDA calculation still yields a gap and not a metallic structure. The electronic bands are dispersive, and there is no indication of any particular localisation of the electrons neither along certain directions nor around certain atoms. The conduction bands are considerably more dispersive than the valence bands, but this is common for a large majority of solids.

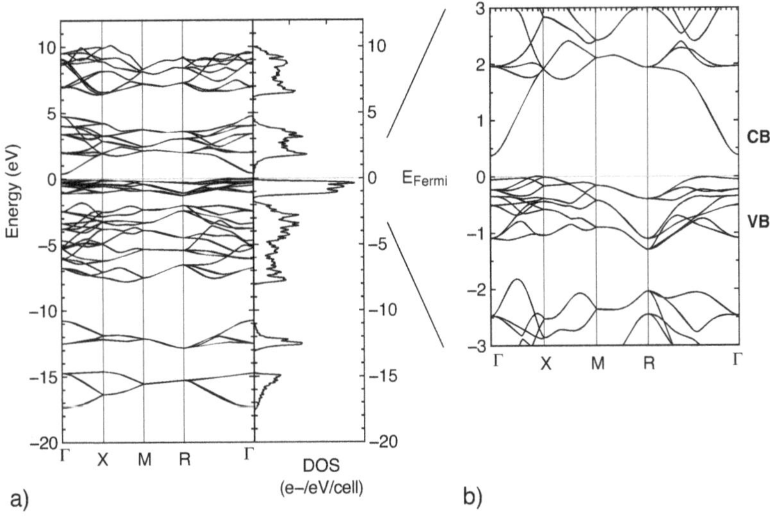

Fig. 4.20   Electronic band structure and the corresponding electronic density of states (DOS) (a) of pyrite, FeS_2. Pure pyrite is a small gap semiconductor with an experimental gap of 0.95 eV. DFT yields a direct gap in Γ of about 0.4 eV (see zoom in b). All the bands are dispersive, as there are no localised electrons along preferential directions in the crystal or around particular atoms. $\Gamma = [0\,0\,0]$; $X = [1/2\,0\,0]$; $M = [1/2\,1/2\,0]$; $R = [1/2\,1/2\,1/2]$.

The A_2B_3 minerals, where A stands for Sb, or Bi, and B stands for S or Se, with minerals such as stibnite, Sb_2S_3, have a characteristic elongated needle-like habitus. These minerals have an electronic band structure characterised by the presence of several band groups well delimited by wide gaps. In increasing order of energy, the DOS peaks correspond to the s states of the A atoms, then to the s states of the B atoms, and then to hybrids of the p states of both A and B atoms. Another peculiar characteristic of the electronic band structure of minerals like stibnite is the presence of flat bands, that is, non-dispersive, along the $\mathbf{c}^*$ direction. Because of the inverse character of the reciprocal space and the direct space, weakly or even non-dispersive bands in the reciprocal space correspond to localised states in the direct space. For example, deep-energy s states that are non-participating in the chemical bonds or hybridisations are highly localised around the atomic nuclei. Consequently, these bands are non-dispersive in the reciprocal space.

In the case of stibnite-type minerals, the non-dispersive character of the bands in the reciprocal space is restricted to the $\mathbf{c}^*$ direction. This implies that the electrons are localised along the $\mathbf{c}$ axis of the real space, which corresponds to the longitudinal axis of the rods forming the structure. The electron density has minima between the rods. This axis is also reflected in the preferred crystalline habitus of this group of minerals, as the cleavage planes are parallel to the c direction and comprise the (100) and the (010) planes. The preferential distribution of electrons along one given direction or within a plane can easily be correlated not only with habitus and morphology but also with the anisotropy of other physical properties of the minerals, like thermal and electrical conductivities.

This pattern of concentrating the electrons along one direction or within a plane corresponds to low-dimensional solids (Box 3.1). We have seen the example of pyroxenes and of A_2B_3 stibnite-like minerals. This feature of the band structures is important, and it is always worth looking for. Oftentimes this can lead to other interesting physical properties, which can be used in day-to-day life. For example, tailored stibnite-like compounds are used for photovoltaic cells and as a thermoelectric. Quasi-two-dimensional molybdenite, MoS_2, when intercalated with layers of alkali metals, becomes a high-temperature superconductor [266, 268]. With first-principles calculations, we have the tools today to proactively investigate such materials and, even more, to propose chemical changes and substitutions that can improve their physical properties. It is enough to look with an attentive eye throughout the mineral realm, there are many gems that can pop up.

For transitional metals, the electronic bands are more dispersed as a direct result of the delocalisation of the valence electrons. Their d states are usually at the top of the conduction bands but cross the Fermi level. Au, represented in Figure 4.15, is an excellent example. The s electrons are transferred to the d orbitals, and the hybridised $6s$ and $6p$ are empty, above the Fermi level.

Under pressure, as the atoms lie closer and closer, their potential wells overlap, and the core electrons begin to delocalise (Figure 3.5). In the large majority of cases, under extreme conditions, the electronic gap between the valence bands and the conduction bands decreases and the bands become more and more dispersive. Figure 4.21 exemplifies this behaviour for jadeite, $NaAlSi_2O_6$, which is a pyroxene stable at high pressures. Jadeite has a DFT-calculated gap of more than 5 eV at ambient pressure

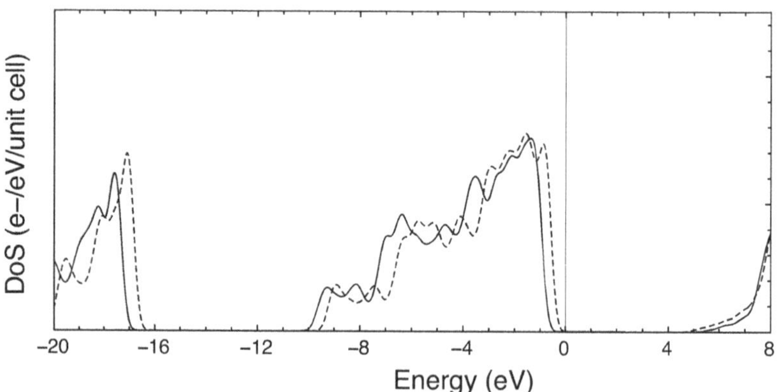

Fig. 4.21 The electronic density of states of jadeite at two different pressures: ambient (solid line) and 20 GPa (dashed line). Over this pressure range, the electronic gap decreases by about 1 eV. The gap reduction, or even closure, is a common behaviour of most minerals under compression.

conditions. Under pressure, as the atoms get closer to each other to accommodate compression, the gap shrinks. In the 0–20 GPa pressure range, it decreases by almost 1 eV.

In extreme cases, the gap can completely close, and the minerals become metallic. This is a pressure-induced metallisation. With a few exceptions, all solids eventually become metallic under enough compression. In the Earth's mantle, FeO is famous for its gap closure under pressure [121, 196]. Actually, describing correctly the electronic properties of FeO is a real challenge for all the DFT codes and for all the exchange-correlation functionals [238].

There are also exceptions where the compression first contributes to an opening of the gap because of a particular geometry of the orbitals. Periclase, MgO, is a good example of a mineral exhibiting such behaviour. The gap opens under compression within the B1 phase, as illustrated in Figure 4.22. However, this is only a transitory state, and under enough compression, the structure ends up becoming metallic. The transition to the higher pressure B2 phase of MgO, at multimegabar pressures, characteristic, for example, for the deep interior of super-Earths ([40]), is associated with a reduction of the gap.

Minerals are never pure chemical phases. They always have various impurities that affect their physical properties, which are discussed at

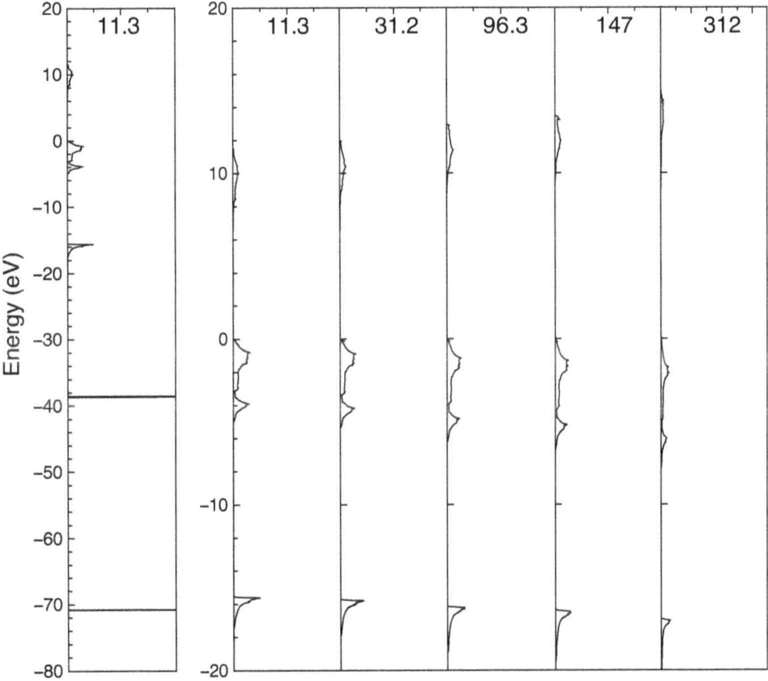

Fig. 4.22 The electronic density of states of periclase, MgO, as a function of pressure. Values of pressure (in GPa) are written explicitly at the top of each panel. At ambient pressure, periclase has a wide gap, on the order of about 7.8 eV (left panel). Under compression, the gap opens to about 12 eV before the transition to the higher-pressure B2 phase.

length in Chapter 5. And many times, they are part of broad solid solutions. Homovalent cations of similar sizes with abundances in nature, like Mg and Fe, or K and Na, or U, Th, W and Zr, and so on, are often exchanged. Performing repetitive calculations of specific physical properties over the entire range of a solid solution can save a lot of time and improve efficiency. Box 4.7 exemplifies the computation of the electronic DOS along the olivine solid solution, going from forsterite, Mg_2SiO_4, to fayalite, Fe_2SiO_4. At each step, one Mg atom is replaced with one Fe atom. These structures are ideal and ordered approximants. The structure used in the example comes from AMCSD and is fixed for the entire solid solution. The presence of iron in olivine reduces the gap, as the d electrons are positioned in the middle of the gap of pure forsterite. But the d electrons of Fe still preserve a small gap, which comes from the $t_g - e_g$ splitting. A small gap is preserved even for pure fayalite, Fe_2SiO_4. The Fermi energy is shifted from the top of the Op bands, in pure forsterite, to the top of the t_g states, in Fe-bearing olivine. This shift is seen in Figure 4.23. Olivine becomes darker and less transparent with increasing Fe content.

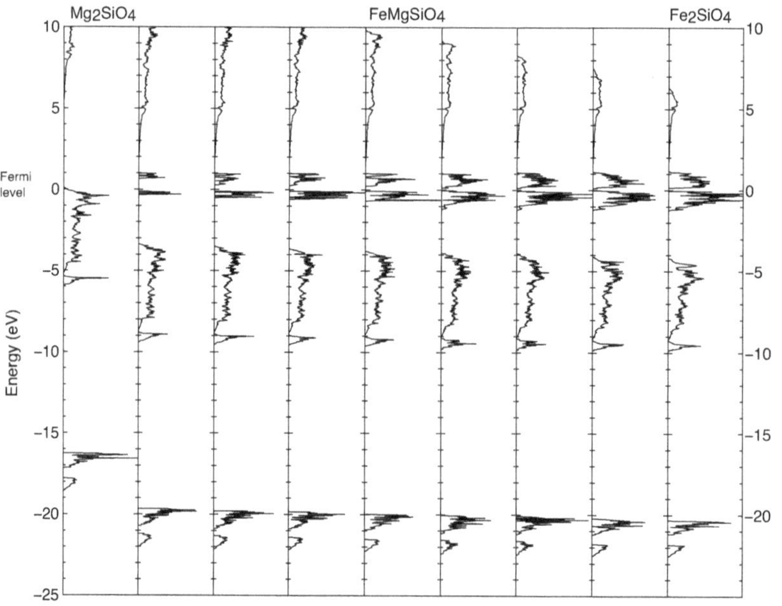

Olivine, $(Me)_2SiO_4$, is not a proper mineral but a complex solid solution where various cations can substitute on the Me site. The most common cations are Mg and Fe. Here, the electronic changes induced by the Mg $<=>$ Fe substitution are clearly seen. Forsterite, Mg_2SiO_4, in the first column, has an electronic gap of about 6.5 eV. The d electrons of Fe that are brought in olivine by the Mg-Fe substitution, starting with the second column, lie high in energy around the Fermi level. This actually makes olivine darker and more conductive with increasing Fe content. However, computing correctly the variation of the electronic properties in olivine as a function of iron content at a given pressure should be done after the structure is permitted to relax to take into account the influence of Fe and Fe/Mg distribution on the residual pressure.

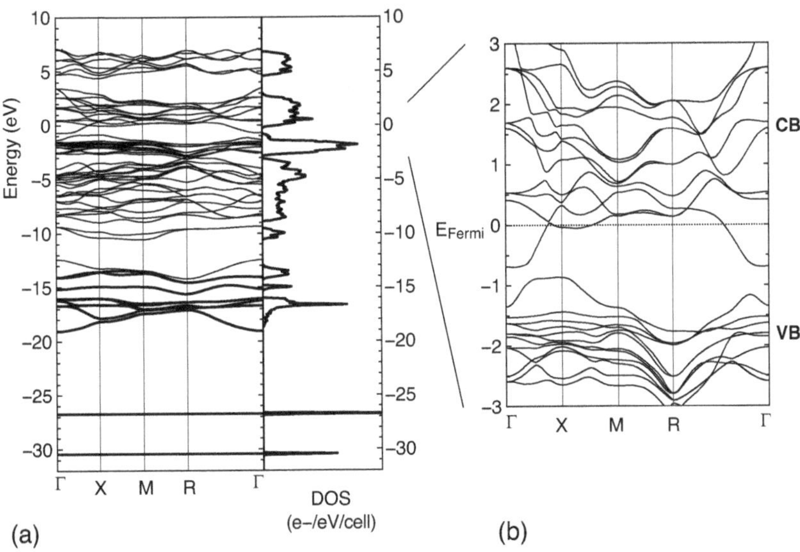

Fig. 4.24 The presence of Cu in the structure of pyrite, FeS_2, closes the gap, as the *d* states of Cu are placed in the former gap of pyrite.

The addition of Cu as a (common) impurity in pyrite, FeS_2, closes the electronic gap (Figure 4.24), in contrast to the pure phase of pyrite (Figure 4.20), which shows a small gap.

While we can go on forever with interesting stories about the electronic structure of various minerals and groups of minerals, for many such minerals, interesting properties are not necessarily found at the level of the electronic band structure, *per se*, but rather in the change of the electronic properties under perturbations.

## 5

Origin of Colour, Optical Constants, and Electronic Spectroscopy

Minerals interact with the electromagnetic waves from the visible spectrum, that is, the light, in various ways. The incoming light on a mineral surface is partly transmitted, absorbed, and reflected. The amount of each depends on the electronic band structure of the mineral, the light frequency, and the incident geometry. At the very fundamental level, the electrons dictate the interaction of the minerals with the external electromagnetic field.

The amount of transmitted light gives the mineral transparency. At the two ends of the scale are the ideal perfectly transparent medium, for 0% absorption, and the ideal black body, for 100% absorption. Real materials lie between these two extremes.

The incoming photon on a surface can transfer its energy and its momentum to the electrons, which will move to an excited state. This light absorption is dependent on the availability of empty electronic bands where the electrons can be excited. The electron excitation can be done by preserving its wavevector, $\mathbf{k}$, or changing it. In this last case, there is a momentum transfer between the incoming photon and the electrons. The electronic excitation has a certain finite lifetime.

The deexcitation of the electron takes place with the release of electromagnetic radiation. The process is reversed with respect to the excitation: The wavevector can change or remain constant, and the electron can fall down directly to its initial ground state position in one or several steps or change its wavevector and fall into another available hole.

The selective interaction between light with different wavelengths and the electronic band structure determines the colour of minerals. In general, if a given mineral is metallic or semi-metallic, it is also opaque, and its colour is a constant that can be used as an identification tool. For transparent minerals, colour is usually variable and depends on impurities. The ratio of reflected and transmitted + absorbed light gives the mineral lustre.

5.1 Interaction of Light with Metals and Insulators

As metals have no electronic gap, there is a continuous distribution of available empty states in the conduction bands where the electrons can be excited. Consequently, the absorption of light is done at all wavelengths.

Hence, metals are opaque. After the electrons are excited, they live for a certain time in the excited state; then, they can de-excite and fall back into the holes previously created on the valence bands. The de-excitation process is also continuous in terms of wavelength, giving the typical metallic lustre [178].

Semi-metals and small-gap semiconductors have similar behaviour to metals. Because their electronic gap is small, the incoming photons of visible light can easily excite electrons and move them in the conduction bands. The absorption spectra would be less intense than for metals and more dependent on the frequency of the incoming light. This gives the typical semi-metallic lustre.

In other cases, the minerals do not interact with the incoming photons, and then they are transparent and colourless. This is typical for insulators, where the energetic gap between the conduction bands and the valence bands is larger than the energy of visible light photons. There are no available empty states, and the electrons cannot be excited. Light passes through the mineral, largely unaffected by the electrons. The result is a transparent mineral. Typically, minerals having fully occupied s and p states, for example, various salts with large electronic gaps, are transparent; they have a glassy lustre. Smaller gap insulators and semiconductors can have partial light absorbance. In this case, the electronic gap is small enough that some photons can be absorbed, and electrons can be excited to the available empty bands. Their lustre is often adamantine or even semi-metallic.

For insulators, the presence of atomic impurities of transitional elements can create real havoc in the electronic structure. The partly occupied d or f states of transitional metal ions are high in energy and lie inside the gap between the valence and the conduction bands. They can serve as an easy pump of electrons to be excited to the conduction bands, and they can act as anchor states for the excitation of the electrons of the host mineral on their way up or for their deexcitation on their way down. The position of these bands will infer the wavelength of the radiation that can be absorbed or that is emitted; their occupancy will act on the intensity of the spectrum.

The complementarity of the frequency-dependent light absorption yields the mineral colour: Minerals absorbing the red light are blue, and minerals absorbing the blue light are red.

5.2 Crystal Field Theory

In an insulating transparent mineral, the partially occupied d or f states of a transitional impurity metal typically lie in the middle of the electronic gap. The incoming photons can excite some of these d electrons that will jump between the occupied and the unoccupied d states. This produces partial absorbance of the white light and colours the transparent mineral. This process also lies at the origin of a variety of luminescent phenomena.

With respect to a Cartesian system of reference, the five d orbitals of transitional metals are decomposed into two groups. Three orbitals lie in the diagonal planes, the d_{xy}, the d_{yz}, and the d_{zx}, and two along the principal axis, the d_{z^2} and the $d_{x^2+y^2}$ ones. Because of their symmetry, the first three are often denoted as the t_3 group and the last two as the e_2. For an isolated atom of a d metal, the five d orbitals are fully degenerate and have the same energy. The degeneracy is preserved when the isolated atom is immersed in a uniform electromagnetic field; only the energy of the orbitals is increased. But in the presence of a non-uniform electrical field, due to the surrounding anionic electronic clouds, there can be a split into two or more groups of bands as the orbitals try to form chemical bonds with these surrounding atoms. The splitting depends on the local environment, that is, the symmetry of the positions of the ligands.

The general rule is that the d orbitals facing neighbouring ligand electrons raise their energy because of the negative-negative repulsion, and the d orbitals facing empty orbitals lower their energy.

In a regular octahedral field, generated, for example, by oxygen atoms, the ligands are positioned along the Cartesian axis, [100], [010], and [001]. The electronic repulsion due to the p electrons of oxygen facing the d orbitals directly along the Cartesian axis increases the energy of the d_{z^2} and the $d_{x^2+y^2}$ orbitals and decreases the energy of the d_{xy}, the d_{yz}, and the d_{zx} orbitals. This results in opening a t_3–e_2 energy gap. The splitting confers the t_3 bands bonding character (lower energy) and the e_2 bands anti-bonding character (higher energy). As the electrons prefer to populate first the lowest-energy bands, the t_3 bands are first occupied, followed by the e_2 bands. If the e_2 bands are partly filled, the electronic excitations would be from the t_3 to the e_2 orbitals.

In a regular tetrahedral field, the situation is reversed, as the p electrons of oxygen face the d orbitals along the diagonals. In this case, the electronic repulsion acts on the diagonal orbitals. As a result, the t_3 group of orbitals, that is, d_{xy}, the d_{yz}, and the d_{zx}, see an increase in energy, and the e_2 group of orbitals, that is, the d_{z^2} and the $d_{x^2+y^2}$, suffer a relative decrease. The e_2 orbitals are the ones to be occupied first, and an e_2–t_3 gap opens. If the t_3 orbitals are partly occupied, then the electronic excitations would be from the e_2 to the t_3 orbitals.

In both these cases described earlier, the cubic parent symmetry of the ensemble is preserved. This is the case with the ideal crystal-field theory. But if the position of the ligands breaks the cubic symmetry, the groups of d orbitals, seeing a larger or a smaller electrostatic repulsion, are further degenerate. This is called the Jahn–Teller distortion. Each structural arrangement has a different distortion. Partial occupancy of the d bands will result in various possible electronic transitions. Each transition can be activated by specific photons, whose wavelengths, hence energy, match the electronic gaps between the various occupied and the unoccupied d states.

For example, chromium replaces the aluminium in the hexagonal structure of corundum. Hence, it is in octahedral coordination. Cr has three available electrons to populate the five d orbitals. An octahedral distortion occurs after Al replacement by Cr, which also depends on the amount of Cr impurity. Moreover, the small trigonal distortion on the Cr site induces further splitting between the t_3 levels. From all the possible combinations of d orbital splitting, the t_3–e_2 splitting is the dominant one, corresponding to green light around 500 nm (20,000 cm^{-1}). The result is the well-known red colour of ruby. Furthermore, as the actual value of the splitting varies with pressure, the wavelength of the luminescent light generated by the de-excitation is used to estimate the pressure in diamond-anvil cell experiments [176, 86].

The colours of most gems originate in the Jahn–Teller effect acting on the partially occupied d or f orbitals of atomic impurities. For example, V^{3+} causes the green colour of tsavorite, V-bearing Ca$_3$Al$_2$Si$_3$O$_{12}$ garnet, Co^{2+} in tetrahedral coordination is typical blue, Cu^{2+} yields blue in azurite, Cu$_3$(CO$_3$)$_2$(OH)$_2$, or green in malachite, Cu$_2$CO$_3$(OH)$_2$, depending on the local distortions, Fe^{2+} generally yields yellow and Fe^{3+} red, Mn yields a variety of colours, like red, green, and pink (as in rhodonite, MnSiO$_3$, or rhodochrosite, MnCO$_3$), or black. The *Mineralogical Applications of Crystal Field Theory* [46] is an old but useful textbook explaining in detail the origin of colour in minerals.

5.3 Electronic Spectroscopies

The basic principle behind electronic spectroscopy is to measure the frequency-dependent absorption or emission of light, as this gives direct information about the position of the electronic bands and the gaps. The interplay between the energy of various electronic bands, the existence of impurities with their partially occupied d or f states, the electronic gap, and the excitation/deexcitation process can be captured in detailed electronic spectra, which can be measured in a variety of spectrometers. Several electronic excitation and deexcitation types are schematised in Figure 5.1.

Absorption spectroscopy uses the frequency-dependent transfer of incoming energy to the excitation of electrons in the sample. The gain in electronic energy in the sample takes place as electronic transitions. As a result, the electrons can be excited from the valence bands into the conduction bands. The excitations can be into different types of bands; for example, p electrons from an occupied band can be excited to empty s bands of higher energy. During the electronic excitation process, only the part of the energy of the incoming beam that matches the gap between the respective conduction and valence bands is absorbed. The result is a broad sampling of the electronic DOS upwards in energy (Figure 5.1). The absorption happens on the timescale of the femtosecond or less.

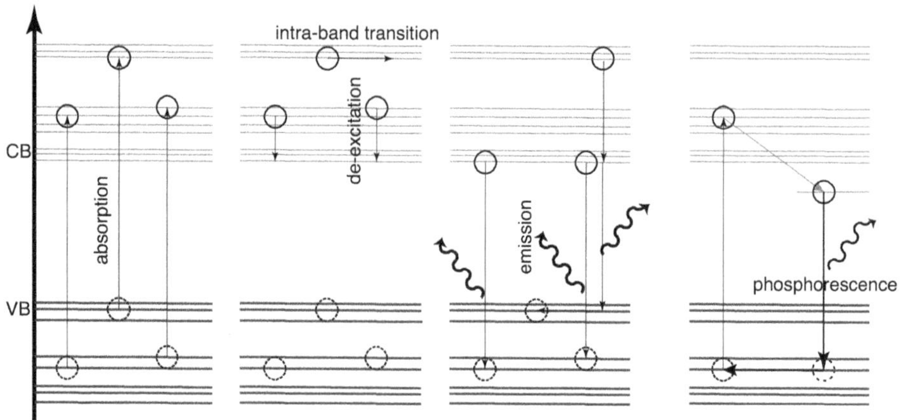

Fig. 5.1　Luminiscence is associated with the de-excitation of electrons. Electrons absorb energy from the incoming light and are excited in the unoccupied conduction bands, leaving holes in the valence bands. They de-excite first to the bottom of the conduction bands in one small jump or after travelling through the conduction bands. The transition from the conduction to the valence bands to annihilate the holes is associated with a strong emission of photons. If transitional metals are present as impurities in various salt minerals, their *d* states are placed in the electronic gap. Then de-excitations can happen with a stop in these bands. The resulting process is phosphorescence.

The transitions in the visible spectrum, described in Section 5.2 lying at the origin of mineral colour, involve much smaller energies and usually within the same types of bands inside the gap.

Inside the conduction bands, the electrons can undergo intra-band or inter-band transitions. In the first case, they preserve their energy but change their moment, that is, their wavevector. In the second case, they suffer several deexcitation transitions between the conduction bands, migrating towards the conduction bands with lower energy. The two processes, intra- and inter-band transitions, can be combined with electrons that de-excite and change their momentum simultaneously. These transitions occur within the first picoseconds following the excitation. Eventually, the electrons fall to the bottom of the conduction bands.

The various de-excitation processes occur with the release of energy in the form of emitted photons, that is, light. The frequency of these photons is measured, and the energy difference between the different bands can be easily inferred. The photoemission spectroscopy is the experimental tool that allows us to measure these processes.

In *angle-resolved photoemission spectroscopy*, the change in momentum is also measured. This is a powerful technique widely employed for metallic systems. It allows the sampling of the electronic bands around the Fermi level and hence provides the direct imaging of the shape of the Fermi surface. In this case, the incoming photons require much smaller energy as the electronic transitions are measured within a narrow energy window around the Fermi level. They typically lie in the X-ray region of the electromagnetic spectrum.

Under a source of (invisible) ultraviolet light, certain crystals doped with selected transitional metals show fluorescence. This can be nicely seen in a dark space and is always an attraction point in mineralogy museums. Calcite, baryte, or fluorite are excellent examples of such fluorescent minerals. The de-excitation typically happens in several steps, with a large part of the emitted light in the visible spectrum.

There can be other sources of energy to induce electronic transitions. In thermo-luminescence, the source of energy for electronic excitation is provided by heating. In chemoluminescence, the source of energy is a chemical reaction.

All these transitions involve the electrons being excited up to the conduction bands and leaving behind electronic holes in the valence bands, into which they fall later. The electrons have a given finite time in the conduction bands. In the valence bands, the electrons recombine with the holes, and the system retrieves its ground state.

5.4 Optical Constants and the Problem of the Electronic Gap

Apart from the beauty of colours and the gemmological value of crystals, the amount and quality of the reflected light on a mineral or rock surface are particularly important in planetology and space exploration. Here, it bears the general name of 'optical constant'. Just as in the case of absorption, which induces colour, the different reflection peaks are specific for the electronic difference between certain electronic states. As such, they can be used to identify various minerals at the surface of various solar system objects. We can only hope that one day the telescope will have enough resolution to identify minerals on extrasolar objects where their suns provide enough incoming light.

An idealised scheme for computing optical spectra involves determining the position and the occupancy of the various electronic bands. Transitions are allowed from an occupied state to an empty state if the appropriate amount of energy is provided to the system. Figure 5.1 illustrates this concept.

But as the electronic transitions happen between occupied and unoccupied states, the key ingredient for realising a correct calculation of optical constants is to accurately obtain the energetic difference between the two sets of bands. This implies calculating the electronic gap between the valence states (all occupied, the source of electrons) and the conduction bands (all unoccupied, the recipient of electrons). And this is where DFT fails on us. As mentioned in Chapters 3 and 4, calculating the electronic gap raises difficulties in DFT, where it is almost always underestimated. This failure comes from poorly estimating the electrostatic interactions between the various charges in the system. These interactions are screened by the presence of the other electrons. As the screening is unknown and there is no analytical formula for it, numerical approximations are the only solution today. Eventually, the DFT results are not reproducing experiments; the gap is underestimated, so all the optical spectra are unreliable. While recent advanced functionals, like the r2scan improve considerably the gap, the only accurate solution is to know or measure the value of the gap and then to split the valence and the conduction bands 'by hand' by the appropriate amount.

It is only natural that a lot of work was dedicated to correcting the electronic gap in DFT. Probably the most efficient correction today comes from the GW method. Here, G stands for Green's functions, which describe single-particle electronic wavefunctions, and W stands for the screened Coulombian interaction. W is written in terms of the dielectric function ϵ and the standard unscreened Coulomb potential, V_C:

$$W = \frac{1}{\epsilon} V_C \tag{5.1}$$

The dielectric function is written in terms of the electronic polarisability χ° as

$$\epsilon = 1 - V_C \chi^\circ \tag{5.2}$$

The electron polarisability is expressed using cross products between all the valence and the conduction bands, including at different wavevectors, like $< \psi_i(k+q)|e^i(q+G))|\psi_j(k)) >$. The GW method greatly improves the values of the gap. Several packages allow performing GW calculations, though the method remains heavy computationally.

5.5 Hubbard U and J Parameters

Wüstite, $Fe_{1-x}O$, is a minor mineral on the Earth's surface, usually found in rust and other weathered environments rich in iron, alongside various other oxides and hydroxides of iron. Chemical measurements have shown that the x parameter is small but never null at ambient conditions. Switch on the pressure, and x becomes smaller and smaller, vanishing eventually as the compression removes vacancies from crystal structures. More importantly, FeO forms a solid solution with MgO, and together, the ferropericlase or magnesiowüstite, (Mg,Fe)O, is the second most important mineral of the Earth's lower mantle. But for DFT calculations, (Mg,Fe)O is a nightmare that few other phases can equal. Experimentally, FeO is an insulator, but in DFT, it turns out to be a metal. FeO belongs to the class of materials called Mott insulators.

The problem, which created so much headache over the last decades, is due to the localised character of the d states of iron because of the strong correlation between these states. As DFT tends to delocalise electrons, the gap between the d states is reduced, and the d states hybridise with the surrounding Op and Fed states. This allows the d electrons of one Fe atom to hop from one site to the next. As long as the hopping probability is larger than the on-site Coulombian repulsion, the electrons can hop, and FeO has a metallic character.

One of the proposed solutions to solve this anomaly is introducing a local potential that prevents this hopping. The potential is explicitly applied on the d sites of the Fe atoms. This on-site electronic potential is the Hubbard U parameter.

The explicit introduction of the Hubbard U parameter offers an arbitrary fix to this problem, artificially opening a barrier to the hopping of the d states. The value of U can be either empirical, increased until the calculated gap matches the experimental one, or can be self-consistently calculated. The most widely used method today to self-consistently determine the U_{eff} is to use the linear response approach [74, 75]. The underlying idea is to search for the energy difference for charged systems when an electron is added to the system, and then one electron is subtracted from the system.

Box 5.1　　　　　　　　　　**How to do an LDA + U calculation?**

The +U correction for FeO in the projector augmented wave (PAW) formalism transforms it into an antiferromagnetic insulator at low pressures. As the pressure increases, the magnetic spin collapses, the gap between the t_{2g} and e_g states widens, and the d electrons pair up. The correction is applied using the few keywords detailed below. In this example, the Hubbard U = 4.2eV and J is 0 eV.

acell	1.0 1.0 1.0	# experimental ambient unit cell
rprim		# rhombohedral supercell
	1.0 0.5 0.5	
	0.5 1.0 0.5	
	0.5 0.5 1.0	
ntypat	2	#two types of atoms
znucl	26 8	#Fe and O
natom	4	#supercell that respects AFM symmetry
typat	1 1 2 2	#Fe_{up} Fe_{down} O O
spinat		
	0 0 1	# Fe spin up
	0 0 −1	# Fe spin down
	0 0 0	#no spin on O
	0 0 0	#no spin on O
usepaw	1	#use +U
lpawu	2 −1	#orbitals on which the +U correction is applied
upawu	4.2 4.2 0 0	# Hubbard U
jpawu	0 0	# Hubbard J

The Hubbard U energy correction, positive for low occupancies and negative for high occupancies, favours a population closer to an integer, making the system insulating (insulators have no partial occupancies). Hence the resulting DFT+U calculation (Box 5.1), for large enough values of U, corrects the metallic character for Mott insulators. This +U correction is typically applied only to highly localised states, like the d and f states. The value of U depends on the local environment, which can change with the actual chemistry, pressure, and so on. For example, the value of U is typically in the 3–5 eV range over 0–120 GPa pressure range [74, 75, 238].

The Hubbard J parameter is similar to U but affects the magnetic system, favouring an energy difference between the spin populations and a net magnetic moment. There is no available scheme to determine J self-consistently, and the varied parameter is an effective $U_{eff} = U - J$. However, so far, most of the simulations corrected for the Hubbard parameters assumed either a J = 0 or a J = 1.

5.6 Beyond and Alternatives to DFT+U

An improvement over DFT+U is the use of a mean-field theory. An example is the dynamic mean-field theory (DMFT), in which the highly correlated atom, the one containing the d or the f orbitals, is treated as an impurity immersed in a potential bath of the surrounding lattice [102, 9, 171, 162]. A DMFT calculation is a combined DFT + DMFT calculation. A double self-consistent loop is put into place. The standard DFT or DFT+U calculation yields the band structure of the entire lattice, including the correlated atom. Then the interaction between the highly correlated atom and the rest of the crystal is self-continuously corrected.

DFT + DMFT is still at its beginnings, partly because of the complicated nature of the theory and some computing time limitations. Several implementations of DFT + DMFT are in use today, but production simulations are rare. Various schemes to avoid electron double counting are still to be verified, and results obtained with different codes are still to be compared. DFT + DMFT can also be used together with the +U corrections.

However, just as in the case of the QMC methods, these techniques are highly dependent on the quality of the input wavefunctions. If the initial LDA or GGA calculation was wrong for whatever reason, the advanced simulations could not really help correct them.

The typical case of a magnetic material starts with a transitional metal with 5 d orbitals but fewer than 10 d electrons. Unless this atom can find some electrons to fill or some host to empty its entire d layer, it ends up in a state where some of its d orbitals are partially occupied. As each electron has an angular moment, its value and orientation define the spin of that electron. This is commonly represented as a vectorial property. Most of the time, it is represented only by its z component, which, in this case, is also named *the spin* and defines its orientation. Consequently, the spins are represented as collinear vectors with the arrows pointing *up* or *down*.

The electrons with opposite spins are placed on the available orbitals of a given atom. The difference between the total number of electrons with spin *up* and the total number of electrons with spin *down* defines the net magnetic moment. In atomic units, its value is also expressed in magneton-Bohrs, where 1 magneton-Bohr = 1 unpaired electron. Having a partial occupancy in a d- or f-orbitals-bearing structure almost automatically results in a net magnetic moment.

In practice, the net magnetic moment of a unit cell is computed as the difference between the integrals of the two spin populations of the electronic density of states in a unit cell. The local magnetic moment is more complicated to obtain. As the electronic states are described using lattice periodic Bloch functions, like every local property, the atomic magnetic moment cannot be obtained directly. Instead, the electronic states must first be projected on atomic orbitals within some, more or less arbitrary, atomic limits (see all the discussion about charges in Section 4.3). In projector augmented wave (PAW) calculations, the atomic limits may be chosen, for convenience, to be spherical, with the radius a scaling of the PAW spheres. The atomic orbitals are already defined from the part of the wavefunctions inside the spheres, and the contribution of the remaining outer electrons is obtained from integrating the outer charge inside the PAW spheres. For standard norm-conserving calculations, it is necessary to perform first projections of the wavefunctions on the atomic orbitals.

In a crystalline lattice, there are magnetic domains, typically on the size of nanometers, in which the magnetic spins are homogeneously distributed. Within a given domain, there are several possible cases allowed by symmetry to arrange the magnetic moments, listed in Table 6.1.

Table 6.1 Types of magnetic ordering.			
Magnetism type	Magnetic moment	Spin ordering	Response to external field
Collinear			
Ferromagnetism	>0	all parallel	aligned with field
Antiferromagnetism	0	all antiparallel	aligned with field
Ferrimagnetism	>0, small	partially antiparallel	aligned with field
Non-collinear			
Paramagnetism	0	random	resistive to field
Spiral	>0	partially parallel	
Non-magnetic			
Diamagnetism	0	no spin	aligned with field

The first three cases in Table 6.1 represent collinear magnetism, that is, the local magnetic spins can be oriented either *up* or *down* parallel to the z axis, depending on the dominant population. In ferromagnetic crystals, all the spins point in the same direction within the same magnetic domain. Consequently, the resultant magnetic moment is the sum of all local magnetic moments within a magnetic domain. In the magnetic domains of the antiferromagnetic crystals, the spins are aligned opposite each other. Within these domains, the resultant net magnetic moment is zero. Finally, in ferrimagnetic crystals, there are two unequal populations of *up* and *down* spins within each magnetic domain.

Hematite, Fe_2O_3 is a special case of antiferromagnet. The iron sublattice is magnetised, and the Fe spins are ordered opposite to each other. But they are also all slightly tilted out of the hexagonal basal plane, along the c axis, such that there is a total net non-zero vertical component. This particular spin orientation is called 'canted' antiferromagnetic.

In the following two cases, the spins are not parallel. In paramagnetic crystals, the local magnetic moments point in random directions of space, yielding a resultant zero magnetic moment. From a magnetic point of view, these materials behave as glasses. Hence, they can also be called *spin glasses*. Local magnetic moments come from partly occupied d or f orbitals.

In spiral magnets, the local magnetic moments are aligned in the structure according to a spiral. These materials are another special case of canted antiferromagnets. This is the case for the rare-earth-based hexagonal magnets, where the out-of-plane orientation of the magnetic spins follows a spiral pattern along the c axis. The horizontal components cancel out, and the vertical components sum up, just like for hematite. Alternating layers of rare Earths atoms and oxides or organics characterise the crystal structure of these magnets. Magnetisation occurs on the rare earth atoms. The type and thickness of the non-magnetic layer can tune the periodicity of the spiral magnetisation. The magnetic spins interact via correlation effects, like those hidden in the exchange-correlation term of the Kohn–Sham equations of the DFT.

Finally, in the last type of magnetism, the diamagnetic crystals, there is no local magnetic moment on any atom. This is the case for insulating crystals like halite or quartz. In these crystals, all the orbitals are fully occupied by two electrons, one with spin up and one with spin down. Consequently, there is no local magnetic moment.

Ferromagnetic structures are metallic; paramagnetic and ferrimagnetic crystals are usually metallic, while antiferromagnetic crystals can be either metallic or insulators. The diamagnetic crystals are insulators, as all the orbitals are doubly occupied.

6.1 Magnetic Hysteresis

When an external magnetic field acts on a material, the result is a variation of its magnetic properties. This variation as a function of the intensity of the applied field is typically represented in a magnetic hysteresis diagram. In a certain way, this represents the inertia of the system with respect to magnetic change.

In ferromagnetic materials, the magnetic spins spontaneously form on each magnetic site in the absence of an external magnetic field. They are aligned within magnetic domains, which are homogeneous but small. If the crystal is an aggregate of small domains with random spin orientations, its resultant magnetisation may be zero. When an external field is applied, the domains migrate, and the spins rotate to align with the external magnetic field. As a result, the magnetisation inside the material starts to increase. At the first application of an external field, the sample follows the dashed line in Figure 6.1. The increase is neither linear nor infinite, as it asymptotically reaches the maximum possible magnetisation, corresponding to having all the local magnetic spins pointing parallel to and in the direction of the external magnetic field.

When the field is reversed, the spins start to reverse. The reverse path is not linear either (Figure 6.1). At zero external magnetic field, there is a finite remaining magnetisation corresponding to the spontaneous magnetisation of the sample. As the field continues to reverse, the magnetisation decreases until it reaches zero. The corresponding field is called a *coercive* magnetic field. As the field amplitude increases, the spins align with the field until the maximum magnetisation is again reached, but with the spins pointing in the opposite direction from point 1. The cycle can be repeated, and the same steps are found again upon reversal of the field. Figure 6.1 shows a typical example of a hysteresis diagram for a ferromagnetic material.

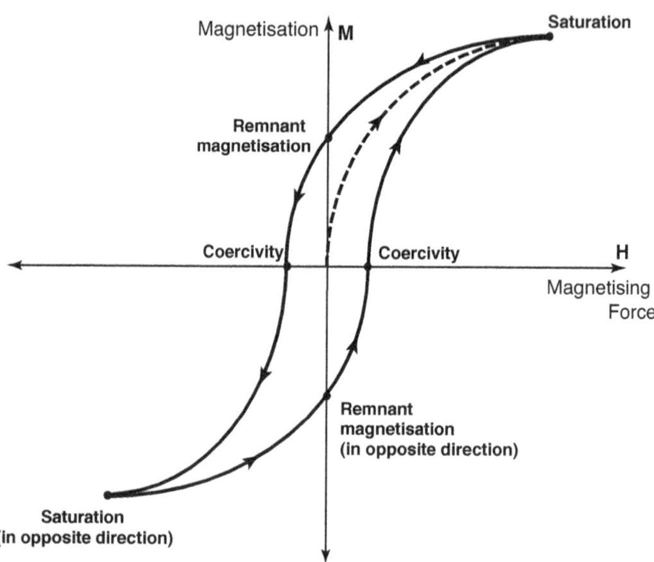

Fig. 6.1 A typical hysteresis curve for a ferromagnetic system. When the external magnetic field is switched on, the magnetic spins in the different domains start to align. If, in the initial state, the total magnetisation of the sample is zero because of disorder, the induced magnetisation of the sample follows the dashed line. The resulting internal magnetisation increases until the saturation, where all spins are aligned to the direction of the external spin. When the external magnetic field is reversed, the spins start to rotate to reorient with the new direction of the field. The process continues until all the spins have opposite orientations, and the sample reaches saturation again. The remnant magnetisation corresponds to the aligned spins after the field application. The coercive field is the field necessary to cancel the resulting internal magnetisation.

The magnetic domains interact with each other. At the boundary between the domains, the spins rotate to change from one orientation to the neighbouring one. The rotation can happen parallel to the domain wall or perpendicular to the domain wall. The energetics of the two processes are not the same, and they are also highly dependent on the material. The reversal of the spin, one way or another, takes place over several atomic layers. When the external magnetic field is applied to the crystal, the boundaries between the magnetic domains migrate, leaving behind an extended single magnetic domain. Antiferromagnetic and ferrimagnetic materials display similar behaviour under a magnetic field, but their response is weaker than for ferromagnetic crystals.

An external magnetic field induces an increasingly large and opposite magnetic moment in diamagnetic crystals. Consequently, diamagnetic crystals are repelled by the external magnetic field. The relation between the two is magnetic susceptibility. In paramagnetic crystals, the same phenomenon happens, but in the opposite sense, as the random spins align, and the induced magnetisation is reversed concerning the applied magnetic field. Paramagnetic crystals are attracted to the external magnetic field.

6.2 Curie and Néel Temperature and the Origin of Palaeomagnetism

At high temperature, because of entropy, the electronic states are excited. Above a certain temperature, the magnetic spins become disordered, and the crystal becomes paramagnetic. This is called the Curie temperature for ferromagnets and the Néel temperature for antiferromagnets. When the temperature decreases below the Curie or the Néel temperature, the magnetic spins start to align, and the sample recovers the well-known ferromagnetic and antiferromagnetic domains. This property has tremendous importance in geology.

Lavas rich in iron erupt at the surface; there are local magnetic moments on the iron atoms. When the lava is molten, the temperature is above the Curie or the Néel temperature. Consequently, there is no preferred alignment of the magnetic spins. As the temperature drops during cooling, when the lava reaches the Curie or Néel temperature, the magnetic moments on the iron atoms start to align. They are aligned according to the local external magnetic field, which is the magnetic field of the Earth. In this way, those lavas capture the orientation of the magnetic field in that particular place at that particular moment on the surface of our planet.

At the mid-ocean ridges, as the tectonic plates slowly spread, basaltic lavas continuously erupt and arrive on the top of the plates. After the eruption, they cool, and the magnetic moments on the iron atoms capture the orientation and the intensity of Earth's magnetic field along the ridge.

Later, oriented samples of these basalts are collected on the two sides of the mid-ocean ridges. In the laboratory, the natural process can be back-engineered. The orientation of the magnetic spins can be measured in the oriented basaltic samples collected in the field. In this way, the spatial orientation of the magnetic field during their eruption can be retraced, more exactly the moment when their temperature went below the Curie or Néel temperature. As the basalts erupt on the two diverging plates and as their age can be reconstructed using the radioactive decay of selected isotopes, the orientation of the plates can be reconstructed relative to the magnetic field (assuming a dipole north-south magnetic field) at different moments in time. Eventually, putting everything on a map reveals the continental drift and tectonic activity over geological time [246]. This method of orienting geological samples to the former magnetic field of the Earth is at the base of the field of paleomagnetism. Simple unpaired d electrons provided major strong evidence for plate tectonics and are used today as a powerful tool for plate reconstruction!

6.3 Contributions of DFT Calculations

Computing the local magnetic moments (Boxes 6.1 and 6.2) for Fe-bearing minerals is notoriously difficult. Describing the correct magnetisation of minerals like wüstite, FeO, or fayalite, Fe_2SiO_4 can quickly become a nightmare. Advanced DFT techniques, like the Hubbard U parameter (Section 3.5), help our community considerably. By comparison, magnetisation measurements on natural samples, even if tedious and expensive, are much easier and more reliable. However, things are different as soon as the pressure increases. This is where the computational community brings a considerable contribution.

Box 6.1	How to perform a collinear spin-polarised magnetic calculation?

For magnetic calculations, the symmetry of the electronic wavefunctions of the two spin populations needs to be broken. This is done by initialising the spins for each magnetic moment. Consequently, the energy minimisation may find a minimum where the two spin populations differ. The initialisation of the spin is done using three numbers representing the projection of the local magnetic spin along the three axes. For non-magnetised atoms, like the ones from the s or p blocks, all values are 0. In this case, there is no distinction between the two spins, whose populations are strictly equal. For collinear spin calculations, the common convention wants the initialisation to be done along the z axis (an arbitrary choice).

For ferromagnetic systems, all values may be positive or negative, but they all need to be the same and with the same sign:

```
natom       4          # an example with four atoms
ntypat      2          # two atomic types
typat       1 1 2 2    # atomic types
occopt      4          # force the system to be metallic to allow for
                       # partial band occupancies
nspden      2          # two spin components of the density
nsppol      2          # two distinct spin populations
spinat      0 0 1      # initialise the local magnetic moment
            0 0 1      # similarly on each atom
                       # one line of three entries per atom
            0 0 0      # non-magnetic atom
            0 0 0      # non-magnetic atom
fixmagmom   4          # fix the total number of unpaired electrons
                       # if unspecified, magnetisation is unconstrained
```

<table>
<tr><td>Box 6.1</td><td>(cont.)</td></tr>
</table>

For antiferromagnetic structures, the initialisation of the spin is done such as to have two distinct spin populations, but of equal amount and related by a magnetic inversion centre of symmetry; it is done with alternatively positive and negative numbers, such as to respect the inversion symmetry:

nspden	1	# one spin density
nsppol	2	# two spin populations
spinat	0 0 1	# initialise the local magnetic moment
	0 0 −1	# inversely on each atom
	0 0 0	# non-magnetic atom
	0 0 0	# non-magnetic atom

The atoms #1 and #2 need to be related by symmetry. The magnetic inversion adds up as a new symmetry operation.

<table>
<tr><td>Box 6.2</td><td>How to perform a non-collinear spin-polarised magnetic calculation?</td></tr>
</table>

For non-collinear calculations, the spin can point in any direction. Consequently, the spinat vectors can be oriented in space, either randomly or according to symmetry, such as in the case of spiral magnetisation. The nspinor value must be set at 2:

nspden	2	# consider the two spin components of the electronic density
nsppol	2	# consider the two distinct spin populations
nspinor	2	# colinear magnetism
spinat	0.1 0.2 0.3	# initialise the local magnetic moment in 3D
	−0.4 − 0.5 − 0.6	# initialise the local magnetic moment in 3D
		# one line of three entries per atom
	0 0 0	# non-magnetic atom
	0 0 0	# non-magnetic atom

In general, there are fewer calculations published in the literature for non-collinear magnetic materials.

Once the U parameter is properly calculated or adjusted so that numerical results fit experimental data, its value can be used to explore the physical properties of these minerals under extreme conditions. First-principles simulations explained the mechanism of spin transition, actually a spin collapse under pressure, for mantle minerals (e.g., Refs. [251, 146, 60]). Moreover, the predictions of the variation of the elasticity of these minerals during the spin transition fit remarkably well with seismic observables and explain a whole series of anomalies and heterogeneities of the Earth's lower mantle [224].

Closer to the surface, the DFT simulations can describe the behaviour of minerals in magnetic aggregates, the formation and migration of magnetic domains, and help interpret magnetic measurements in soils and outcrops.

The electric dipole is the product of the charge and the distance. It is a vectorial property. Polarisation is the sum of the dielectric dipoles in a crystal.

In the classical theory of polarisation from the old textbooks, the polarisation of a crystal was represented as a sum over the charge dipoles that occur in that crystal. As atoms come close to each other to form bonds, the resulting electrical bond dipoles sum up to generate the macroscopic polarisation:

$$\mathbf{P}_{\mathrm{mac}} = \sum p = \sum qr \tag{7.1}$$

where p are the electrical dipoles defined as $p = q\mathbf{r}$ occurring between charges q with opposite signs, separated by a distance $\mathbf{r}$, and the sum runs over all atomic pairs. Diatomic molecules with two identical atoms, such as N_2, H_2, and so on, have no net resulting dipole. The water molecule has one O atom and two H atoms, disposed at about 104.5 degrees. Each hydrogen atom is charged with a positive charge $+q$, and it induces on the oxygen atom a corresponding charge $-q$. Consequently, there are two dipoles, $p_{H-O} = qd_{H-O}$, which sum up as two standard vectors to define the dipole of the water molecule. However, what is the charge q? This question is so difficult to answer (see Section 4.3) that for years we knew how to calculate polarisation's variations and derivatives. Still, we did not know how to calculate the polarisation itself.

The definitions of static electric charges, based on the gradient of the electronic density as in the case of Bader charges, were already discussed (Section 4.3). But even if absolute and correct values for the charges q could be obtained following such an approach, the charges would still be delocalised in space. Then, an alternative is to try to write an integral over the charge density in the entire sample. For a crystal, this is written as:

$$\mathbf{P}_{\mathrm{mac}} = \frac{1}{V_{sample}} \int_{sample} q\rho(\mathbf{r})d\mathbf{r} \tag{7.2}$$

In this formulation, the integral must be done over the entire sample, which is finite. Charges arising on the boundaries or surfaces are hard or impossible to calculate. If the boundaries are neglected, and the polarisation is normalised to one unit cell, the integral then becomes:

$$\mathbf{P}_{\mathrm{cell}} = \frac{1}{V_{cell}} \int_{cell} q\rho(\mathbf{r})d\mathbf{r} \tag{7.3}$$

This brings the problem of the origin and the shape of the unit cell, which are both arbitrary. An average over all possibilities vanishes.

This set of approximations and alternative formulations concludes that it is extremely hard or even impossible to formulate an accurate universal formulation of the polarisation in a crystal.

7.1　The Berry Phase and the Value of Polarisation

In a crystal, the polarisation can be decomposed into two contributions, one coming from the ions and the other from the electrons, $\mathbf{P} = \mathbf{P}_{ion} + \mathbf{P}_{el}$. The ionic contribution is straightforward to calculate as the sum over point-like charged ions (Equation 7.1).

The electronic contribution is the most difficult to calculate, as discussed earlier. The solution to compute the polarisation in the framework of electronic structure calculations in methods like the DFT is to use the Berry phase [255, 257, 214]. The central idea behind this treatment is that the electrical polarisation can be related to a transient polarisation current. In quantum mechanics, the currents are related to the phase of the wavefunctions and their variations. The wavefunctions have a real and imaginary part in all the points of the reciprocal space except for Γ, where the imaginary part vanishes.

Then the electronic polarisation can be written in terms of the change of the phase of the wavefunction along a path in the reciprocal space that traverses the Brillouin zone as:

$$\mathbf{P} = \frac{e}{(2\pi)^3} Im \sum_n \int <u_{n\mathbf{k}}|\nabla|u_{n\mathbf{k}}> d\mathbf{k} \tag{7.4}$$

Because the reciprocal space is always sampled on a discrete grid of $\mathbf{k}$ points, the integral corresponds to extracting the phase from a product between wavefunctions over neighbouring points, which describe chains spanning the entire Brillouin zone:

$$... <u_{k1}|u_{k2}><u_{k2}|u_{k3}><u_{k3}|u_{k4}> ... \tag{7.5}$$

Each term of this product corresponds to the overlap between the wavefunctions on two neighbouring $\mathbf{k}$-points. This is equivalent to the projection of the wavefunction in $\mathbf{k_i}$ to the wavefunction in its neighbouring point $\mathbf{k_{i+1}}$. The projection of two wavefunctions with the same norm (as the number of electrons remains constant) corresponds to a change of phase. Consequently, the calculation requires a particularly dense grid of k points along the direction of the polarisation vector. The summation needs to be done over all fully occupied bands. The procedure is not valid for metals. The parameters of the actual calculation are exemplified in Box 7.1.

| Box 7.1 | How to compute the spontaneous polarisation using the Berry phase? |

Like many other codes, ABINIT allows the calculation of the polarisation using the Berry phase. Moreover, the polarisation can be decomposed on the electronic bands. There are two available implementations, each with its own specificities. They both require a dense grid of **k** points along the direction in which the polarisation is computed. In the case of macedonite, $PbTiO_3$, with a tetragonal unit cell with a = 3.895 and c = 4.171 angstroms, the polarisation develops along the c axis. It is accompanied by displacements of all atoms. Using the experimental parameters for the atomic positions, the input file contains the following:

acell	3.895 3.895 4.171	angstroms
angdeg	90.0 90.0 90.0	
xred	0.0 0.0 0.0	
	0.0 0.0 0.54	
	0.5 0.5 0.118	
	0.5 0.0 0.621	
	0.0 0.5 0.621	
natom	5	
ntypat	3	
typat	1 2 3 3 3	
znucl	82 22 8	
toldfe	1.0d-10	# hard convergence
ecut	40	
kptopt	2	# computes all the k points
nshiftk	1	# on one grid
ngkpt	4 4 4	# with a high density along the c axis
shiftk	0 0 0	# no need for shift
nstep	100	

The polarisation itself is computed if the berryopt variable is not zero. For a value of 1, the sequential version of the code needs to be used. This is valid for norm-conserving s. In this case, the associated variables are:

berryopt	1	# initial implementation
nberry	1	# no. of Berry calculations
bdberry	1 18	# limit the calculation to a specific
		# subset of electronic bands
kberry	0 0 1	# the direction(s) of polarisation

At the end of the calculation, the output file contains the value of the polarisation along the two directions of the reciprocal space, [0 0 1]* and [1 0 0]*. Because of symmetry, the second polarisation is zero, as the polar axis is the c axis. When using the PAW flavour of DFT, simply put berryopt −1 in the input file. Of course, as always, the full description of the variables and their values can be found in the help file of the package.

7.2 Ferroelectricity

The symmetry of the electrical polarisation and the corresponding classification of the different types of insulators follow a similar pattern to magnetisation in magnetic crystals. There are, however, several exceptions.

For dielectric crystals, polarisation is present only during the action of an external electric field, which distorts the electronic clouds. Once the action of the field is stopped, the polarisation disappears, and the electronic clouds return to their initial non-perturbed state. Halite, NaCl, is an excellent example of such a crystal. The large majority of insulating minerals is dielectrics. They are repelled by the external electric field. Note that halite is also a diamagnetic material. As a general rule, the dielectrics are also diamagnetic.

In paraelectric crystals, local dielectric dipoles can form, and they point randomly in all directions of space, the resultant net polarisation being zero. Under the action of the external electric field, these dipoles start to align, they orient against the field, and hence the paraelectric crystal is attracted to the field. High-pressure, high-temperature water ice VII is an example of such a crystal.

For crystals that have local dipoles but are centrosymmetric, the net resultant polarisation is zero. All the local dipoles that can form would mutually cancel each other. To each dipole that forms, the centre of symmetry makes it correspond to another dipole of exactly opposite orientation. These are antiferroelectric crystals. High pressure, low temperature water ice VIII is an example of such a crystal [65].

For non-centrosymmetric crystals, things are much more interesting but also more complicated. In a non-centrosymmetric crystal, there is always a net resultant dipole, however small, because the dipoles do not cancel each other. Consequently, crystals that have one singular axis of symmetry are called polar crystals. They were also called pyroelectric in the beginning. Because of the existence of a net resultant internal dipole inside each unit cell, the entire crystal has a total electric dipole equal to the number of unit cells times the dipole in each unit cell. This dipole induces the presence of charges on the fresh ideal crystal surfaces. For real crystals, these surface charges are compensated by the presence of ions. Under heating, the crystal deforms, the dipole changes, and part of these external ions evaporates. The result is a change of the charge on the surface following a change of temperature, hence the name pyroelectrics. The pyroelectric crystals are also piezoelectrics. One of the most famous pyroelectric crystals that is also piezoelectric is quartz.

Polar (pyroelectric) crystals that are switchable are called ferroelectric crystals. According to the pure definition based on symmetry, in this category of crystals, the electric dipoles are all oriented in the same direction. But the action of an external electric field results in reversing the orientation of the local dipoles. Switching the field again switches the orientation

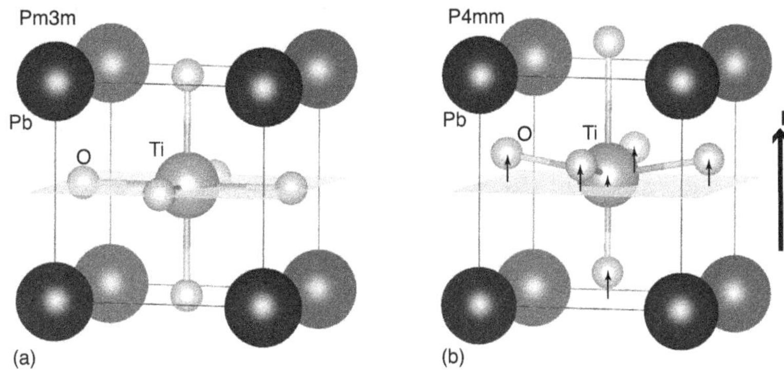

Fig. 7.1 (a) The cubic parent structure of the ABO_3 perovskite minerals, exemplified by the mineral macedonite, $PbTiO_3$, is stable at high temperatures. It is a cubic non-polar structure. (b) Upon lowering the temperature, a phase transition develops that sees all the atoms being displaced along the z axis by different amounts and in different directions. The centre of symmetry is lost during the phase transition, and the resulting structure, with tetragonal symmetry, is polar, exhibiting a net resultant spontaneous polarisation, which can be computed using the Berry phase approach described in this chapter. (c) Reverting the direction of the external electric field induces an opposite displacement, which results in the formation of a reversed dipole.

of the dipoles, the whole phenomenon describing a hysteresis loop with the electric field and polarisation similar to the magnetic hysteresis loop (Figure 6.1).

A typical case, extremely studied over the last decades, is the ferroelectric perovskite oxides. This group of ABO_3 perovskites has the mineral perovskite, cubic $CaTiO_3$ with $Pm3m$ symmetry as the archetype. The structure of the ABO_3 perovskites has a three-dimensional network of BO_6 octahedra that are connected by their corners by the O atoms. The A atoms lie in between the octahedral network. In the high-temperature cubic phase, both the A and the B atoms occupy cubic sites (Figure 7.1(a)). In some perovskites, an instability may develop at lower temperatures, which results in all the atoms being displaced along one of the Cartesian axes by different amounts, but in opposite directions (Figure 7.1(b)). Consequently, the centre of symmetry is broken, and the structure becomes tetragonal, $P4mm$, and non-centrosymmetric. An electric dipole develops between the sublattice of O atoms and the sublattice of cations. The resultant spontaneous polarisation points along the direction of the tetragonal distortion, which transforms the crystal into a polar one.

Switching a dipole requires moving charges around, which translates into a relative displacement of the cations and anions. Reversing the external field, as in Figure 7.1(c), results in opposite displacements of the B and O atoms. This leads to the formation of an opposite dipole.

Macedonite, the $PbTiO_3$ mineral with a P4mm perovskite structure, has one of the largest recorded spontaneous polarisations in the mineral world, with a computed value of 0.2 C/m^2. The value is obtained

Table 7.1 Types of electrical polarisation ordering.

Ordering type	Resulting dipole moment	Ordering of local dipoles
Ferroelectric	>0, all parallel, switchable	
Antiferroelectric	0	all antiparallel
Ferrielectric	>0, small	partially antiparallel
Paraelectric	0	random
Dielectric	0	no local dipole

using the parameters from Box 7.1. The measured values of the polarisation can go up to 0.8 C/m^2, depending on the temperature [132]. In all perovskites, under a strong enough external electric field, the displacement of the B atoms can be inverted, and therefore, the polarisation is switched.

The ferroelectric property is highly looked after by industrial applications, where external sensors can read the orientation of the electric field. The two states can be equated to the two binary states of the Boolean algebra. A dipole pointing up means an entry of one, and a dipole pointing down means an entry of zero. In this way, zeros and ones can be stored using local dipoles. The persistent polarisation of each state and the possibility of both switching and reading its orientation with an external electrical field is the basis for using these materials as ferroelectric memories in computers today.

Finally, for ferrielectric crystals, the dipoles are parallel, but like in ferrimagnetic structures, there are more dipoles oriented in one direction than in the opposite direction. The result is a net though small polarisation. Again, the ferrielectrics need to be switchable by an external electric field; otherwise, they are 'just' polar or pyroelectrics (Table 7.1).

7.3 Piezoelectricity and Piezomagnetism

Piezoelectric crystals exhibit a coupling between mechanical strain and electrical charge. Piezoelectricity occurs only in polar crystals, which do not exhibit an inversion of symmetry. Electric charge can accumulate on the faces of the crystal. This charge produces an electric field, which can be measured by external electronic devices. Any change in the distribution of the electric charge couples with the atomic positions, which results in a mechanical distortion of the lattice. Then a change in the electric field induces a change in the electric charge, which results in a change in the dimensions of the crystal. The reverse is also true: Applying a strain to the structure results in atomic displacements and thus a redistribution of the charge in the crystal, a change of the charge on the surface, and eventually, the occurrence of an electric field.

Table 7.2 Piezoelectric tensor of quartz.

−2.3	2.3	0	−0.67	0	0
0	0	0	0	0.67	4.6
0	0	0	0	0	0

Values in 10^{-12} m/V [49]

All the insulating crystals that are polar exhibit piezoelectricity. Of course, the amount of piezoelectricity is highly material dependent. Consequently, piezoelectricity is hugely important for practical applications because it allows to convert between two types of energy: electric field and mechanical deformation.

The piezoelectric tensor (third-rank tensor) connects the strain (second-order tensor) to the electric field (a vector, thus a 1st order tensor). It is typically represented by a reduced $3 \times 3 \times 3$ tensor, which, like all the other physical properties of crystals, obeys the symmetry rules of the point group. The calculation of the piezoelectric response can be done either using finite differences or the linear response. While the second method is described further in the book, the first is straightforward. The electrical polarisation needs to be computed as a vector for various amounts and orientations of the strain using the Berry phase. The resultant relation between the two yields the piezoelectric tensor. Examples of the piezoelectric tensor of quartz are given in Table 7.2.

Similarly, piezomagnetic crystals exhibit at the same time a distortion of the electronic wavefunctions, which results in the presence of magnetisation and distortion of the lattice, which results in mechanical deformation of the structure. However, because magnetic circuits are less common than electric ones, piezomagnetic crystals received considerably less attention than their piezoelectric counterparts.

7.4 Multiferroics

The presence of vectorial properties on atomic sites, like magnetisation and electric polarisation, defines a specific class of physical properties, all related to the geometrical arrangement of these vectors. This general class of properties is called *ferroic* properties. The ferroic materials are extremely useful in industry. There is an extensive range of objects from day-to-day life made of such materials, ranging from electronic watches to oil pressure sensors in cars and from fridge magnets to sophisticated electronic circuits. In geology, the ferroic properties of minerals help with their identification and separation and may serve as geological markers.

But the modern-day holy grail of ferroic crystals is the multiferroics. These crystals exhibit multiple ferroic properties; for example, they are ferroelectric and antiferromagnetic. The magnetism needs to be antiferromagnetic. Otherwise, the structure would be metallic, so it would not show ferroelectricity.

What is the interest in such materials? The answer lies in the multiplicity of responses. When an external magnetic field acts on such a multiferroic, it couples with the magnetic moment and distorts the structure by displacing the atoms and modifying the unit cell parameters. This distortion induces a change in the charge distribution, eventually leading to a change in the electric field. In short, the magnetic field is the input, and the electric field is the output. Similarly, an external electric field is transformed into a magnetic response. This coupling between various external fields and crystal responses can be used in sophisticated electronic circuits where there is a need to change different types of energy.

The key to having a good ferroic composite material is to ensure the best possible couplings between successive layers. In practice, most of the multiferroics realised at the industrial scale today are obtained as multi-layered materials: one layer is ferroelectric and the other is ferromagnetic. The coupling between the different layers is purely mechanical, as they are piezoelectric and piezomagnetic, respectively.

Alternatively, a few materials, like $BiFeO_3$ [261, 230], intrinsically exhibit both ferroelectricity and magnetism.

## 8

Mechanical Properties

Calculating the mechanical properties of crystals is one of the most straightforward operations. Its usefulness in Earth sciences cannot be underestimated, as this relates directly to the structure at the atomic level with large-scale observations. The increase in density with pressure is the most important parameter that determines the functioning of the interior of any celestial body (temperature is second). Seismic waves are acoustic waves, so their propagation directly depends on the density of the materials they pass through and their elastic properties. They offer a direct way of sampling the interior of the Earth. And finally, the rheological behaviour of the minerals constituting the interior determines the large-scale structures, including convection and deformation.

8.1 The Stress Tensor and the Pressure

The ratio between the force exerted on a surface and its area determines the pressure or *the stress*, the two terms being interexchangeable. Pressure is measured in Pascals, where 1 Pa = 1 N/m^2; pressure inside the Earth goes up to several megabars, where 1 bar = 100 kPa and 1 GPa = 10 kbars. The Earth's atmospheric pressure is 1.013 bars, the pressure at the core-mantle boundary is 137 GPa, and at the centre of the Earth's core, it is about 360 GPa [93]. For comparison, it is estimated that the pressure at the centre of Jupiter is around 4,000 GPa, depending on the equations of state used for modelling [181].

Because the force is vectorial, one can define a normal stress, corresponding to the component perpendicular to the surface, and a shear stress, corresponding to the component parallel to the surface. The force can be expressed as the derivative of the energy with respect to distance. The stress or pressure can be expressed as the derivative of the energy with respect to volume P = dE/dV. The stress is a second-order symmetric tensor, $\sigma_{ij} = \sigma_{ji}$, with components along the three directions of the space:

$$\begin{pmatrix} \sigma_{xx} & \sigma_{xy} & \sigma_{xz} \\ \sigma_{xy} & \sigma_{yy} & \sigma_{yz} \\ \sigma_{xz} & \sigma_{yz} & \sigma_{zz} \end{pmatrix} \tag{8.1}$$

where the first index represents the direction of the force, and the second index is the axis normal to the plane.

Pressure is then defined as the trace of the stress tensor:

$$P = (\sigma_{xx} + \sigma_{yy} + \sigma_{zz})/3 = (\sigma_1 + \sigma_2 + \sigma_3)/3. \tag{8.2}$$

The Einstein simplified notation helps reduce the number of indices that are needed to describe the components of the stress tensor:

$$\begin{aligned}
\sigma_{11} &= \sigma_1 \\
\sigma_{22} &= \sigma_2 \\
\sigma_{33} &= \sigma_3 \\
\sigma_{23} &= \sigma_4 \\
\sigma_{13} &= \sigma_5 \\
\sigma_{12} &= \sigma_6.
\end{aligned} \tag{8.3}$$

Then the matrix notation becomes:

$$\begin{pmatrix} \sigma_{11} & \sigma_{12} & \sigma_{13} \\ \sigma_{12} & \sigma_{22} & \sigma_{23} \\ \sigma_{13} & \sigma_{23} & \sigma_{33} \end{pmatrix} = \begin{pmatrix} \sigma_1 & \sigma_6 & \sigma_5 \\ \sigma_6 & \sigma_2 & \sigma_4 \\ \sigma_5 & \sigma_4 & \sigma_3 \end{pmatrix}. \tag{8.4}$$

If the three components are equal and the off-diagonal components are null, then the pressure is hydrostatic. This is the case for homogeneous fluids. In solids, deviations may occur, representing non-hydrostatic or deviatory stresses. They are obtained by subtracting from the stress tensor the hydrostatic value; the result can be either off-diagonal components or different values of the uniaxial stresses. The off-diagonal stresses are also called shear stresses. The consequence of the deviatoric stresses may be crystal deformation and preferentially oriented texture. The textures found in metamorphic rocks are formed mainly under the influence of deviatoric stress.

In calculations, the pressure can be computed by monitoring the change in energy due to a volume change. In practice, as the number of plane waves depends on the unit cell volume and the kinetic energy cutoff, keeping the volume constant, a variation of the number of plane waves is equivalent to a change in volume (Figure 8.1). Then a convenient way for the plane wave codes to determine the stress tensor is by computing the variation of the total energy induced by a change of the actual number of plane waves. This facilitates the entire procedure, and in most of the ab initio codes today, the calculation of the stresses is implicit and does not require any special input keyword. The result is typically printed at the end of the self-consistent loop after the energy is converged (Box 3.4). In some instances, like some of the QMC codes, where the stresses (nor the forces) are not directly computed, several calculations of the total energy need to be done as a function of volume; then the stress is obtained as the energy derivative with respect to volume.

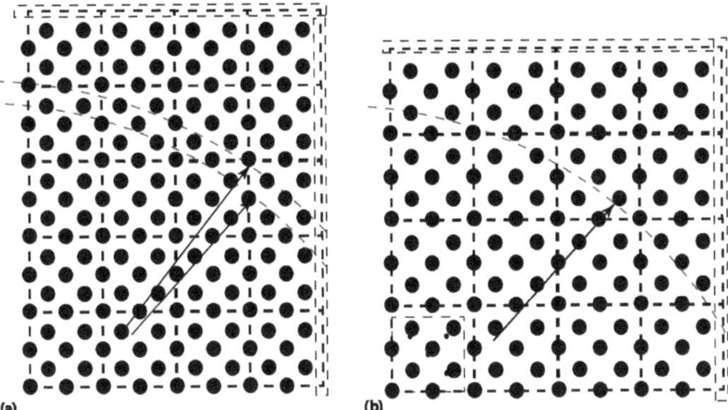

Fig. 8.1   A variation of the volume induces a change in the distance between the **k**-points. Consequently, the number of planewaves whose eigenvectors fall on the **k**-points also changes. This induces a change in energy, which can be related to pressure. On the other hand, a variation of the kinetic energy cut-off at fixed volume also implies a change in the number of planewaves, and, thus, in energy. The two approaches, (a) infinitesimal variation of the kinetic-energy cutoff or (b) infinitesimal variation of the volume, yield a change in the energy. The resulting variation of the energy is equivalent and equal to the pressure: $dE/d(\# \text{plwv}) = dE/d(\text{cut-off}) = dE/dV = P$.

8.2 Structural Relaxations

The solids that are not constrained mechanically by uniaxial or shear stresses will relax towards hydrostatic pressures and minimum forces on the atoms. This procedure lies at the base of the structural determinations in *ab initio* calculations. Because the atomic forces correspond to the energy derivative with respect to the atomic positions, the minimum energy is obtained when all the atoms lie at the bottom of their potential wells, which corresponds to null forces (Figure 2.1). The same is true for the stresses. The minimum energy is obtained at pure hydrostatic stresses, at the bottom of the potential wells of the energy with respect to volume deformations.

The theoretical structures are obtained by displacing the atoms until the forces are null and by deforming the unit cell until the stresses are purely hydrostatic. These are coupled processes, as a mechanical deformation of the crystal induces a change in atomic positions, which induces a change in interatomic forces. Similarly, a change in atomic positions induces a change in stresses. The exception is found when the reduced positions of the atoms are fixed by symmetry.

In practice, the structural relaxations are done using two minimisation cycles. The geometry is changed in the outer structural cycle according to the computed forces, stresses, and energies. Once the new geometry is set, the inner electronic cycle computes the forces and the stresses. Convergence is achieved when the interatomic forces are null, the stresses are hydrostatic, and the energy is at a minimum. The result is the theoretical

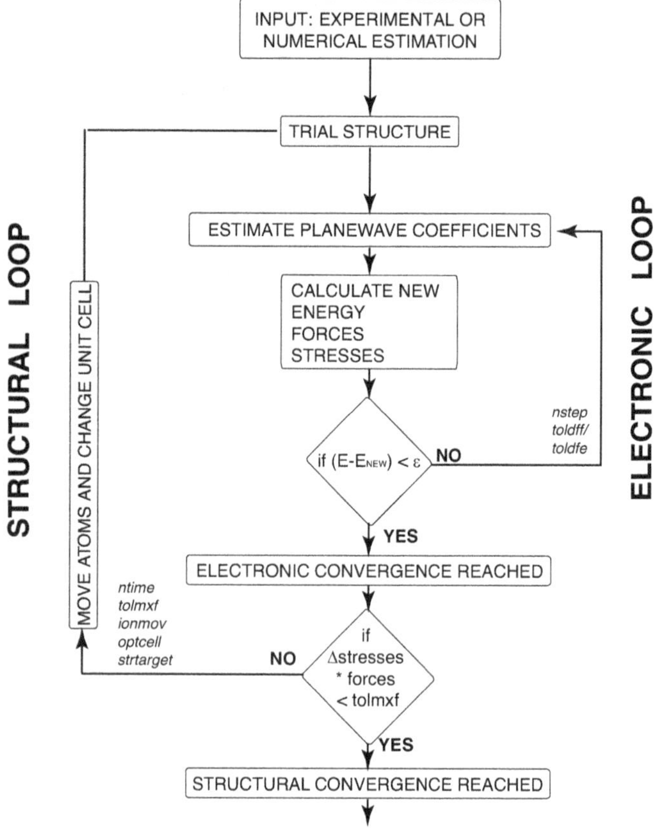

Fig. 8.2 Simplified scheme of the standard relaxation calculations. The structural parameters, that is, the reduced atomic positions and the size and shape of the unit cell, are adjusted in the outer loop. The adjustments are made based on the forces and stresses computed in the inner loop, electronic. Convergence is reached when the forces are null, and the structure is hydrostatic. In abinit, there is a criterion to estimate the quality of the structural result based on a combination of forces and stresses. The different abinit keywords needed for the relaxation are shown in Box 8.1. Note that if a theoretical structure is correctly relaxed, it lies at the absolute energy minimum for those thermodynamic conditions.

structure. Figure 8.2 illustrates the two cycles, electronic and structural, and Box 8.1 details the workflow of a standard relaxation calculation.

The refinement of the atomic positions and the lattice parameters employs minimisation algorithms that can estimate the best combinations of geometrical changes to attain the global energy minimum. When the energy hypersurface, that is, the value of the energy as a function of structural degrees of freedom, is smooth, efficient algorithms, [220], can reach the energy minimum in less than a couple of dozen steps. If the minimisation takes longer, the structural relaxation should be done in several steps. The unit cell parameters should be fixed, and the internal degrees of freedom are first relaxed. Then the entire geometry optimisation can be restarted using the relaxed internal atomic positions.

Box 8.1   **How to perform a structural relaxation?**

In these simulations, the atoms move to minimise the interatomic forces, and the unit cell changes size to obtain only hydrostatic stresses. Symmetry constraints on the cell parameters and the atomic positions are preserved. The relaxation can be done at a given pressure or a fixed volume. Because the volume changes during a structural relaxation, the number of planewaves that enter the computation of the total energy varies. Consequently, the energy values are not directly comparable between the different relaxation steps. The solution to this problem is to add a spherical shell around the $\mathbf{k} + \mathbf{G}$ sphere, where the weight of the $\mathbf{k} + \mathbf{G}$ points fades away. Some codes add a further constraint on the maximum volume change allowed between consecutive relaxation steps, for example, needed to prevent memory overflows.

The following table gives typical keywords of *abinit* needed in structural relaxations.

ionmov	2	# atomic relaxation
optcell	2	# full unit cell relaxation
ecutsm	1	# additional spherical shell that
		# preserves the energy consistency
		# in terms of number of planewaves
dilatmx	1.2	# limits the volume change
ntime	60	# forces exit of the relaxation loop
tolmxf	1.0d-05	# convergence on forces & stresses
toldff	1.0d-06	# convergence on forces in Ha/Bohr
strtarget	100/29421.033	# target pressure.
		The ratio transforms Ha/Bohr^3 into GPa

The convergence of the forces needs to be one order of magnitude better than the mixed forces and stresses. For symmetric systems, like in the case of halite, NaCl, or diamond, C, where the interatomic forces cancel each other, the convergence of the energy should be used as the exit criterion for the inner cycle.

There might be even cases when reaching a very flat or a very irregular energy hypersurface makes finding the energy minimum a daunting task. In these cases, BFGS exhibits an oscillatory behaviour and might even completely fail to find the minimum. This is a nightmarish scenario, especially when trying to confirm experimental data, which are produced *on-the-fly* by synchrotron users. In this case, alternative minimisation schemes, like damped molecular dynamics, might be more successful. In this procedure, the atoms are moved in small steps, following the direction of the interatomic forces, until the latter become minimal. This has a big chance of succeeding but at a high computational cost.

When searching for convergence, the criterion is highly relevant to the quality of the result and the time spent on relaxation. Figure 8.3 shows the evolution of various parameters during the relaxation of quartz, SiO_2.

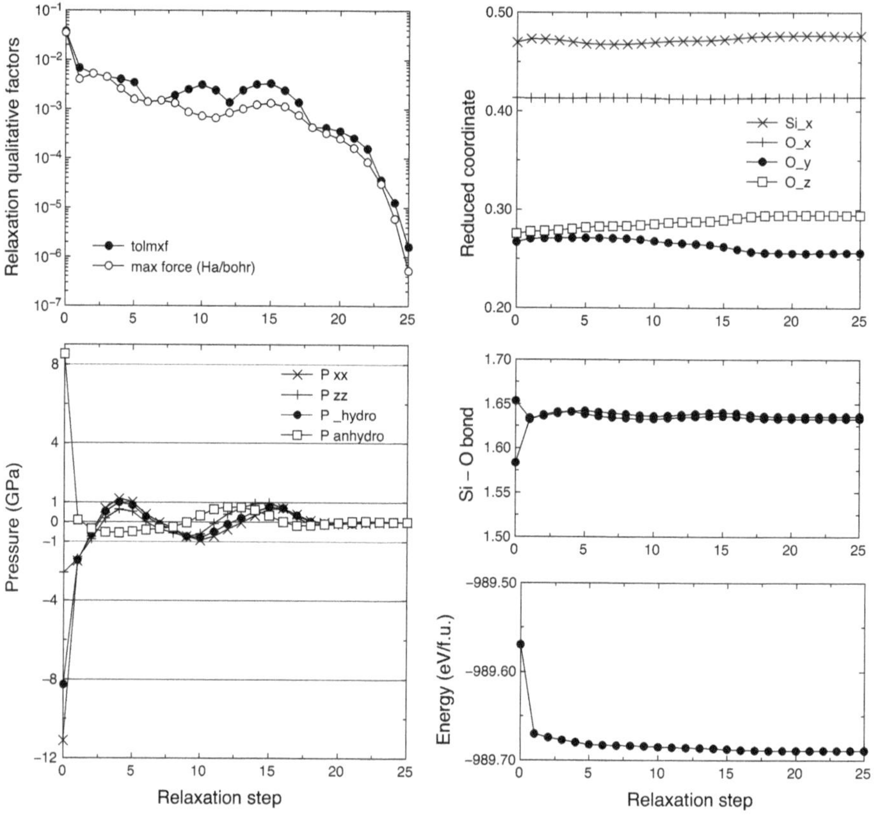

Fig. 8.3 Simplified scheme of the standard relaxation calculations. The structural parameters, that is, the reduced atomic positions and the size and shape of the unit cell, are adjusted in the outer loop. The adjustments are made based on the forces and stresses computed in the inner loop, electronic. Convergence is reached when the forces are null, and the structure is hydrostatic. In abinit, there is a criterion to estimate the quality of the structural result based on a combination of forces and stresses. The different abinit keywords needed for the relaxation are shown in Box 8.1. Note that if a theoretical structure is correctly relaxed, it lies at the absolute energy minimum for those thermodynamic conditions.

The process starts with experimental values for the internal degrees of freedom of the structure. The relaxation is done using the parameters listed in Box 8.1.

In abinit, the quality of the relaxation is given by the ratio between the stresses and the forces, represented by the *tolmxf* keyword. In other codes, the energy difference between successive relaxation steps is monitored. At convergence, these approaches yield similar, if not identical, structures. For the relaxation of the structure of quartz, the tolmxf factor gains almost one order of magnitude after the first step, and the energy decreases by 0.1 eV/formula unit. The following six relaxation steps mark another order of magnitude gain in tolmxf. The following steps show an oscillation of the tolmxf parameter. The maximum force on the atoms is always about one order of magnitude smaller than tolmxf. After about 20 steps, the structure does not change significantly anymore. The tolmxf parameter reaches a value of about 1.0E-04, which is a good indication of structural convergence. At this level, the maximum force is also on the order of 1.0E-04. In this example, we pushed the convergence down to tolmxf = 1.0e-05. Unless specific cases, non-hydrostaticity of less than 1 GPa should be considered good enough. In general, the interatomic forces should be less than $5 * 10^{-5}$ Ha/Bohr or less than $2.5 * 10^{-3}$ eV/$\mathring{A}$.

In most cases, the relaxation yields theoretical structures that deviate by only a few per cent from the experimental ones. In Section 3.2, the exchange-correlation plays an important role in the quality of the relaxation. In general, LDA yields bond lengths underestimated by 1–2% with respect to the experimental values and interatomic angles underestimated by about one degree. On the contrary, GGA yields bond lengths that are overestimated by 1–2% and angles that are overestimated by one degree. These values depend highly on the materials studied and the thermodynamic condition.

As a rule of thumb, crystal structures with relatively homogeneous electronic densities are better refined in *ab initio* calculations than structures with large heterogeneities. For example, diamond-like structures, B1 or B2 structures, and so on, are very well reproduced by calculations. But the relaxations of micas or zeolites fail, sometimes by more than 5% compared to the experiment, because DFT has difficulties solving localised charges and large void spaces in the structure. In other isolated cases, like FeSi, both LDA and GGA fail, as both underestimate the lattice parameters [56]. In some cases, more sophisticated exchange-correlation formalisms, like meta-GGA or hybrid functionals, yield better results, especially for molecular solids, though they come with a heavy computational cost (see the discussion in Section 3.2).

Differences between theoretical results and experimental data also come from temperature. As DFT is a ground-state theory, static simulations are performed at 0K, with no electronic excitations considered. There are numerical techniques to approximate the Fermi–Dirac distribution of electrons with temperature [180]. They can solve some of the discrepancies, such as those related to the thermal pressure of electrons. But when comparing structural results at 0 K with experimental data at room temperature, differences arise from the bare thermal dilatation of the structures, which adds to the inherent DFT errors.

Finally, the structural relaxations allow for crystal structure prediction. This is a field under great expansion with enormous potential. The user typically chooses the chemistry for a given mineral or material, some external parameters, like the pressure or the density, or target values for specific physical properties, like magnetisation or polarisation. Then more or less automatic algorithms generate random or educated initial guesses for the structures. The *ab initio* methods refine these structures under the given constraints. The theoretical structures with the lowest enthalpies, or even better, the lowest Gibbs free energies, are the most promising candidates for a polymorph to adopt that structure at those conditions. Several software packages available today solve these problems using various methods and implementations. They employ *ab initio* packages to compute the forces, stresses, and energies of the various theoretical structures. The *ab initio* random structure searching [208] starts from the initial random positions of atoms and tries to find the global energy minimum; various educated guesses help the search. The Calypso code uses the particle-swarm method to find the optimal structure; the search follows and improves the local minimum [262]. The Uspex code uses genetic algorithms to reach the global energy minimum [195].

8.3 Compressibility and the Equation of State

Under hydrostatic pressure, the equation of state (EOS) relates the pressure and the volume. This simple and fundamental relation is extremely useful for modelling the structure of planetary interiors.

There are numerous formulations for the EOS circulating in the literature today. Some of the widely used are obtained from expansions of the energy or pressure in the Taylor series as a function of deformation. There are two possible mathematical descriptions for the deformation: the Eulerian strain, where the deformation of a given volume is expressed relative to a fixed external reference, and the Lagrangian strain, where the deformation is defined relative to the initial internal coordinates. At small stresses, the two expressions are equivalent [152, 151].

Using the Eulerian description, with f as the notation for the strain:

$$f = \frac{1}{2}[(V_0/V)^{2/3} - 1] \tag{8.5}$$

the Taylor expansion of the Helmholtz energy, H, becomes:

$$H = H_0 + af + bf^2 + cf^3 + df^4 + \dots \tag{8.6}$$

Starting from this formulation, with the pressure, P = dH/dV, for a homogeneous solid, Birch [33] and Murnagham [188] proposed several EOS, which eventually resulted in the Birch–Murnagham (BM) EOS. The BM EOS was designed to describe the behaviour of isotropic solids under hydrostatic pressure. It is widely used today for describing the behaviour under compression of most Earth and planetary materials. The full fourth-order form is given as:

$$P = 3Kf(1+2f)^{5/2}\left[(1+\frac{3}{2}(K'-4)f + \frac{3}{2}(KK'' + (K'-4)(K'-3)\right.$$
$$\left. +\frac{35}{9})f^2\right] \tag{8.7}$$

Most of the time, K'' is neglected. With $K'' = 0$, only the terms in K and K' are preserved. This simplified form is called the third-order BM (BM3) EOS. When there are not enough data points, it is quite common practice to fix $K' = 4$, and only fit V_0 and K. This defines the second-order Birch–Murnagham (BM2) EOS. From its construction as a Taylor series, the BM EOS works best for isotropic crystals and becomes less and less accurate for highly anisotropic solids.

Many alternative EOS have been proposed, some better adapted to specific conditions. Amongst those used in mineral physics, the Vinet EOS is particularly appropriate in the high-pressure regime:

$$P = 3K \left(\frac{1 - \eta}{\eta^2} \right) exp(3/2(K' - 1)(1 - \eta) \tag{8.8}$$

where $\eta = (V/V_0)^{1/3}$ is the relative deformation from the initial conditions.

The EOS formulations above are all athermal. The thermal part can be added in terms of volume dilatation. Several forms of thermal EOS have been proposed. A thermal EOS can be obtained either from interpolating athermal EOS fit along different isotherms or from a general fit to the entire P - V - T set of points.

The EoSfit software [8] is a good option for fitting various EOS with a wide variety of thermal and athermal formulations. Also, their documentation is an excellent tool for finding out about other forms of EOS.

All the EOS yield a series of characteristic material parameters, like the equilibrium volume at 0 GPa, V_0, and the pressure derivatives with respect to volume up to n^{th} order. This last parameter defines the order of the EOS. The first-order derivative is the bulk modulus, $K = dP/dV \mid_V$, which is expressed in units of pressure and is a measure of compressibility: how fast the pressure increases with squeezing volume at the beginning of the compression. Table 8.1 lists the bulk moduli for the minerals constituting the original Mohs scale for hardness, discussed in Section 8.4. The largest bulk modulus of a crystalline material recorded so far belongs to diamond, which has a value of 430 GPa. Higher values were obtained only for polycrystalline samples of nano-scale diamonds, whose values go beyond 442 GPa [87]. Bulk moduli values higher than 300 GPa qualify for *extrahard* or *ultrahard* materials. There is a continuous quest to find such materials due to their wide-ranging applications in industry. Higher-order derivatives of the bulk modulus, $K' = dK/dP$, $K'' = dK'/dP$, and so on, express the behaviour of the material at high compression.

Table 8.1 Several scales for mineral hardness. Frederich Mohs designed a first scale (1822) based on relative values. The absolute scale is based on Vickers microindentation measurements [260]. The bulk modulus stems from the equation of state [26]. The scales are non-linear. Bulk modulus is in GPa.

Mineral	Formula	Mohs	Vickers	Bulk modulus
Talc	$Mg_3Si_4O_{10}(OH)_2$	1	1	41
Gypsum	$CaSO_4.2H_2O$	2	2	42
Calcite	$CaCO_3$	3	14	73
Fluorite	CaF_2	4	21	86
Apatite	$Ca_5(PO_4)_3(OH^-,Cl^-,F^-)$	5	48	80*
Orthoclase	$KAlSi_3O_8$	6	72	54
Quartz	SiO_2	7	100	38
Topaz	$Al_2SiO_4(OH^-,F^-)_2$	8	200	167
Corundum	Al_2O_3	9	400	253
Diamond	C	10	1,500	430

* Values for apatite vary with the chemistry.

The typical calculation of an EOS curve consists of relaxations of the structure at several volumes in the relevant pressure range. At least six different pressure–volume points are needed to improve the numerical fit. To save time, the calculations can be iterative, that is, the output of one calculation can be used as the input of the next. Compressing or expanding the structure in consecutive small steps improves the convergence to the final desired pressure. Box 8.2 gives an example of a recursive relaxation calculation of SiO_2 quartz up to 12 GPa.

Box 8.2	How to calculate the equation of state?

For an EOS, you will need several pressure–volume points. The sequence of pressure-density points is obtained by performing a series of structural relaxations targeting a given pressure. In the multi-dataset mode of abinit, you can perform all the structural relaxations in one chained calculation. The advantage is to use the structure relaxed at the previous pressure step as a starting guess for the new simulation. This saves considerable computing effort. For a structural relaxation between 0 and 100 GPa in 10 GPa steps, the abinit input file should contain the following:

ndtset	11	# number of data points
strtarget:	3*0.0/29421.033 3*0.0	# target pressure at 0 GPa
strtarget+	3*-10.0/29421.033 3*0.0	# target increase in 10 GPa steps
ionmov	2	# atomic relaxation
optcell	2	# full unit cell relaxation
ecutsm	1	# additional spherical shell that
		# preserves the energy consistency
		# in terms of number of planewaves
dilatmx	1.2	# limits the volume change
ntime	60	# forces exit of the relaxation loop
tolmxf	1.0d-05	# convergence on forces and stresses
toldff	1.0d-06	# convergence on forces, Ha/Bohr
getxred	−1	# use previously parameters
getcell	−1	# use previous cell

Note the use of atomic units for pressure, where 1 GPa $= 29421.033$ HaBohr^3. The pressure increment is negative because of the sign convention in abinit.

Do not forget to leave the variables related to the structural relaxations unchanged: optcell, ionmov, tolmxf.

The computed values of the pressure and the volume are needed for the EOS fit. For this, use the *grep* function of Linux. Then use software that can fit non-linear equations and where you can define the type of equation. There are several options available, some as free software, some licensed. Alternatively, you can write a script to fit the EOS. EosFit software (www.rossangel.comtext_eosfit.htm) is a standalone application that can fit in various ways pressure–volume data. Moreover, EosFit can even analyse temperature-dependent data to yield thermal EOS.

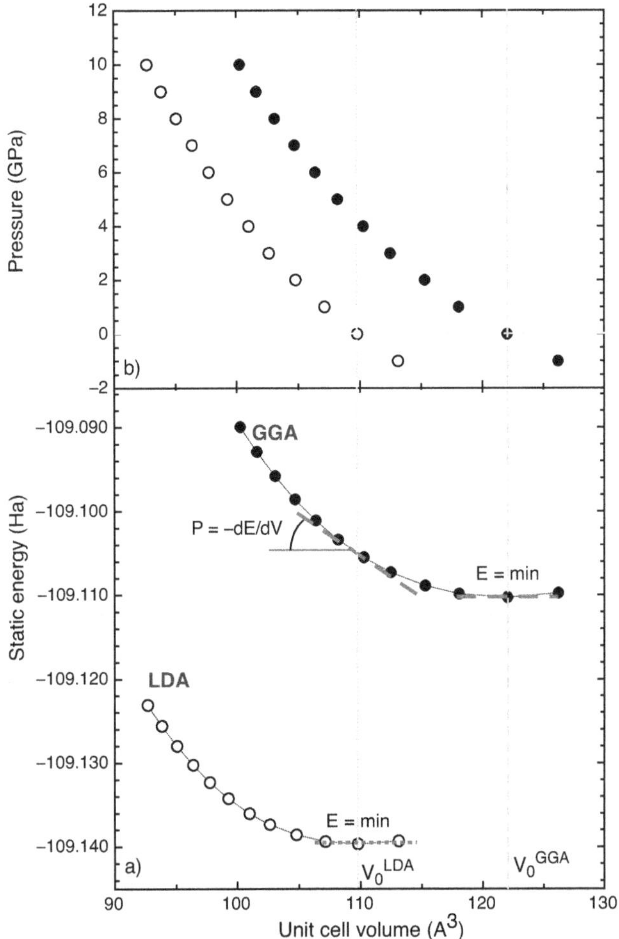

Fig. 8.4 Compressibility curve of SiO_2 quartz. See Box 8.2 for the model input files used to perform the calculations. LDA underestimates and GGA overestimates the experimental equilibrium volume (113.3 Å^3).

Figure 8.4 shows the variation of the computed static (T = 0) energy of SiO_2 quartz as a function of unit cell volume, obtained using the parameters in Box 8.2. The negative tangent to the energy curve at any point gives the pressure. The minimum of the energy corresponds to the horizontal tangent at the volume, V_0, which is the equilibrium volume at ambient pressure, P = 0 GPa. The energy increases with decreasing volume and increasing density. At small volumes, the tangent becomes steeper as the pressure increases. At higher volumes than V_0, the pressure becomes negative. At first sight, this is counterintuitive, especially from an experimental perspective: Negative pressure does not occur in experiments! However, this is nothing else than tension, a common constraint in materials science. At the atomic level, negative pressure can be achieved in the case of intergrowths. This has already been observed, for example, in feldspar intergrowths. Because of the entropy, there is a continuous solid solution between K and Na feldspars at high temperatures. As the temperature decreases, an immiscibility gap opens, and two phases separate, each rich in one of the end-member cations. This is known as the perthitic texture. Albite intergrown within a high sanidine matrix exhibits longer Si-O bonds, translated as a larger b cell parameter than at ambient conditions, indicating negative pressure [155]. In ab initio simulations, it is common practice to go at volumes slightly larger than the equilibrium one at ambient conditions. This helps better constrain the equilibrium conditions, fit the EOS, and explore phase diagrams at higher temperatures by offering educated guesses of the possible structures at larger volumes.

8.4 Static Transition Pressure

The relaxation calculations can be performed at various thermodynamic conditions. They yield the theoretical structures at different pressures (and temperatures) and their energies. In the static case, for which the temperature is 0 K, there is no thermal component to the free energy. From the calculations, the enthalpy is obtained as $H = E + PV$ (Equation 1.2), where E, P, and V are, respectively, the internal energy, the pressure, and the volume.

Box 8.3 illustrates the general procedure for extracting the enthalpy from a series of calculations at different pressures. All the necessary values are contained in the output files. In abinit (and many other codes), all properties are given in atomic units and are normalised to the content of one unit cell.

Comparing the enthalpies of various polymorphs as a function of pressure directly gives the relative stability of the different crystal structures. From the definition of enthalpy, it results that the enthalpy variation with pressure is quasi-linear. The standard procedure is to perform a series of structural relaxations on a sequence of pressures for all polymorphs and compare the results. Box 8.4 shows how to compute the enthalpy for two polymorphs of halite, NaCl, the B1 and the B2 phases, as a function of pressure. The absolute values of the enthalpy for the two structures are similar, so their crossing is hard to observe in a standard plot (Figure 8.5). A better representation is to show the difference in enthalpy between the two phases (Figure 8.5) as a function of pressure. This gives an immediate answer to where the transition pressure lies.

This extremely useful exercise offers an idea about the relative stability of different polymorphs. For many silicate and oxide minerals, the thermal contribution in the range from absolute 0 K (corresponding to the pure static calculation of the enthalpy) to the ambient temperature of 300 K is limited. Consequently, the results of static calculations can be easily compared to actual data of high-pressure experiments performed at ambient temperature. Of course, there are exceptions that require special care.

| Box 8.3 | How to calculate the transition pressure? |

Once you have performed the calculations for the EOS for at least two polymorphs of the same mineral, you can extract and compare their enthalpies. The most efficient way to extract the different components of the enthalpy from the output file is to use the Linux/Unix function *grep*:

grep Etotal file_polymorph1.output

grep Pressure file_polymorph1.output

grep ucvol= file_polymorph1.output

Only the last value is relevant.

Then do the same with the second polymorph. Note that in the output of abinit, the energy is expressed in Hartrees, the pressure in GPa, and the volume in Bohr cubes. Use appropriate transformations to express the enthalpy, for example, in eV/formula unit (also check how many formula units, or molecules, you have in the unit cell):

$$H \text{ (eV/fomula unit)} = (\text{Etotal} + \text{Pressure} \times \text{Volume}/29421.033)/\text{no.fomula.units} \times 27.2116.$$

Sometimes, it is helpful to transform the enthalpy units from eV/formula unit into kJ/mol. You can solve that exercise for fun.

Note that the number of formula units per unit cell depends on the crystallographic setting. Inner-centred and single-face-centred lattices have double the number of formula units between the conventional and the primitive lattice. This number differs by a factor of four for face-entered lattices.

| Box 8.4 | How to calculate the B1–B2 transition pressure for NaCl? |

First, you need to perform the structural relaxation on a regular grid of pressures in a given pressure range, for example, every 10 GPa in the 0–100 GPa pressure range. The relaxation must be done for both the B1 and the B2 structures. Use the following lines in your input file:

optcell	2	# modify the unit cell
ionmov	2	# displace the atoms
dilatmx	1.2	# change volume
ecutsm	1	# smooth number of planewaves
tolmxf	1.0d-04	# total convergence
toldfe	1.0d-09	# convergence on energy
ndtset	11	# number of pressure points
strtarget:	3*0.0/29421.033	# initial pressure
	3*0.0	
strtarget:	3*-10.0/29421.033	# external pressure
	3*0.0	

Note that the increment of the pressure is again negative because of the sign convention for external versus internal pressure in abinit. The exit criterion for the electronic convergence loop was changed to *toldfe* from *toldff*, because the atoms are fixed by symmetry, and the forces are strictly 0.0E+00.

Then, once the calculations are performed, the *grep* function extracts the values of the relevant thermodynamic properties directly from the output file. *grep Pressure* gives all the Pressure values. *grep Etotal =* gives all the internal energy values.

Finally, the *grep ucvol =* gives all the volumes obtained during all the structural relaxations. As stated before, only the last value of each relaxation is relevant.

Then take all the values extracted from the output file and use simple spreadsheet software to compute the enthalpy for both the B1 and the B2 phases. Finally, use the B1 as the reference phase and plot the difference in enthalpy ΔH, between the B2 and the B1 phases: $\Delta H = H_{B1} - H_{B2}$. The result is shown in Figure 8.5 for both LDA and GGA calculations.

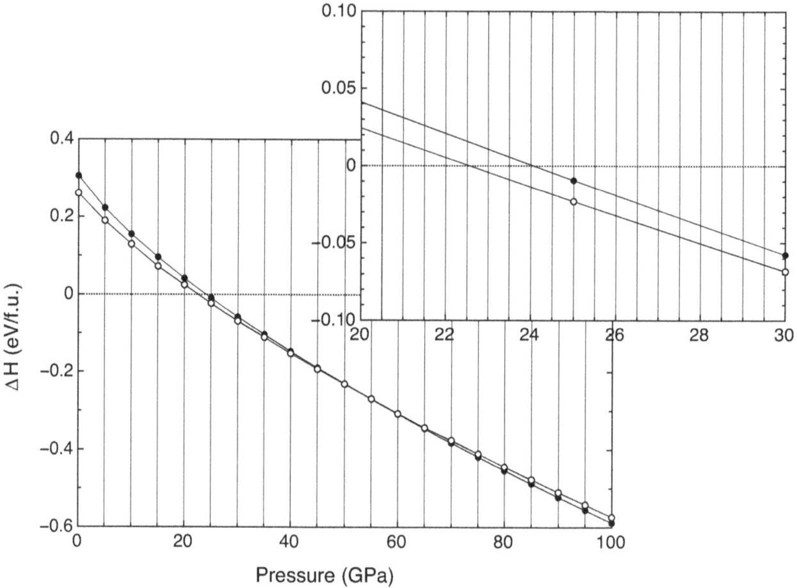

Fig. 8.5   Enthalpy difference between the B2 and the B1 phases of halite, NaCl, as a function of pressure. At 0 K, both the LDA (solid symbols) and the GGA (open symbols) underestimate the transition pressure with respect to the experimental observations at 300 K [168]. LDA yields a transition pressure of 24 GPa and GGA of 22.2 GPa, compared to the experimental values at 26.8 GPa. The main cause of the difference is the temperature.

For example, the OH groups present in a mineral often undergo an order-disorder transition at temperatures below ambient values. It is the case for brucite [247], gypsum, hydrohalite [53], and so on. Many of these minerals are present on the surface or in the shallow parts of the icy worlds of the outer solar system. In these cases, the relative stability of different polymorphs at ambient temperature may be quite different than at low temperatures and static conditions. At high temperatures, the enthalpy difference is less reliable for defining relative stability fields, and the Gibbs free energy needs to be specifically computed and compared.

8.5 Hardness

In the real life of a field geologist, the bulk modulus is reflected in hardness, a physical property highly relied on for mineral identification. Hardness is defined as the resistance of the mineral to scratches. In 1822, Frederich Mohs, a German mineralogist and naturalist, established an empirical relative scale of hardness for minerals, ranging from 1 to 10, with 1 being the weakest and 10 being the hardest (Table 8.1). Talc, the softest mineral on the Mohs scale, is a phyllosilicate in which parallel sheets of polymerised AlO_6 and SiO_4 are loosely held together by weak perpendicular van der Waals bonds. Because of these weak bonds, talc is easily breakable, that is, it has perfect cleavage, and in nature it is always found as an aggregate. Scratching it with the nail is easy as the sheets slide one on top of the other, and the entire aggregate can be easily reduced into a powder. The following mineral is gypsum, with ionic and van der Waals bonds. Several salts, where cations and anionic groups are held together by ionic bonds, follow next. Starting from the middle of the scale, there are several polymerised aluminosilicates, like quartz, orthose, and topaz. Corundum, a mineral with strong covalent bonds, is second last. The scale tops with diamond, the polymerised form of carbon with sp^3 orbital hybridisation.

But the Mohs hardness scale suffers from two major drawbacks: relative and non-linear. To overcome these problems, in metallurgy and, more generally, in materials science, there are different ways to quantify hardness. One dynamic measurement is the indentation method, in which a pyramid of diamond with an angle between opposite faces of $136°$ and under a lightweight is released apex down to penetrate the surface of a material. This is called the Vickers hardness, or sometimes micro-hardness, because of the reduced load. Table 8.1 compares the Mohs and the Vickers scales for hardness with the bulk modulus obtained from the EOS measurements [260]. The relation between the Mohs and Vickers scales and the bulk moduli are proportional but not linear.

8.6 Elastic and Compliance Tensors

The deviation in shape and size of a crystal from its equilibrium unit cell at hydrostatic conditions represents the deformation, characterised by the spatial *strain* tensor:

$$\begin{pmatrix} 1 - \epsilon_{xx} & \epsilon_{xy} & \epsilon_{xz} \\ \epsilon_{xy} & 1 - \epsilon_{yy} & \epsilon_{yz} \\ \epsilon_{xz} & \epsilon_{yz} & 1 - \epsilon_{zz} \end{pmatrix} \tag{8.9}$$

where, just like for the stress, the first index is the direction of the deformation, and the second is the axis of the normal to the plane. Typical strain values in natural systems are on the order of the per cent.

Figure 8.6 shows the mechanical response of a mineral under deformation, that is, strain. This diagram has two main regions: the linear and the non-linear response. In the linear part and the beginning of the non-linear part of the diagram, once the deviatoric stress is released, the deformation vanishes, and the crystal retrieves its initial shape and size. The linear portion of this region, from the origin to the yield point, defines the linear elasticity and is of fundamental importance for Earth sciences, as it corresponds to the domain of seismic wave propagation. The highest stress at which the crystal reverses to its original shape is called yield stress. Beyond this point, the deformation is partly irreversible. In this region, the elasticity is non-linear. The amount of non-linear elasticity depends on many factors and is highly material dependent. The remnant deformation after stress release corresponds to the plastic behaviour of the structure.

The Hooke law states that the elasticity tensor, **C**, gives the linear stress response of a mineral under small *strain*. As both the stress and the strain are second-order tensors, elasticity is a fourth-order tensor whose components satisfy:

$$\sigma_{ij} = \sum_{kl} C_{ijkl} \cdot \epsilon_{kl} \tag{8.10}$$

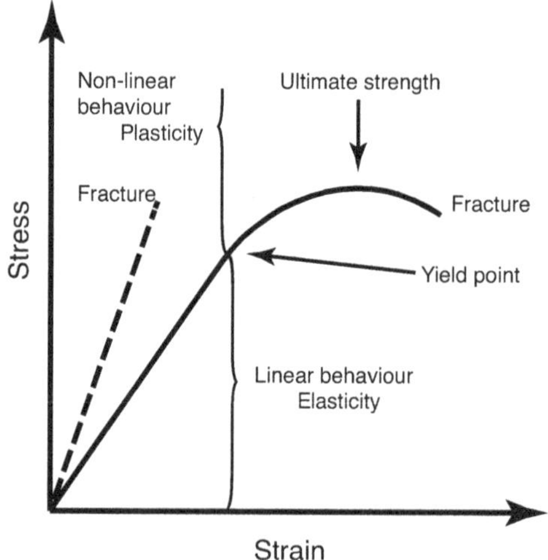

Fig. 8.6 Stress-strain relation for ductile materials (solid line) and for brittle materials (dashed line).

The inverse of the elasticity tensor is the compliance tensor, $\mathbf{S} = \mathbf{C}^{-1}$, which relates the strains to the applied stresses:

$$\epsilon_{ij} = \sum_{kl} S_{ijkl} \cdot \sigma_{kl} \tag{8.11}$$

All the elastic tensors are symmetric:

$$C_{ijkl} = C_{jikl} = C_{ijlk} = C_{klij} = C_{klij} = C_{lkij} = C_{klji} = C_{lkji} \tag{8.12}$$

A maximum of 21 values need to be determined to characterise the elastic properties of a material. But both the elasticity and the compliance tensors preserve the symmetry of the point group. This helps reduce the number of independent elastic constants (Table 8.2). For example, the orthorhombic symmetry, adopted by many minerals of the deep Earth, requires only the computation of nine independent constants, while the cubic structures require only three.

There are two ways to compute the elastic constants tensor: finite differences and linear response. In the finite differences approach, small strains are applied to a structure, the atoms are allowed to relax within the new unit cell, and the residual stresses are computed. The system of linear equations between the strain tensor and the residual stress tensor yields the elastic constants. This procedure is illustrated in detail in Box 8.5. Then, Box 8.6 gives a further example for computing the elastic constants tensor of an orthorhombic mineral, like olivine, $(Mg,Fe)_2SiO_4$, or bridgmanite, $MgSiO_3$.

The linear response procedure is discussed in Section 15.2.

8.7 Seismic Wave Velocities

The elastic constants, together with the crystal density, determine the propagation of the seismic waves, which are elastic waves. They represent three-dimensional displacive perturbations, eventually affecting the atomic positions. They correspond to a stress field propagating through the material along a certain direction.

The elastic waves can be divided relative to the propagation wavevector into one longitudinal or compression wave, denoted with the 'P' index, and two transversal, or shear, waves, denoted with the 'S' index. The two shear waves can be separated into one *horizontal* and one *vertical* wave, depending on how they are polarised along the seismic path. For an average Earth mineral, the wavelengths of the seismic waves are on the order of several kilometres, which makes them about 12–14 orders of magnitude larger than the size of the unit cell.

Table 8.2 Symmetry of the elastic constants tensors.

Symmetry	Independent elastic constants Equivalence with the other constants
Isotropic	C_{11}, C_{12} $C_{11} = C_{22} = C_{33}, C_{12} = C_{13} = C_{23},$ $C_{44} = C_{55} = C_{66} = (C_{11} - C_{22})/2$
Cubic 23, m$\bar{3}$, 432, 43m, m$\bar{3}$m	C_{11}, C_{12}, C_{44} $C_{11} = C_{22} = C_{33}, C_{12} = C_{13} = C_{23},$ $C_{44} = C_{55} = C_{66}$
Hexagonal 6, $\bar{6}$, 6/m, 622, 6mm, 6m2, 6/mmm	$C_{11}, C_{12}, C_{13}, C_{33}, C_{44}$ $C_{11} = C_{22}, C_{13} = C_{23}, C_{44} = C_{55}$ $C_{66} = (C_{11} - C_{12})/2$
Trigonal 3, $\bar{3}$	$C_{11}, C_{12}, C_{13}, C_{14}, C_{15}, C_{33}, C_{44}$ $C_{11} = C_{22}, C_{13} = C_{23}$ $C_{24} = -C_{14} = -C_{56},$ $C_{15} = -C_{25} = C_{15}, C_{44} = C_{55}$ $C_{66} = (C_{11} - C_{22})/2$
32, 3m, $\bar{3}$m	$C_{11}, C_{12}, C_{13}, C_{14}, C_{33}, C_{44}$ $C_{11} = C_{22}, C_{13} = C_{23}$ $C_{24} = -C_{14} = -C_{56},$ $C_{44} = C_{55}$ $C_{66} = (C_{11} - C_{22})/2$
Tetragonal 4, $\bar{4}$, 4/m	$C_{11}, C_{12}, C_{13}, C_{16}, C_{33}, C_{44}, C_{66}$ $C_{11} = C_{22}, C_{13} = C_{23}, C_{16} = -C_{26}, C_{44} = C_{55}$
4mm, $\bar{4}$2m, 422, 4/mmm	$C_{11}, C_{12}, C_{13}, C_{33}, C_{44}, C_{66}$ $C_{11} = C_{22}, C_{13} = C_{23}, C_{44} = C_{55}$
Orthorhombic 222, mm2, mmm	$C_{11}, C_{12}, C_{13}, C_{22}, C_{23}, C_{33}, C_{44}, C_{55}, C_{66}$
Monoclinic 2, m, 2/m	$C_{11}, C_{12}, C_{13}, C_{16}, C_{22}, C_{23}, C_{25},$ $C_{33}, C_{35}, C_{44}, C_{45}, C_{55}, C_{66}$
Triclinic 1, $\bar{1}$	$C_{11}, C_{12}, C_{13}, C_{14}, C_{15}, C_{16},$ $C_{22}, C_{23}, C_{24}, C_{25}, C_{26},$ $C_{33}, C_{34}, C_{35}, C_{36},$ $C_{44}, C_{45}, C_{46}, C_{55}, C_{56}, C_{66}$

In an isotropic medium, the propagation of the elastic waves takes place with the same velocity in all directions. But in a crystal, the elastic wavefront propagates at different speeds along different directions according to symmetry. Starting from Hooke's law, the Christoffel equation [189]

Box 8.5   **How to calculate the elastic constants tensor?**

The general form of the elastic constants tensor, using the Einstein reduced notation (Equations 8.3 and 8.4), is:

$$\begin{matrix}
C_{11} & C_{12} & C_{13} & C_{14} & C_{15} & C_{16} \\
C_{12} & C_{22} & C_{23} & C_{24} & C_{25} & C_{26} \\
C_{13} & C_{23} & C_{33} & C_{34} & C_{35} & C_{36} \\
C_{14} & C_{24} & C_{44} & C_{44} & C_{45} & C_{46} \\
C_{15} & C_{25} & C_{45} & C_{55} & C_{55} & C_{66} \\
C_{16} & C_{26} & C_{46} & C_{66} & C_{66} & C_{56}
\end{matrix}$$

For calculating the elastic constants, one needs to apply a set of small uniaxial and shear strains to the relaxed unit cell. The residual stresses are computed in this deformed cell after the full relaxation of only the atomic positions (relaxation of the entire cell would result in the cancellation of the deformation). Examples of pure uniaxial strain along the x axis and pure shear strain around the x axis are respectively:

$$\sigma_{uniaxial} = \begin{matrix} 1+\epsilon & 0 & 0 \\ 0 & 1 & 0 \\ 0 & 0 & 1 \end{matrix} \quad \text{and} \quad \sigma_{shear} = \begin{matrix} 1 & 0 & 0 \\ 0 & 1 & \epsilon/2 \\ 0 & \epsilon/2 & 1 \end{matrix}$$

For finite-differences calculations, using the linear stress-strain relations, these strain matrices need to multiply the basis of the crystallographic vectorial space at the right (and NOT at left):

$$\sigma = rprim * \sigma_{uniaxial} \text{ and respectively } \sigma = rprim * \sigma_{shear}.$$

Using the computed stresses and the applied strains, we obtain a system of linear equations, like

$$\sigma_1 = C_{11}\epsilon_1 + C_{12}\epsilon_2 + C_{13}\epsilon_3 + C_{14}\epsilon_4 + C_{15}\epsilon_5 + C_{16}\epsilon_6$$
$$\sigma_2 = C_{12}\epsilon_1 + C_{22}\epsilon_2 + C_{23}\epsilon_3 + C_{24}\epsilon_4 + C_{25}\epsilon_5 + C_{26}\epsilon_6$$
$$\cdots\cdots$$
$$\sigma_6 = C_{16}\epsilon_1 + C_{26}\epsilon_2 + C_{36}\epsilon_3 + C_{46}\epsilon_4 + C_{56}\epsilon_5 + C_{66}\epsilon_6$$

In practice, you can apply the strains one by one, which considerably simplifies the above equations, with many terms cancelling out. For a pure uniaxial strain along the x axis, the above set of equations simplifies to:

$$\sigma_1 = C_{11}\epsilon_1$$
$$\sigma_2 = C_{12}\epsilon_1$$
$$\cdots\cdots$$
$$\sigma_6 = C_{16}\epsilon_1$$

The values of the strains should be on the order of 1–2% or less, with sets of both positive and negative values. The residual stresses are obtained after subtraction of the hydrostatic pressure of the undeformed cell. Linear fits to the individual stress-strain relations yield the elastic constants.

There are nine independent elastic constants for orthorhombic structures. Their calculation can be fully automated using the multi-dataset features in *abinit*. An *abinit* input file to compute the C_{11}, C_{12}, and C_{13} elastic constants (datasets 2–5) and the C_{44} pure shear constant (datasets 6 and 7) of an orthorhombic crystal should contain:

ionmov	2	# atomic relaxation
optcell	0	# do not relax the strained cell
ntime	60	# forces exit of the relaxation loop
ndtset	7	# number of data points
rprim1	1 0 0 0 1 0 0 0 1	# reference P calculation
rprim2	0.98 0 0 0 1 0 0 0 1	# -2% strain along x
rprim3	0.99 0 0 0 1 0 0 0 1	# -1% strain along x
rprim4	1.01 0 0 0 1 0 0 0 1	# +1% strain along x
rprim5	1.02 0 0 0 1 0 0 0 1	# +2% strain along x
rprim6	1 0 0 0 1 0.01 0 0 1	# reference P calculation
rprim7	1 0 0 0 1 0.02 0 0 1	# reference P calculation

The output file will have seven sets of stress tensors. To apply the formalism in Box 8.5, subtract the hydrostatic stress obtained in dataset 1 from the stresses obtained at each dataset 2 to 7. The results from dataset 1 can be obtained in a separate stand-alone simulation. However, the advantage of performing it here once more is to have the reference stress tensor of the unperturbed crystal in the same output file as all the other calculations.

yields the velocity of the propagation of the elastic waves as a function of direction in the interior of a given crystal:

$$det(\rho v^2 \delta_{ik} - C_{ijkl}n_j n_l) = \begin{vmatrix} \Gamma_{11} - \rho v^2 & \Gamma_{12} & \Gamma_{13} \\ \Gamma_{21} & \Gamma_{22} - \rho v^2 & \Gamma_{23} \\ \Gamma_{31} & \Gamma_{32} & \Gamma_{33} - \rho v^2 \end{vmatrix} = 0 \quad (8.13)$$

where, n_i represents the direction of the propagation of the wavefront, and $\Gamma_{ik} = \sum_{jl} C_{ijkl}n_j n_l$ are the Kelvin–Christoffel stiffness coefficients.

Figure 8.7 exemplifies the dependence of the seismic shear wave velocities with the crystal orientation for a crystal of C-doped hcp Fe at inner core pressure conditions [62]. The presence of ordered interstitial carbon in iron breaks the hexagonal symmetry down to trigonal.

The difference between the maximum and the minimum velocities yields the seismic anisotropy. This is defined for both the compressional and the shear waves. The absolute difference between the average propagation of the two shear waves, $|V_{SH} - V_{SV}|$, defines the shear wave splitting, which can be directly measured from seismograms.

When analysing geological bodies, we mostly deal with rocks, which are mineral aggregates. The propagation of the seismic waves through an aggregate is determined by averaging over the propagation through

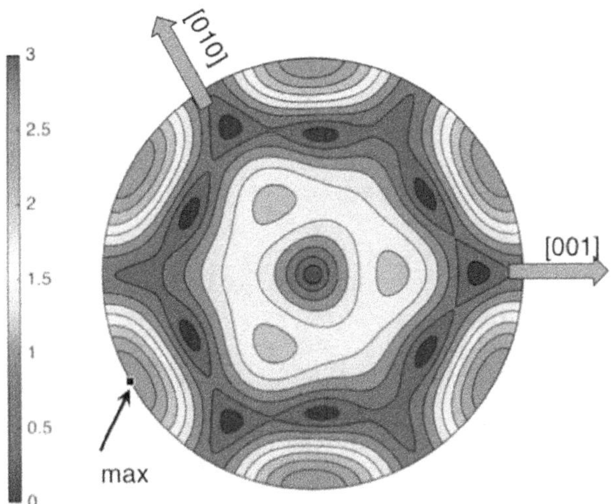

Angular dependence of the propagation of the shear waves in ordered trigonal FeC$_{0.125}$ crystals at inner core pressures in the basal plane, perpendicular to A$_3$. One recognises the trigonal symmetry. (after Ref. [62])

each constituent crystal. Thus the result corresponds to the average elastic constants, which are the elastic moduli of the aggregate. Several elastic moduli stem from various definitions. The bulk modulus K is a measure of crystal deformation under uniform hydrostatic compressional stress. The bulk modulus from the elastic constants tensor should have the same value as the one in the EOS and the compressibility equations because they express the same response of the crystal to hydrostatic deformation. The shear modulus, G, is a measure of the response of the crystal under non-hydrostatic deformation. The averaging for these two moduli can be expressed either by considering uniform propagation of the stress field, in the Reuss (R) formalism, or uniform propagation of the strain field, in the Voigt (V) formalism. The Voigt formalism uses the elastic constants tensor, and the Reuss formalism uses the stiffness constants tensor to build their corresponding averages. The resulting moduli in both formalisms have units of pressure. General formulas for the average seismic wave velocities were derived for monomineral homogeneous aggregates of crystals of trigonal and hexagonal classes, as well as for orthorhombic, tetragonal, and cubic crystals. In these formulations, the surface effects of the crystals are neglected. Despite numerous efforts, there is no analytical derivation yet for the monoclinic and the triclinic crystals.

The bulk (K) and shear (G) moduli for hexagonal crystals, in the Voigt and Reuss formalisms, are, respectively:

$$K_V = \frac{2(C_{11} + C_{12}) + 4C_{13} + C_{33}}{9} \tag{8.14}$$

$$G_V = \frac{C_{11} + C_{12} + 2C_{33} - 4C_{13} + 12C_{44} + 12C_{66}}{30} \tag{8.15}$$

$$K_R = \frac{(C_{11} + C_{12})C_{33} - 2C_{12}^2}{C_{11} + C_{12} + 2C_{33} - 4C_{13}} \tag{8.16}$$

$$G_R = \frac{5}{2} \frac{[(C_{11} + C_{12})C_{33} - 2C_{12}^2]C_{55}C_{66}}{[2(C_{11} + C_{12}) + 4C_{13} + C_{33}]C_{55}C_{66}/3 + [(C_{11} + C_{12})C_{33} - 2C_{12}^2](C_{55} + C_{66})} \tag{8.17}$$

For orthorhombic crystals, the two bulk moduli, in the two formalisms, are:

$$K_V = \frac{(C_{11} + C_{22} + C_{33}) + 2(C_{12} + C_{13} + C_{23})}{9} \tag{8.18}$$

$$G_V = \frac{(C_{11} + C_{22} + C_{33}) - (C_{12} + C_{13} + C_{23})}{15} + \frac{3}{15}(C_{44} + C_{55} + C_{66}) \tag{8.19}$$

$$K_R = \frac{1}{(S_{11} + S_{22} + S_{33}) + 2(S_{12} + S_{13} + S_{23})} \tag{8.20}$$

$$G_R = \frac{15}{4(S_{11} + S_{22} + S_{33}) - 4(S_{12} + S_{13} + S_{23}) + 3(S_{44} + S_{55} + S_{66})} \tag{8.21}$$

From these relations, it is easy to derive the formulas for the tetragonal systems, knowing that $C_{11} = C_{22}$, $C_{12} = C_{13}$, and $C_{33} = C_{44}$. However, there is a particular class of tetragonal crystals, with 42 symmetry, that contain a non-zero C_{16} constant (Table 8.2). In this case, the elastic tensor must first be rotated to vanish the C_{16} constant [54], and then the averaging can be applied as in the general tetragonal case.

Finally, cubic crystals benefit from high symmetry, and the corresponding elastic moduli are further simplified as follows:

$$K_V = K_R = \frac{C_{11} + 2C_{12}}{3} \tag{8.22}$$

$$G_V = \frac{1}{5}(C_{11} - C_{12} + 3C_{44}) \tag{8.23}$$

$$G_R = \frac{5(C_{11} - C_{12})C_{44}}{4(C_{44} + 3(C_{11} - C_{12}))} \tag{8.24}$$

In real aggregates of crystals, the actual propagation of the seismic waves lies between the Voigt and Reuss end-member terms. Hill [139] proposed a formalism that corresponds to the arithmetic average of the two previous moduli:

$$K_H = \frac{K_V + K_R}{2} \tag{8.25}$$

$$G_H = \frac{G_V + G_R}{2} \tag{8.26}$$

Then the average compressional (Vp) and shear (Vs) seismic wave velocities for propagation in homogeneous aggregates are obtained from the elastic moduli and the density as:

$$v_P = \sqrt{\frac{K + 4G/2}{\rho}} \tag{8.27}$$

$$v_S = \sqrt{\frac{G}{\rho}} \tag{8.28}$$

Further relations between the elastic constants define a series of moduli, ratios, and parameters useful in particular circumstances. For example, the Young modulus, E, gives a measure of the uniaxial deformation under uniaxial stress along a given direction in a material. It is expressed in terms of the bulk and shear moduli as [139]:

$$\frac{1}{E} = \frac{1}{3G} + \frac{1}{9K} \tag{8.29}$$

The Poisson ratio, v, is a measure of the amount of shear deformation under uniaxial stress. It is defined as the negative ratio between the transversal and the longitudinal strains, occurring under a given stress. Its relation to other elastic moduli is given by:

$$v = \frac{1}{2}(1 - \frac{3G}{3K + G}) \tag{8.30}$$

Its values are larger than -1 and smaller than 0.5, with most materials in the $0 - 0.5$ range, that is, their cross-section thins under elongation. Notably, the Poisson ratio of the inner core is used as an additional constraint for determining the core composition. Its measured value estimated from seismic data at 0.44 is hard to attain only with standard hcp-like Fe-based alloys [210].

8.8 Rheology and Deformation

Any stresses larger than the yield stress (Figure 8.6) result in permanent deformation of the structure, even after release. The yield stress marks the beginning of the plastic regime.

Beyond this point, if the material deforms continuously, the deformation process is called creep or cold flow. It is due to nucleation and displacement of point defects, planar defects, dislocations, and grain boundaries. The actual creep and the creep rate depend on many factors, including temperature, pressure, point defect concentration and their diffusion rate, and the general deformation rate of the material. Metals and some polymers may exhibit necking at the highest deformation, where the transversal section in a part of the sample starts to decrease under extensive traction. Often polycrystalline materials exhibit recrystallisation to accommodate the deformation rate as grains rotate and deform. Metals and insulators behave differently for obvious electronic reasons. In metals, the delocalisation of the electrons allows for an easy structural rearrangement even under extensive deformation. In insulators, where the atoms are often charged, the ionic repulsion between displaced atoms prevents the preservation of the crystal integrity at large deformations.

Fracture occurs at the ultimate point of the deformation, where the mechanical integrity of the sample is lost. To make things more complicated, some materials exhibit discontinuities related to partial fragmentation in the plastic range called cataclastic flow. The materials that exhibit extended plastic regimes are ductile, plastic, or viscoelastic, while the ones with limited plasticity and where fracture occurs rather rapidly are called brittle.

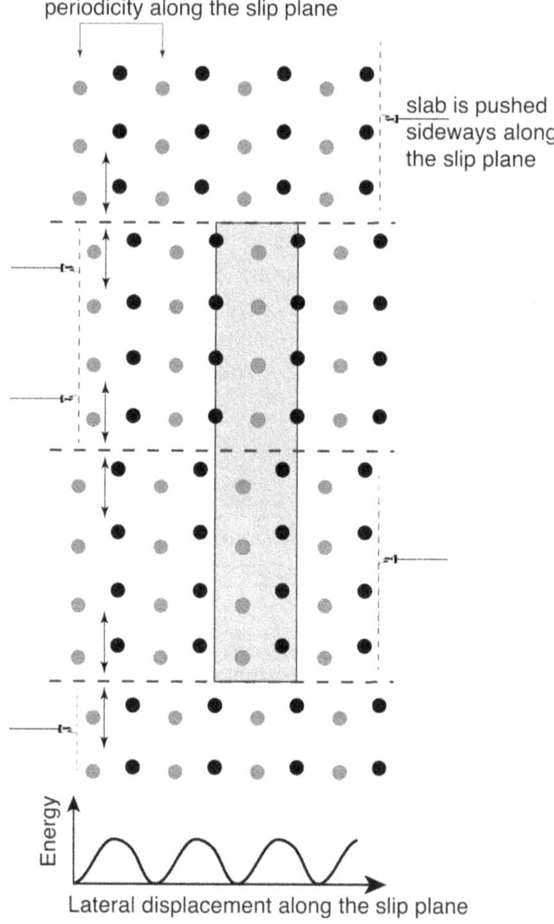

Fig. 8.8   Structural model of the setup for a stacking fault (γ surface) calculation. The simulation box covers the supercell outlined in grey. Slabs of equal width glide on top of each other in discrete small steps. Each simulation box contains two slabs. Atoms are allowed to relax their position only perpendicular to the slip plane.

The phenomena that happen in the plastic regime are complex and not always predictable. Most of these are statistical phenomena that occur at meso-scale. Several in-depth discussions and explanations of the deformation of Earth materials already exist ([151]), and we will not enter into more details here.

Because of the scale difference, the plastic behaviour of minerals cannot be directly modelled using ab initio techniques. This remains the domain of atomistic potentials, whose simplicity allows for large-scale simulations with tens of millions of particles that are appropriate to study, for example, fracture propagation [123, 217, 21].

However, some aspects can be modelled at the atomistic scale using first-principles simulations ([45]). These simulations follow the displacement of the support planes of the dislocations according to the Peierls–Nabarro theory ([68]). First, the elastic constants are necessary to model the elastic interaction between the undeformed regions of the crystal separated by dislocations. Then the energy of the slip systems is needed to constrain the position, the spread, and the dynamics of the dislocations.

Box 8.7   **How to calculate the γ surface?**

We consider the slip plane (001). The supercell, as drawn in Figure 8.8, contains $N = 6$ unit cells stacked as $1 \times 1 \times 6$. Each unit cell has two atoms (a blue one and a red one). The total number of atoms is $natom_{total} = N \times natom_{unitcell}$. The γ surface is constructed from the energy difference between the configuration with the displaced slab by {dx,dy} and the non-distorted supercell: $E_{dx,dy}$-E. The corresponding abinit input file that will do the relaxation at the step {dx,dy} should contain:

acell	1 1 N	#N = number of stacked cells (odd)
xred	x_1+dx y_1+dy z_1/6	# First slab shifted by {dx,dy}
	x_2+dx y_2+dy z_2/6	# x and y are shifted
	x_1+dx y_1+dy 1/6+z_1/6	# by dx and dy
	x_2+dx y_2+dy 1/6+z_2/6	# z takes into account
	x_1+dx y_1+dy 2/6+z_1/6	# the vertical sequence
	.....	
	x_1 y_1 3/6+z_1/6	# The second slab is not shifted
	x_2 y_2 3/6+z_2/6	
ionmov	2	# atomic relaxation
optcell	0	# do not relax the strained cell
ntime	60	# forces exit of the relaxation loop
natfixx	$natom_{total}$	# all atoms are fixed along x
iatfixx	1 2 3 ... $natom_{total}$	
natfixy	$natom_{total}$	# all atoms are fixed along y
iatfixy	1 2 3 ... $natom_{total}$	

For each different simulation, the {dx,dy} values have to be changed by hand.

This energy can be easily obtained from ab initio simulations, for example, modelling the energetics of stacking faults, also called γ surfaces. This technique pushes slabs of equal width on top of each other. The movement is discretised, and at each step, the atoms are allowed to relax only perpendicular to the slip plane (Figure 8.8). Box 8.7 illustrates the corresponding input file for calculating the γ surface of the slip system from Figure 8.8.

## PART III

DYNAMICAL PROPERTIES

## 9

Lattice Dynamics

In Chapters 3 to 8 of this book, the atoms were treated as static entities. Chemical bonds were allowed to form by redistributing valence electrons, but the actual position of the atomic nuclei did not change. They occupied the same reduced positions in a vectorial space where they were initially put. The only allowed movements were during structural relaxations, whose sole purpose was to bring the atoms to their equilibrium positions of zero forces, from where they would not move anymore. However, in real crystalline lattices, in glasses, and even more so in melts and fluids, the atoms are not static. They do not reside in one specific place, as they are sometimes shown in crystallography books. Instead, they are dynamic entities. The position of every atom in a lattice changes continuously. In crystalline lattices, the atoms gravitate around their equilibrium positions; there is a constant back-and-forth movement, but over long periods, the atoms do not describe any net displacement. This movement has a vibrational character. In fluids, there is a non-zero net displacement of the atoms that increases over time. During these movements, the atoms may continuously exchange electrons. In fluids, old bonds are broken, and new bonds are formed. Now, we will start addressing various aspects and properties stemming from this state of constant movement.

9.1 Potential Energy Wells

The image of a static lattice is similar to the classical representation of potential energy as a landscape where a balloon falls down a slope, stays temporarily in a local potential minimum until it is kicked out, and finally falls and remains in the global energy minimum. In a way, atoms do the same. At equilibrium, they lie at the bottom of potential energy wells. When they are out of equilibrium forces, push them towards the bottom of these potential wells, just like the balloon.

In Section 8.3, the atoms were moved, under lattice symmetry constraints, until the total energy of the crystal was brought to a minimum and the forces on the atoms vanished. At this state of relaxation, each atom would lie at the bottom of its local potential well. Once relaxed, the atoms were not allowed to exit those locations.

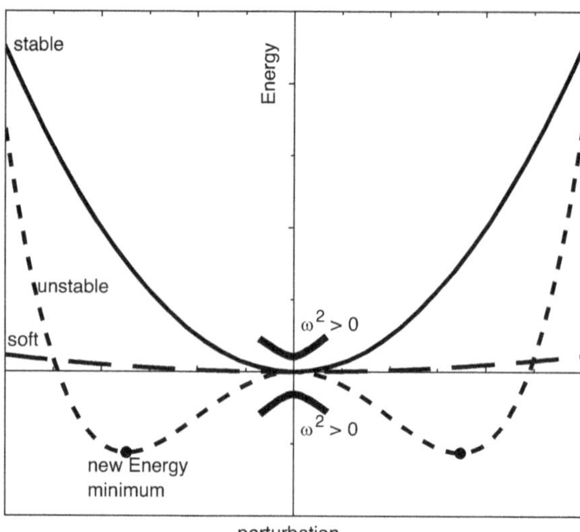

Fig. 9.1 Idealised potential curves describing the possible pattern of the change in energy during a given perturbation of the crystal lattice. The perturbation can be anything like temperature that induces an atomic displacement, an external electric or magnetic field, stress, and so on. If the perturbation is an atomic displacement, then the unperturbed lattice has the atoms lying at the $x = 0.0$ position. In this case, the second-order derivative of the energy with respect to the displacements corresponds to the eigenvalue of the dynamic matrix (Equation 12.10), which yields the square of the frequency of the vibration, omega. The crystal structure is stable with respect to the atomic displacements if the curvature of the energy curve is positive when omega squared is positive. In this case, the atoms oscillate around their initial equilibrium position. For negative curvatures, omega squared is negative, and omega becomes imaginary. In this scenario, the atoms move away at the slightest perturbation, fall into a new minimum of energy, and oscillate around new equilibrium positions, represented by the big black dots.

In this part of the book, infinitesimal displacements are applied to each atom. Then the total energy of the system is monitored as a function of the amplitude and direction of these displacements. There are three possible outcomes, schematised in Figure 9.1 by lines with different patterns [24]. The scheme covers atomic displacements along one arbitrary direction of the space.

The first outcome is that the energy increases (the solid line in Figure 9.1). This signifies that the initial position was at the bottom of an energy well. When the atom is released, it will want to return to its initial state, as this was an energy minimum. The applied displacement is translated in terms of a gain in energy. At maximum amplitude, the total energy is potential, but as the atom approaches the minimum, part of this energy is transformed into kinetic energy. In an ideal system without friction, the result is that the atom will pass by the equilibrium position, and its kinetic energy will propel it to the other side of the potential well, just like a balloon in the bottom of a well or as an elastic resort that is released after being compressed or extended. On the other side of the well, the atom slows down as more and more of its kinetic energy is transformed into potential energy, and at a certain moment, it will arrive at a halt. Then reverse, it will come back, pass again by the equilibrium position, and go back on the other side. As the system is ideal, with no friction, this back-and-forth movement never stops. This is an atomic vibration; all the atoms vibrate around their equilibrium positions. Because the atoms always remain around their equilibrium positions in a crystalline solid, the average structure remains unchanged. This crystalline arrangement is then stable.

As with all oscillatory movement, this back-and-forth movement is characterised by an amplitude and a frequency directly related to an energy gain, as explained in detail later.

If the potential well is steep, then at equal amplitude, the rise in potential energy is high, and consequently, the frequency of oscillation is high. For example, as the atoms approach each other in the case of compression, the potential wells that develop between them become steeper. In this case, the frequencies of the oscillations increase.

On the contrary, if the potential well opens up, the movement of atoms becomes more sluggish. In large, wide-open wells, the atoms require a long time to reserve during a vibration. As a result, the corresponding frequency decreases. The process of decreasing frequency when the potential well opens up is called *softening*.

Eventually, the potential well opens to such an extent that it becomes virtually flat. This is the second outcome of the problem in Figure 9.1, illustrated with the long-dashed line. In this case, there is no driving force to bring the atoms back, and there will be no oscillation, no vibration. If the atoms start to move in this direction, they can stop anywhere, as the energy curve is flat. The structure is then unstable, becoming disordered, as the atoms end up lying anywhere in this energy landscape. This is the typical case of order-disorder transitions.

Finally, the third outcome is that the potential well keeps opening until it eventually changes shape, bends downwards, and transforms into a double-well potential. This is the short-dashed line in Figure 9.1. Parts of it become negative on both sides of the former equilibrium positions. In this case, as soon as the atoms are displaced away from their initial positions, the energy of the system starts to decrease. The atoms require energy to come back to their initial positions. But as this energy is unavailable, the result is a permanent distortion of the lattice. The atoms keep moving away from their initial positions until they reach the bottom of the double well on one of the two sides. As they remain trapped in this well, these become their new equilibrium positions. Thus, the atoms vibrate around these new equilibrium positions. During this process, the crystalline structure changes; therefore, a new crystalline phase forms – this process marks a phase transition.

The potential well represents the energy variation as a function of perturbation. To build it, we take as an example the conventional cell of ZnS wurtzite; we displace one of the Zn atoms along the z axis by equal amounts in both positive and negative directions, varying the Zn-S distances and the interatomic angles. The corresponding input file should look like this:

acell	3.82 3.82 6.26	angstroms # values from webmineral.com
angdeg	90.0 90.0 120.0	# hexagonal angles
natom	4	# number of atoms
ntypat	2	# number of types of atoms
typat	1 1 2 2	# types of atoms
znucl	30 16	# Z of each atom type
ndtset	21	# several dataset
		# in a chained calculation
xred:	1/3 2/3 -0.050	# initial positions
	2/3 1/3 0.5	
	1/3 2/3 0.3748	
	2/3 1/3 0.8748	
xred+	0.0 0.0 0.005	# increment along z of Zn atom
	9*0.0	# all other atoms fixed
ecut	40	# kinetic cutoff energy
ngkpt	8 8 4	# grid of k point
kptopt	1	# k-points option
nshiftk	1	# no. of k-points grids
shiftk	0.0 0.0 0.5	# hexagonal shift
toldfe	1.0d-08	# SCF exit criterion
nstep	100	# max. no. SCF steps

With this input, the resulting potential well is shown in Figure 9.2. We start the calculation from a negative perturbation of -0.05 along the z-axis, and we go up to a positive value of $+0.05$ along the z-axis. This allows for building a realistic potential well, which is asymmetric.

In practice, the potential well is built by applying displacements along all directions of the space. In the cases discussed earlier, the atomic displacements are the perturbations of the system. In this case, the potential wells are obtained by moving the atoms along the three Cartesian axes in both negative and positive directions. Box 9.1 gives the solution for building a potential well when moving one of the Zn atoms along the diagonal axis in the sphalerite, ZnS, structure. This displacement corresponds to the stretching of the ZnS bond. The result is shown in Figure 9.2. In some cases, it is also interesting to study the energy variation under the action of other perturbations, like electric fields, lattice deformations, or specific structural distortions, for example, to allow the O-H distance to vary in

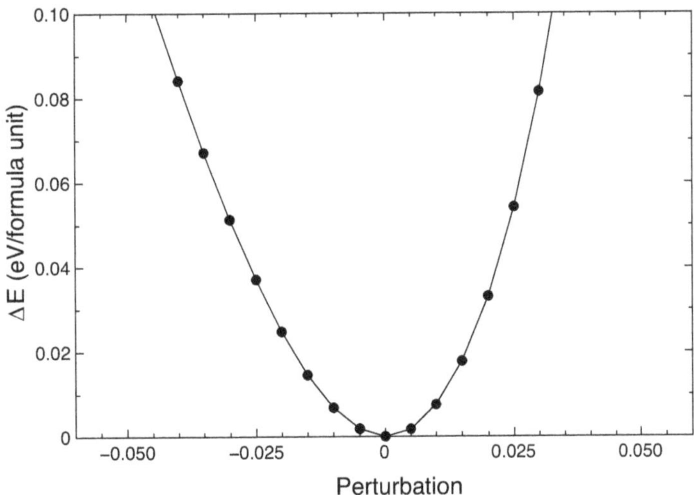

Fig. 9.2 Variation of the energy during the displacement of one Zn atom along the diagonal [111] axis of sphalerite, which plays the role of a perturbation. The potential well is often asymmetric, as is the case here for wurtzite. We discussed this behaviour at length in Chapter 2.

hydroxides or the H-O-H angle in water-based molecular structures. The technique is the same; similar potential wells would eventually be obtained.

Further examples of potential wells show that some are highly asymmetric, like Zn-S or Zn-O in wurtzite, ZnS, and zincite, ZnO (Figure 9.3a), respectively. As the Zn-X distance varies, where X = S or O, the energy increases. But the S or O atoms face different electronic environments towards the Zn atom or away from it. Consequently, the change in energy is slightly different on the two sides, translated into an asymmetry of the potential well. This will be discussed later in Section 15.4, together with solutions to improve its description. However, the potential well of Na-Cl in halite is perfectly symmetric because of the symmetry (Figure 9.3b).

9.2 Taylor Expansion of the Energy

The variation of the total crystal energy as a function of atomic displacements, represented in Figure 9.1, can be expressed using a Taylor expansion. This technique is a common practice in condensed matter physics to explain and resolve the variations of functions under infinitesimal perturbations. The Taylor series rewrites the value of a function f at the point r as a sum of consecutive derivatives with respect to the perturbation x:

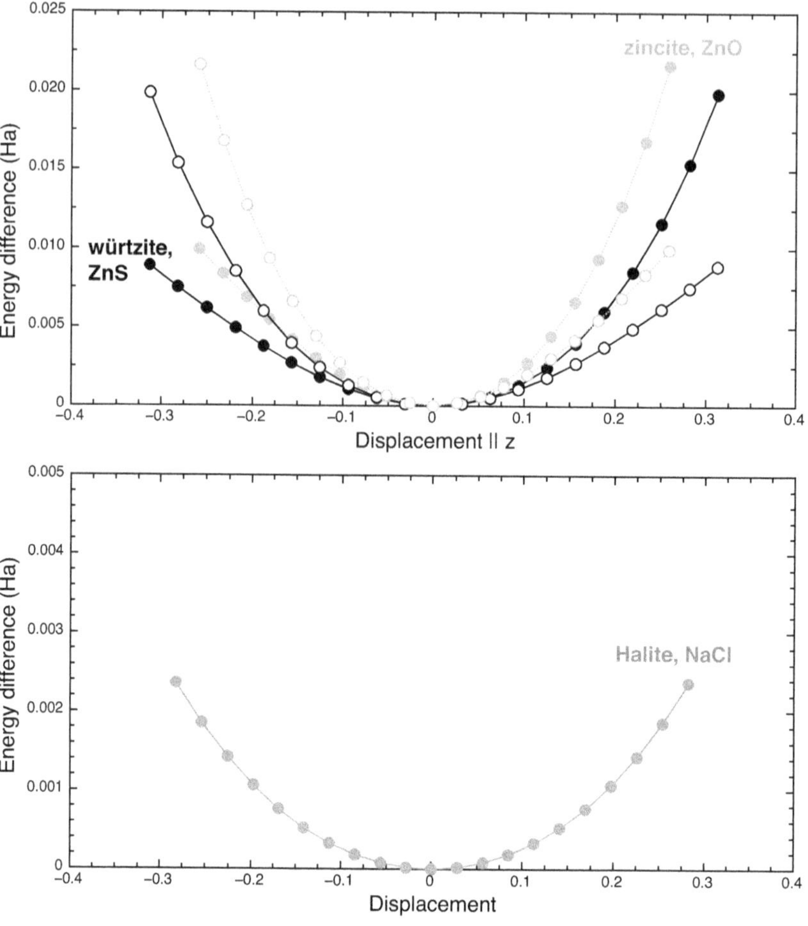

Fig. 9.3 (a) Potential wells for the Zn-S and the Zn-O bonds in isosymmetrical wurtzite, ZnS, and zincite, ZnO. As the S atoms come closer to the Zn atoms, the repulsion from the Zn electrons increases, leading to a faster energy increase. As the S or O atoms move further away, the attraction takes over, but with a smaller energy increase. The computed potentials are shown with solid circles. The empty circles show the symmetrised wells. There are clear deviations from the symmetric wells. (b) In halite, NaCl, the potential wells are symmetric as the electronic environments around Na and Cl atoms are symmetric.

$$f(x) = \sum_{n=0}^{\infty} \frac{\partial^n f(r)}{\partial x^n} \frac{(x-r)^n}{n!} \tag{9.1}$$

or, explicitly, this gives:

$$f(x) = f(r) + \frac{\partial f(r)}{\partial x}(x-r) + \frac{1}{2!}\frac{\partial^2 f(r)}{\partial x^2}(x-r)^2 + \frac{1}{3!}\frac{\partial^3 f(r)}{\partial x^3}(x-r)^3$$
$$+ \frac{1}{4!}\frac{\partial^4 f(r)}{\partial x^4}(x-r)^4 + ... \tag{9.2}$$

As the perturbations are small, the high-order derivatives and the $(x-a)^n$ terms quickly tend to zero. Consequently, most of the high-order terms of the Taylor expansion can be neglected for practical purposes.

9.3 Quasiharmonic Approximation

In the case of a crystal lattice, the internal energy can be treated as the function f from Equation 9.2. Depending on the driving force, there are different types of possible perturbations, x. The atomic displacements are the most straightforward ones, occurring, for example, due to temperature. Lattice distortions result from external stresses, either uniaxial pressure or shear stress, or a combination thereof. Electronic polarisation may occur due to external electrostatic fields and magnetisation due to external magnetic fields.

With a generalised notation for the perturbation as λ, Equation 9.2 for the crystal energy E can be rewritten as:

$$E[\lambda] = E^{(0)} + \sum_i \frac{\partial E}{\partial \lambda_i} d\lambda_i + \frac{1}{2} \sum_{i,j} \frac{\partial^2 E}{\partial \lambda_i \partial \lambda_j} d\lambda_i d\lambda_j$$

$$+ \frac{1}{6} \sum_{i,j,k} \frac{\partial^3 E}{\partial \lambda_i \partial \lambda_j \partial \lambda_k} d\lambda_i d\lambda_j d\lambda_k + \dots \tag{9.3}$$

In Equation 9.3, the i, j, k indexes denote different possible types of perturbations. Let us first analyse in detail what happens when the atomic displacements are the perturbations of the crystal energy. In this case, λ is replaced in the general formulation with τ. The displacements concern all the atoms α, β, γ, and so on. Here we adapt the same notations as in references [118, 119, 120]. Other studies, for example, Ref. [24], perform a similar development but with slightly different notations. Then the Taylor development for the crystal energy becomes:

$$E[\tau] = E^{(0)} + \sum_\alpha \frac{\partial E}{\partial \tau_\alpha} d\tau_\alpha + \frac{1}{2} \sum_{\alpha,\beta} \frac{\partial^2 E}{\partial \tau_\alpha \partial \tau_\beta} d\tau_\alpha d\tau_\beta$$

$$+ \frac{1}{6} \sum_{\alpha,\beta,\gamma} \frac{\partial^3 E}{\partial \tau_\alpha \partial \tau_\beta \partial \tau_\gamma} d\tau_\alpha d\tau_\beta d\tau_\gamma + \dots \tag{9.4}$$

The first term in Equation 9.4 represents the static ground-state energy, where the atoms are lying fixed at positions $\mathbf{r}$. This is the energy computed in Section 4.1; it will always be written in the output files of the density functional theory (DFT) package at the end of the ground-state calculation.

The second term is the first energy derivative to an atomic displacement. This represents the force acting on the atom. If the crystal structure is relaxed before the lattice dynamics calculation, the forces acting on all atoms are null. This is the case when the atoms lie at the bottom of the potential wells, but it is also true when the atoms lie at the saddle points of the energy landscape. In all these cases, the tangent to the energy surface is zero. Consequently, this term cancels out the sum. This cancellation is why the correct way of doing a phonon calculation is to perform a structural relaxation first.

The third term is the second energy derivative, and its treatment deserves a little more attention. This corresponds to the curvature of the energy function around the initial equilibrium positions. A positive derivative means an upward-oriented energy curve; this corresponds to the first case discussed in Figure 9.1, where the atoms lie and vibrate around the bottom of the potential well. A negative derivative implies a downward-oriented energy curve; this corresponds to the third case outlined earlier, where the atoms keep departing from their initial equilibrium positions. A zero derivative corresponds to the flat horizontal energy curve.

The following terms in Equation 9.3 are all higher-order derivatives of the energy multiplied by the atomic displacement raised to the power of the order of the derivative. As the displacement is infinitesimal, this prefactor becomes increasingly small as the order of the derivative is high. As mentioned earlier, it is common practice to neglect all the terms with energy derivatives higher than two. This approximation enormously simplifies the problem that we are trying to solve. Even more interesting, this simplification has a profound physical meaning, as shown in the following.

With the simplifications outlined earlier, the series development of the energy can be written in terms of atomic displacements:

$$E[\lambda] = E^{(0)} + \frac{1}{2} \sum_{\alpha,\beta} \frac{\partial^2 E}{\partial \tau_\alpha \partial \tau_\beta} d\tau_\alpha d\tau_\beta \qquad (9.5)$$

This simplified form of the Taylor expression represents a development as a second-order polynomial function of the atomic displacements. This is a symmetric function, meaning that $f(x) = f(-x)$. The curves in Figure 9.1 were built to be all symmetric precisely for this reason; the dashed curves in Figure 9.3 are symmetrised to fit a quasi-harmonic potential.

The treatment of the potential wells is similar to that of the oscillatory movements of two bodies connected by an elastic resort. Hooke's law tells us that the elastic force between such two bodies connected by a resort is proportional to the distortion of the distance between them and that the proportionality constant is k, the elastic constant:

$$F(x - x_0) = -k(x - x_0) \qquad (9.6)$$

where x represents the perturbation, the atomic displacement away from the equilibrium distance x_0. At equilibrium, $x = x_0$, the elastic force is zero. The energy of the elastic system is:

$$E(x - x_0) = k(x - x_0)^2 \qquad (9.7)$$

In the case of a rigid body attached to an ideal resort without damping and friction, once a perturbation affects the system, like a displacement of that body, this will keep oscillating indefinitely (Figure 9.4). The energy continuously transforms between potential and kinetic energy as the body oscillates. The total energy of the elastic system is conserved, and it

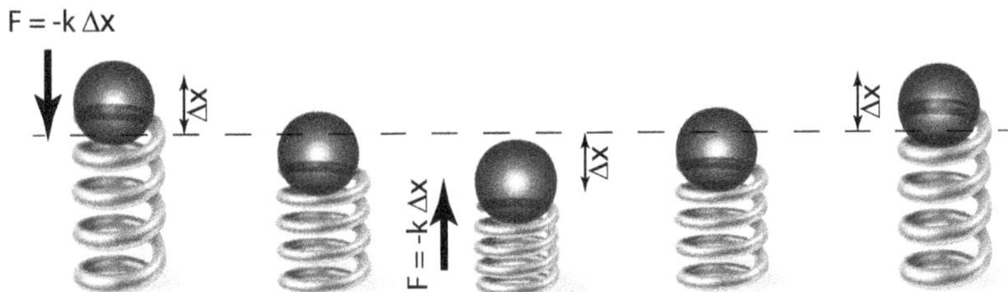

Fig. 9.4 In a deformed elastic resort, an elastic force is directly proportional to the deformation via the elastic constant **k**. For an ideal resort, without friction or energy damping, once deformed and released, it will oscillate harmonically, the potential energy of the maximum deformation being fully transformed into kinetic energy when the resort passes by its undeformed state. (The resort is made after an image designed by macrovector/Freepik)

depends on the amount of initial perturbation as given by the formula of Equation 9.7.

At this point, a nice parallel can be drawn between the two oscillatory systems, atomic and elastic. Using the notation $\lambda = (x - x_0)$, the two expressions of the variation of the energy can be equated. This gives:

$$E[\lambda] = E^{(0)} + \frac{1}{2} \sum_{\alpha,\beta} \frac{\partial^2 E}{\partial \tau_\alpha \partial \tau_\beta} d\tau_\alpha d\tau_\beta = E^{(0)} + \sum_{\alpha,\beta} k_{\alpha,\beta} d\tau_\alpha d\tau_\beta \qquad (9.8)$$

From this results that:

$$\frac{1}{2} \sum_{\alpha,\beta} \frac{\partial^2 E}{\partial \tau_\alpha \partial \tau_\beta} d\tau_\alpha d\tau_\beta = \sum_{\alpha,\beta} k_{\alpha,\beta} d\tau_\alpha d\tau_\beta \qquad (9.9)$$

The $1/2 \partial^2 E / \partial \tau_\alpha \partial \tau_\beta$ term plays the role of the elastic constant of a fictitious resort connecting every two atoms α and β of the crystal. For this reason, this term is called the interatomic force constant.

The approximation that is illustrated in Eq. 9.8 is called the *quasiharmonic approximation* because this describes the oscillatory movement of the atoms during vibrations as if they were connected by ideal (=quasi) elastic (=harmonic) resorts. Compared to the initial development in the Taylor series of the energy, its simplicity is a huge gain and thus made its use widespread. It is the basis of decades of calculations of vibrational spectra, phonon dispersions, thermodynamics, and predictions of phase transitions.

9.4 Coupled Perturbations

Often external driving forces induce coupled perturbations. For example, applying a strain implies displacing atoms; applying an electrostatic field induces atomic displacements. The atomic displacements cause a reorganisation of the electronic density, which in turn leads to the appearance of polarisation and hence of an electrostatic field. In all these cases, the expansion of the energy, of the Hamiltonian, or of the eigenvalues needs to be done in terms of all the perturbations. For example, under the action of two external driving forces, λ_1 and λ_2, the Taylor expansion for the perturbed external potential (Equation 9.3) becomes:

$$v_{ext}(\lambda_1\lambda_2) = v_{ext}^{(0)} + \sum_{i=1,2} \lambda_i v_{ext}^{(1,\{\lambda_i\})} + \sum_{i,j=1,2} \lambda_i\lambda_j v_{ext}^{(2,\{\lambda_i\lambda_j\})} +$$

$$\sum_{i,j,k=1,2} \lambda_i\lambda_j\lambda_k v_{ext}^{(3,\{\lambda_i\lambda_j\lambda_k\})} + \sum_{i,j,k,l=1,2} \lambda_i\lambda_j\lambda_k\lambda_l v_{ext}^{(4,\{\lambda_i\lambda_j\lambda_k\lambda_l\})} + \ldots \quad (9.10)$$

with the notation:

$$v_{ext}^{(n,\{\lambda_i,\lambda_j,\ldots\})} = \frac{1}{n!}\frac{\partial^n X}{\partial\lambda_i\partial\lambda_j\ldots} \quad (9.11)$$

9.5 Energy Derivatives as Physical Observables

This concept of the Taylor expansion of the energy lies at the base of the construction of tensorial responses of the crystal to various driving forces, illustrated by Figure 1.10 from Chapter 1. We will concentrate in the following on the case when two different perturbations affect the system and will take the case of an insulator under a constant electric field. The field interacts with the electronic clouds around the atoms. Consequently, the electronic clouds polarise as they adapt to the presence of the field to reduce the total energy of the system. During this adaptation, they distort and drag the nuclei along, which adapt and change their initial positions. The same happens when the nuclei are displaced from their equilibrium positions, for example, by changing temperature. As the nuclei drag along the electronic clouds, the latter suffer polarisation, generating an electric field. This is the response of the lattice to a coupled perturbation, where the electric field and the atomic displacements interact. But this response is built of several components, and each component is represented by a derivative of the energy.

Today, most of the calculations of lattice dynamics employ the Taylor expansion up to the third order of the energy in the presence of two perturbations, atomic displacements, and electric fields. Many physical properties can be retrieved from solving this equation, and a large amount of effort was put over the last two decades into implementing the solutions of this equation. The Taylor expansion of the electronic energy, E_{el}, as a function of the electric field, $\mathcal{E}$, and the atomic displacements, τ, is:

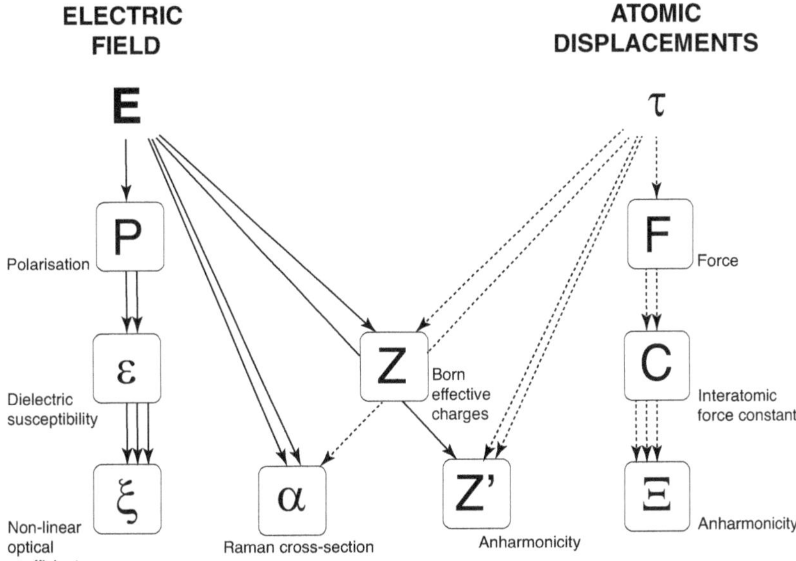

Fig. 9.5 The Taylor expansion of the energy is realised using a series of derivatives of the energy with respect to the electric field (solid arrows) and atomic displacements (dashed arrows). Each arrow symbolises one energy derivative, two arrows mark a double derivative, and three arrows a triple derivative. The result of each such operation is a tensorial physical property.

$$E_{el}[\mathcal{E}, \tau] = E^{(0)} +$$
$$\frac{\partial E_{el}}{\partial \mathcal{E}} + \frac{\partial E_{el}}{\partial \tau} +$$
$$\frac{\partial^2 E_{el}}{\partial \mathcal{E}^2} + \frac{\partial^2 E_{el}}{\partial \mathcal{E} \partial \tau} + \frac{\partial^2 E_{el}}{\partial \tau^2} + \tag{9.12}$$
$$\frac{\partial^3 E_{el}}{\partial \mathcal{E}^3} + \frac{\partial^3 E_{el}}{\partial \mathcal{E}^2 \partial \tau} + \frac{\partial^3 E_{el}}{\partial \mathcal{E} \partial \tau^2} + \frac{\partial^3 E_{el}}{\partial \tau^3}$$

Each term in Equation 9.12 represents a measurable physical property described by a tensor. The rank of the tensors increases as you go down the scheme. Figure 9.5 shows the construction of this equation step by step as combinations of energy derivatives relative to the electric field and atomic displacements.

The analytical derivatives can be obtained up to any given order by building on the lower-order derivatives using the perturbation theory (Chapter 10). But sometimes, for numerical reasons or when implementations are unavailable, the higher-order derivatives can be obtained by deriving the lower-order physical properties directly instead of the energy itself. For example, the Raman tensors can be obtained directly as energy derivatives. But they can also be obtained as derivatives of the Born effective charges under field or of the dielectric tensors under atomic displacements. In these alternative formulations, the physical properties are obtained from derivatives of the properties at previous levels.

10.1 Energy Derivatives from Finite Differences

In Chapter 9, the effect of perturbations on the energy of the crystalline structure was described using potential wells. They are characterised by the Taylor series. Their construction can be done in two major ways. The first approach uses finite differences. For example, we can express the second-order energy derivatives as:

$$\frac{1}{2}\partial^2 E/\partial\tau_\alpha\partial\tau_\beta = \frac{1}{2}\frac{(\Delta E)^2}{\Delta\tau_\alpha\Delta\tau_\beta} \tag{10.1}$$

The α and β are every pair of atoms in the crystal. They may belong to the same or different unit cells. The energy variation is then calculated in a corresponding supercell in the direct space that is big enough to contain the two unit cells. The two atoms are displaced by infinitesimal amounts along each of the x, y, and z Cartesian directions, just like for the construction of potential wells in Chapter 9. The energy variation is monitored as a function of the two atomic displacements. Finally, second-order polynomials can be fit to these energy variation curves. The second-order derivation yields the interatomic force constant between the two atoms, $k_{\alpha,\beta}$, from Equation 9.9. This method of displacing the atoms by small finite amounts is called the *frozen phonon* technique. There are several disadvantages to this approach. The first is that coupling to other perturbations, like electric fields, is absent from the calculation. The second is the need for large supercells as the atoms become further away from each other. With simulations that scale as $O(N^3)$, where N is the number of particles in a system, such a finite difference approach quickly becomes a real computational burden.

10.2 Energy Derivatives from Perturbation Theory

The alternative is to compute directly the energy derivatives that describe the potential wells. This can be done in the density-functional perturbation theory [23, 114, 108, 117, 119, 118, 24], which is briefly described in the following.

We start with the standard Hamiltonian formulation that contains several terms, as in the Kohn–Sham equations (see Section 3.1):

$$H = T + v_{Ha} + v_{xc} + v_{ext} \tag{10.2}$$

where, in order, the terms represent the kinetic energy, the electrostatic potential, the exchange-correlation term, and the last term regroups the electron-ion and the ion-ion electrostatic terms and any external potentials.

An external perturbation can be an ionic displacement, τ, or an external electric field, $\mathcal{E}$. The expansion of the perturbed external potential can be written as:

$$v_{ext}(\lambda) = v_{ext}^{(0)} + \lambda v_{ext}^{(1)} + \lambda^2 v_{ext}^{(2)} + \lambda^3 v_{ext}^{(3)} + \lambda^4 v_{ext}^{(4)} + \dots \tag{10.3}$$

where $v_{ext}^{(n)}$ is the n-th term of the Taylor expansion:

$$v_{ext}^{(n)} = \frac{1}{n!} \frac{\partial^n X}{\partial \lambda^n} \tag{10.4}$$

The perturbation of the external potential induces changes in the physical properties. But any quantity that is affected by the perturbation can be expressed using an expansion in a similar way:

$$X(\lambda) = X^{(0)} + \lambda X^{(1)} + \lambda^2 X^{(2)} + \lambda^3 X^{(3)} + \lambda^4 X^{(4)} + \dots \tag{10.5}$$

where X^n is the n-th term in the Taylor expansion. X can be the total electronic energy, the electronic wavefunctions, $\psi_\alpha(r)$, the electronic density n(r), the electron eigenenergies $\epsilon_\alpha(\mathbf{k})$, or the Hamiltonian H. In all these expansions, the unperturbed terms, $X^{(0)}$, are known, as they characterise the ground state. The next terms can be obtained analytically as energy derivatives.

The following demonstration explains how we can obtain various higher-order derivative terms for energy, wavefunctions, and Hamiltonian, building on the known or previously computed lower-energy derivative terms. This part is typical for a quantum mechanics textbook, and you may skip it if it looks too scary and go directly to Chapter 11.

The bra-ket notation of the operators and functions from quantum mechanics is particularly useful for this exercise because it is much more simplified than the full expansion with integrals. Remember from Section 1.7, Equation 1.20 that:

$$< g|H|f >= \int_r^\infty g(r)^* H(f(r)) dr \tag{10.6}$$

Back to the derivation, the first step is to go back and write the unperturbed Schrödinger equation as:

$$H^{(0)}|\psi^{(0)} >= E^{(0)}|\psi^{(0)} > \tag{10.7}$$

The first-order derivative gives:

$$H^{(1)}|\psi^{(0)} > + H^{(0)}|\psi^{(1)} >= E^{(1)}|\psi^{(0)} > + E^{(0)}|\psi^{(1)} > \tag{10.8}$$

Next, we multiply the two sides of the equation on the left by $<\psi^{(0)}|$ and obtain:

$$<\psi^{(0)}|H^{(1)}|\psi^{(0)}> + <\psi^{(0)}|H^{(0)}|\psi^{(1)}> =$$
$$<\psi^{(0)}|E^{(1)}|\psi^{(0)}> + <\psi^{(0)}|E^{(0)}|\psi^{(1)}> \quad (10.9)$$

The unperturbed $E^{(0)}$ are the eigenvalues of the unperturbed Hamiltonian. Then the integrals are reordered as:

$$H^{(1)} <\psi^{(0)}|\psi^{(0)}> + E^{(0)} <\psi^{(0)}|\psi^{(1)}> =$$
$$E^{(1)} <\psi^{(0)}|\psi^{(0)}> + E^{(0)} <\psi^{(0)}|\psi^{(1)}> \quad (10.10)$$

With the renormalisation relation $<\psi^{(0)}|\psi^{(0)}> = 1$, it results that the first-order derivative of the energy can be obtained as:

$$E^{(1)} = <\psi^{(0)}|H^{(1)}|\psi^{(0)}> \quad (10.11)$$

This last equation is known as the Hellman–Feynman theorem.

The first-order Hamiltonian depends on the perturbation via the external potential. Then, as Equation 10.7 is a system of equations applying to all the bands α, the first-order derivative of the energy can then be expressed as the sum over occupied bands:

$$E^{(1)} = \sum_{\alpha}^{occ.} <\psi_{\alpha}^{(0)}|v_{ext}^{(1)}|\psi_{\alpha}^{(0)}> \quad (10.12)$$

As the first term of the perturbation of the external potential is known, according to the assumptions we made earlier, we can thus obtain the first-order derivative of the energy.

Equation 10.8 can be rewritten as a function of occupied states as in the Sternheimer equation [232]:

$$(H^{(0)} - E_{\alpha}^{(0)})|\psi_{\alpha}^{(1)}> = -(H^{(1)} - E_{\alpha}^{(1)})|\psi_{\alpha}^{(0)}> \quad (10.13)$$

Then the first-order derivative of the wavefunctions is obtained from the Sternheimer equation by expressing them as linear combinations of the unperturbed wavefunctions.

At this point, knowing the first-order derivative of the energy and the wavefunctions, the derivatives are continued to the second order as:

$$H^{(2)}|\psi^{(0)}> + H^{(1)}|\psi^{(1)}> + H^{(0)}|\psi^{(2)}> =$$
$$E^{(2)}|\psi^{(0)}> + E^{(1)}|\psi^{(1)}> + E^{(0)}|\psi^{(2)}> \quad (10.14)$$

After multiplying to the left by $\psi^{(0)}$ and rearranging the terms, the second-order derivative of the energy is:

$$E^{(2)} = \sum_{\alpha} <\psi^{(0)}|H^{(1)}|\psi^{(1)}> + \sum_{\alpha} <\psi^{(0)}|H^{(2)}|\psi^{(0)}> \quad (10.15)$$

with $H^{(1)} = v_{ext}^{(1)}$ and $H^{(2)} = v_{ext}^{(2)}$

In practice, a better, more accurate variational expression leads to the second-order derivatives of the energy as:

$$E^{(2)} = \sum_{\alpha} [< \psi^{(1)} | H^{(0)} - \epsilon_{\alpha}^{(0)} | \psi^{(1)} > + < \psi^{(0)} | H^{(2)} | \psi^{(0)} > +$$

$$+ < \psi^{(1)} | H^{(1)} | \psi^{(0)} > + < \psi^{(0)} | H^{(1)} | \psi^{(1)} >] \quad (10.16)$$

with the constraints that:

$$< \psi_{\alpha}^{(0)} | \psi_{\beta}^{(1)} > = 0 \text{ and } < \psi_{\alpha}^{(0)} | \psi_{\alpha}^{(1)} > + < \psi_{\alpha}^{(1)} | \psi_{\alpha}^{(0)} > = 0 \quad (10.17)$$

Similarly, all the derivatives of the energy and wavefunctions can be obtained step by step up to any order. The 2n+1 theorem [114] shows that one can obtain the (2n+1)-th order derivative of the energy knowing the n-th order derivative of the wavefunctions.

The same procedure is also applied to coupled perturbations. For example, the second-order derivative of the energy to two mixed perturbations is expressed as:

$$E^{i,j} = \frac{1}{2} \frac{\partial^2 E}{\partial \lambda_i \partial \lambda_j} \quad (10.18)$$

This term is obtained from the earlier equations as:

$$E^{i,j} = \sum_{\alpha} < \psi_{\alpha}^{\lambda_j} | v_{ext}^{\lambda_i} | \psi_{\alpha}^{(0)} > + \sum_{\alpha} < \psi_{\alpha}^{0} | v_{ext}^{\lambda_i,\lambda_j} | \psi_{\alpha}^{(0)} > \quad (10.19)$$

In Equation 15.22, the first-order derivatives of the wavefunctions with respect to λ_i, $|\psi_{\alpha}^{\lambda_i} >$ are not needed. However, the earlier equation can be rewritten as a function of $|\psi_{\alpha}^{\lambda_i} >$. There are also more accurate variational expressions that can be used when performing a calculation.

Dielectric Properties

The first physical properties obtained from the energy derivatives are those related to the electric field. The electric fields considered here are continuous or low frequency, where the frequency is low compared to the energy of a typical electronic excitation. The derived physical properties are called dielectric in relation to the dielectric tensor. They are present only in insulators and semiconductors. Applying a finite electric field in these materials leads to a polarisation of the electronic clouds and a change in interatomic bonding. They are absent in metals, where the application of an external electric field induces an electric current. Various tensors relate this deformation of the electronic clouds to the applied field and thus generate a whole series of dielectric properties that we will study later.

11.1 Polarisation as Energy Derivative

When a static external electric field is applied to a crystal, the immediate effect is a deformation of the electronic clouds. This leads to the appearance of an opposite field, such as the resultant field through the crystal is the sum of the two. The potential induced inside the crystal by the external field varies from one unit cell to another, depending on its depth from the surface. Consequently, such a field always breaks the translational symmetry inside the crystal.

Moreover, the external electric field needs to be small enough not to destabilise the electronic gap between the valence and the conduction bands. If the field is too large, it can excite electrons, promote former valence states into conduction, and effectively close the gap. Equation 7.4 of the polarisation are not valid anymore. In practice, such a reaction can be irreversible, and the material may suffer irreparable damage. This is why the materials used for electrical insulating layers, like the plastic protection polymers of electric wires, are wide-gap insulators.

At the macroscopic level, the polarisation corresponds to the first term of the Taylor expansion (Equation 9.12):

$$\mathcal{P}_\alpha(\mathcal{E}) = -\frac{1}{\Omega_0}\frac{\partial E_{el}}{\partial \mathcal{E}} \tag{11.1}$$

This polarisation contributes to the macroscopic displacement induced by the electric and the polarisation fields:

$$\mathcal{D}_{mac}(\mathbf{r}) = \mathcal{E}_{mac}(\mathbf{r}) + 4\pi\mathcal{P}_{mac}(\mathbf{r}) \tag{11.2}$$

In practice, the derivation is done with respect to a longitudinal perturbation potential, which has a wave vector $\mathbf{q}$ tending to $\mathbf{0}$ or Γ. Then the derivation of the energy is done in two steps. First, the derivatives of the wave functions are computed with respect to their wave vector:

$$u_{m\mathbf{k}}^{k_\alpha} = \frac{du_{mk}}{dk_\alpha} \tag{11.3}$$

In the second step, this derivative of the wavefunctions is used to express the energy derivative with respect to electric fields. The heavy, detailed formulas and derivations can be found elsewhere, for example, Ref. [118].

11.2 Born Effective Charges

We have previously seen several possibilities for defining the atomic charges (Section 4.3). The basis of all those formalisms was to analyse the distribution of electron density and decide on the number of electrons that belong to various atoms. The results yield a static image.

The density functional perturbation theory (DFPT) allows defining the atomic charges differently. Because polarisation is related to the deformation of the electronic clouds, computing the changes of the polarisation during the atomic vibrations offers a dynamic perspective to the electronic density. This allows us to define a new type of charge – dynamical charges. These charges are also called Born effective charges and bear the notation $Z*$.

The Born effective charges are obtained from the derivative of the polarisation, obtained at the previous step (Equation 11.1), to atomic displacements:

$$Z_{\kappa\kappa'}^{*\alpha} = \frac{\partial P_\kappa}{\tau_{\kappa'}^\alpha} = \frac{1}{\Omega_0}\frac{\partial^2 E}{\partial\mathcal{E}_\kappa\partial\tau_{\kappa'}} \tag{11.4}$$

where P_k is the polarisation of the unit cell along the direction κ, τ is the displacement of the atom α, Ω_0 is the unit cell volume, E is the energy, and $\mathcal{E}$ is the electric field. The right-hand side of the second equation corresponds to the second term of the second-order derivative of the energy to two perturbations (see Equation 9.12).

Thus, the Born effective charges are the mixed derivatives of the energy with respect to the electric field along direction k, that is, the $\partial E/\partial\mathcal{E}_\kappa$ part, and the atomic displacements along direction κ', that is, the $\partial/\partial\tau_{\kappa'}$ part. This is the practical way of computing these charges.

But the best way to grasp the physical meaning of the Born charges is to use the description from the first part of Equation 11.4. The Born effective charges show how the polarisation of the crystal changes when individual

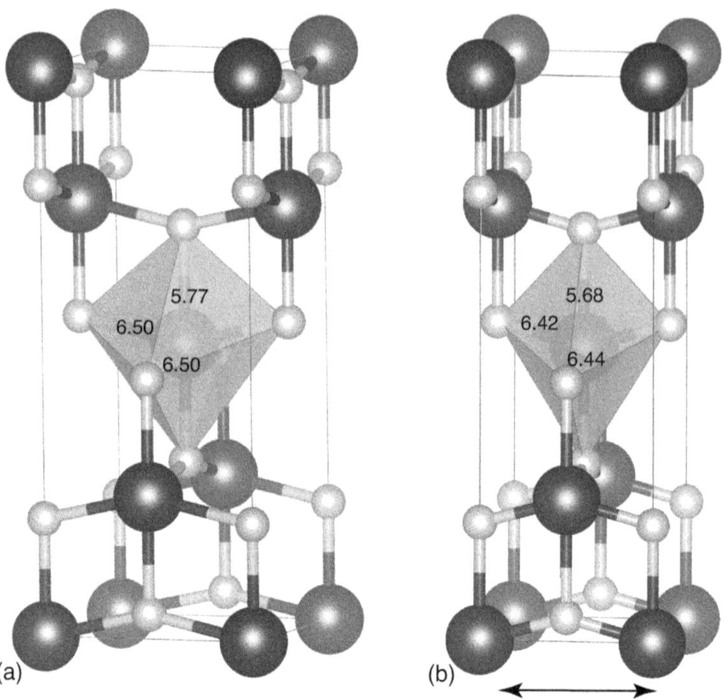

Fig. 11.1 (a) Born effective charges of Ti in anatase, TiO$_2$ along the Ti-O directions. Anatase has a tetragonal structure, and Ti is sixfold coordinated by O. (b) Under uniaxial strain along the x direction, the charge tensor loses part of the initial symmetry. The values calculated here are 'clamped', meaning the atoms are not allowed to relax under strain, which partly explains the decrease in absolute values of the charges.

atoms are displaced. The eigenvalues of the tensors can be compared to the nominal charges from chemistry.

The Born effective charges are 3×3 tensors, they are symmetric, that is, $Z^{*\alpha}_{\kappa\kappa'} = Z^{*\alpha}_{\kappa'\kappa}$, and they can be diagonalised. They depend on the symmetry of the local bonding environment around each atom, but they are independent of the symmetry of the lattice.

As an example of symmetry dependence, in tetragonal anatase, there are two long and four short Ti-O bonds in each TiO$_6$ octahedron. The Ti Born effective charge tensor reflects this environment, with one smaller longitudinal component along the long Ti-O bond and two larger transverse components along the shorter Ti-O bonds. Under uniaxial stress along the x axis, the tetragonal symmetry is broken, and the four Ti-O bond lengths are split into two pairs. Figure 11.1 shows the Born effective charges of anatase, TiO$_2$, and their loss of symmetry under uniaxial stress.

Because the macroscopic polarisation of a crystal is invariant under rigid-body displacement of the crystal, the Born effective charges must obey the charge neutrality of the unit cell. The sum of all Born effective charge tensors needs to be zero, along each of the directions k and k':

$$\sum_\alpha Z^{*\alpha}_{\kappa\kappa'} = 0 \tag{11.5}$$

Table 11.1 Born effective charge of Mg in MgO computed from the response to the electric field as in the density-functional perturbation theory as a function of pressure. While the charges are 3×3 tensors, because of symmetry, only the diagonal compounds are non-null. Their values are equal to each other: $Z^*[1,1] = Z^*[2,2] = Z^*[3,3]$. The charges of Mg and O have opposite signs and very similar values: $Z^*_{Mg} \approx Z^*_O$. The discrepancy comes from numerical reasons associated with the discreteness of the grid of $\mathbf{k}$ points and can be solved during post-processing.

Pressure (GPa)	Z^*_{Mg}	Z^*_O
0	1.953	−1.961
10	1.931	−1.939
20	1.913	−1.920
30	1.898	−1.905
40	1.884	−1.892
50	1.873	−1.881
60	1.863	−1.871
70	1.853	−1.861
80	1.844	−1.853
90	1.837	−1.845
100	1.829	−1.838

Deviations from this rule occur because of the discreteness of the sampling of the electronic density in terms of both the grid of $\mathbf{k}$ points and the number of planewaves. These deviations are a numerical artefact; their amplitude can be reduced by increasing the density of the grid of $\mathbf{k}$ points and the number of planewaves. As explained in Chapter 12, they can also be forced to obey the sum rule, which can be applied in the post-processing part.

For ionic crystals, the values are typically similar to the nominal charges. For example, for magnesite, MgO, the Born effective charges for the Mg and the O atoms are 1.961 and −1.969, respectively. Their values are very close to each other but of opposite signs, and they are also close to the nominal values from chemistry. They were computed for a unit cell parameter of 4.203 Å, within the range of experimental values at ambient conditions. Pressure shifts these values as the bonding evolves with increasing compression. The pressure dependence is not linear, with a faster decrease at the beginning of the compression. Their increased deviation with pressure from the nominal values marks the increased covalent bonding character as the atoms are pushed closer to each other. Table 11.1 lists the values of the Born effective charges for MgO in the $0 - 100$ GPa pressure range.

For covalent crystals, the values of the Born effective charges can be different from the nominal values. In this case, the Born effective charges are called anomalous. The charges of α-quartz are listed in Table 11.2. Their values lie below the nominal charges of +4 and −2 for the Si and O atoms, respectively [115]. The deviation is proof of the covalent Si-O bond. They are also not isometric, as the local bonding environment differs in different directions.

Table 11.2 Born effective charge of Si and O in α-quartz computed from density-functional perturbation theory [115]. The charges are 3×3 tensors, relating the change in polarisation as a function of electric field direction and atomic displacements. Note that not all charge components of Si are positive, nor all charge components of O are negative. However, the sum is zero to ensure the charge neutrality of the cell.

	E $\parallel$ x	E $\parallel$ y	E $\parallel$ z
Si atom			
disp. $\parallel$ x	3.016	0	0
disp. $\parallel$ y	0.0	3.633	0.282
disp. $\parallel$ z	0.0	−0.324	3.453
O atom			
disp. $\parallel$ x	−1.326	0.429	0.222
disp. $\parallel$ y	0.480	−1.999	−0.718
disp. $\parallel$ z	0.298	−0.679	−1.726

Sometimes the anomalous character is very pronounced. The ABO_3 ferroelectric perovskites show some of the largest anomalous Born effective charges. This group of technological materials of fundamental importance has its archetype in the mineral perovskite, $CaTiO_3$, which has a cubic structure. The ABO_3 materials are obtained by exchanging the Ca+Ti cations with various other pairs, like $Me^{II}Me^{IV}$ or $Me^{III}Me^{III}$. For example, in cubic $BaTiO_3$, the computed values of the dynamical charges of Ba are 2.7–2.8, which are relatively similar to the nominal charges of 2. Instead, Ti has Born effective charges on the order of 7.2, which are considerably larger than the nominal charge of 4. The charges of both Ba and Ti are isotropic, as both Ba and Ti occupy sites with cubic symmetry, and all coordinating O atoms are disposed at the same distance to the central cation. But the Born effective charges of the O atoms have a longitudinal component, equal to −5.7, along the O - Ti bonds, and two transversal components, equal to −2.15, along the O-Ba bonds. Their anomalous character compensates for the large positive charges on Ti. The symmetry change associated with the cubic-to-tetragonal phase transition in perovskites (Figure 7.1) leads to breaking the symmetry of the Born effective charge tensors, too, as the local bonding environment changes. The band-by-band analysis of the charges shows a strong electronic charge transfer between the $3d$ orbitals of Ti and the $2p$ orbitals of O [104]. This charge transfer yields the remarkable dielectric properties of $BaTiO_3$.

There are several ways to compute the dynamical charges according to the different equivalent formulations (Equation 11.4). Before the development of the DFPT, the standard way was to use finite difference calculations. The changes of the macroscopic polarisation were computed for a series of discrete atomic displacements along the three directions of the space. The polarisation can be calculated using the Berry phase (Section 7.1). The linear fit between the components of the polarisation and the displacements of the atoms yields the Born effective charge tensor. As with all finite difference calculations, this method was cumbersome and required many calculations.

An elegant alternative widely available today is to directly determine the energy derivatives with respect to the two perturbations, that is, electric field and atomic displacements. You can start computing these charges for insulators using the model from Box 11.1, illustrating the case of quartz. As the computation requires obtaining derivatives, it is important to use a sufficiently dense grid of **k**-points and large kinetic energy cutoffs, even a bit larger than in standard static calculations. Table 11.3 shows the dependence of the Born effective charges of Mg and O with the density of the grid of **k** points. For the low-density grids, like $2 \times 2 \times 2$ and $4 \times 4 \times 4$, the results are so bad that the charges are not even symmetric. The values converge for denser grids (Table 11.3).

The analysis of the Born effective charges is, unfortunately, quasi-unknown in mineralogy today, despite its usefulness.

11.3 Dielectric Tensor and Refractive Index

Contrary to the Born effective charges, the dielectric tensors, properly named dielectric permittivity tensors, are intensively used in various places in Earth sciences. They relate the applied external electric field to the electric field induced inside a sample, called electrical induction or displacement field. The induced field is responsible for the displacement of the atoms as a result of the polarisation of their electronic clouds. The two electric fields are vectors, so the dielectric tensor is a 3×3 tensor. The relation between them is:

$$\mathcal{D}_{mac,\kappa} = \sum_{\kappa} \epsilon_{\kappa\kappa'} \mathcal{E}_{mac,\kappa'} \tag{11.6}$$

Then, the dielectric tensor can be extracted either as the ratio between the two electric fields or, using the polarisation relation from Eq. 11.1, as the change in polarisation due to the external field (see Eq. 29 in [119]):

$$\epsilon_{\kappa\kappa'}^{\infty} = \frac{\partial \mathcal{D}_{mac,\kappa}}{\partial \mathcal{E}_{mac,\kappa}} = \delta_{\kappa\kappa'} + 4\pi \frac{\partial \mathcal{P}_{mac,\kappa}}{\partial \mathcal{E}_{mac,\kappa'}} \tag{11.7}$$

This formulation corresponds to the limit of low frequencies of the applied electric field, where the incoming radiation does not interact with the electronic structure and does not couple or excite the electronic states. This represents only the contribution of the electrons from the ground state to the energy and is denoted with an ∞ superscript, as in $\epsilon_{kk'}^{\infty}$. As the polarisation is the linear response of the energy to an external electric field, these expressions can be eventually related to the derivative of the energy with respect to two electric fields (Box 11.2):

$$\epsilon_{\kappa\kappa'}^{\infty} = -\frac{\partial^2 E}{\partial \mathcal{E}_{\kappa} \partial \mathcal{E}_{\kappa'}} \tag{11.8}$$

Box 11.1

How to compute the Born effective charges?

For insulators, the second-order derivative of the energy with respect to atomic displacements and electric fields yields the Born effective charges, $Z^{*\alpha}_{kk'} \sim \partial^2 E / \partial \mathcal{E}_k \partial \tau^{\alpha}_{k'}$. Because they involve atomic displacements, like during vibrations, these are also called *dynamical* charges. Their computation is done in two steps, as illustrated in the table below. First, the derivatives of the wave functions are computed with respect to their wave vectors (rfelfd 2). Second, these derivatives (getddk 2) are used to compute the derivatives of the energy with respect to the electric field (rfelfd 3 and rfdir 1 1 1). The two derivatives can also be calculated in one dataset with rflelfd 1. Note that the grid of **k** points should be denser than in standard ground-state calculations. As a general rule, the higher the order of the derivative, the denser the grid needs to be to decrease the numerical errors.

#Structure		#Actual Calculation	
acell	3*4.203 angstroms	#Actual Calculation	
rprim	0.0 0.5 0.5	ndtset	3
	0.5 0.0 0.5	#Ground-state calcu-lation	
	0.5 0.5 0.0	prtden1	1
xred	0.0 0.0 0.0	prtwf1	1
	0.5 0.5 0.5	toldfe1	1.02-10
natom	2		
ntypat	2	#Compute du/dk	
typat	1 2	getwfk2	1
znucl	12 8 # Mg O	rfdir2	1 1 1
#Parameters		rfelfd2	1
ecut	24	tolwfr2	1.0d-18
pawecutdg	72		
ngkpt	8 8 8	#Compute d2E/dτ d$\mathcal{E}$	
nshiftk	4	getwfk3	1
shiftk	0.5 0.5 0.5	getddk3	2
	0.5 0.0 0.0	rfdir3	1 1 1
	0.0 0.5 0.0	tolvrs3	1.0d-10
	0.0 0.0 0.5	rfelfd3	2
kptopt	2		

Sometimes the computation of the wave functions at the ground-state level on a highly dense grid of **k** points can be accelerated by performing a first self-consistent calculation on a limited set, for example, $4 \times 4 \times 4$ grid for MgO, and then a non-self-consistent calculation on a dense grid, for example, $8 \times 8 \times 8$ grid.

Table 11.3 Convergence of the Born effective charges of Mg and O in MgO as a function of the grid of **k** points.

grid of **k** points	Z^*_{Mg}	Z^*_O
$2 \times 2 \times 2 \times 4$	1.926	-2.326
$4 \times 4 \times 4 \times 4$	1.952	-1.979
$6 \times 6 \times 6 \times 4$	1.953	-1.962
$8 \times 8 \times 8 \times 4$	1.953	-1.961

Here the energy is normalised to the unit cell.

The full dielectric tensor can be determined only after the phonons are computed, as their frequencies and corresponding atomic displacement patterns are needed to calculate the coupling with the electric response. This coupling plays a central role in obtaining the infrared spectra of minerals (discussed in detail in Chapter 13).

In mineralogy, the dielectric tensor is extensively used for mineral identification. The refractive index of a mineral is the square root of the ϵ^∞. Thus, the image formed in a petrological microscope is due to the dielectric tensor. The thin section defines the cross-section through the refractive tensor, and the polarised light samples the different components of the tensor.

11.4 Non-linear Optical Properties

Higher-order energy derivatives as a function of the electric field yield other properties. The limiting factor for computing them is related to the available implementations. Many DFT codes have started to make available the computation of even higher responses of the energy. Amongst these, the third-order derivative of the energy with respect to the electric field opens the gate towards studying and predicting non-linear optical properties. While they might sound a bit far from the standard mineralogical interest, they participate in a better description of the response of a lattice under the action of an electromagnetic field. This has a direct technological interest in manufacturing optical materials, for example, optical fibres.

A third-order response only adds to the induced polarisation by the external field. Only non-centro-symmetric crystals can have a non-zero, non-linear optical coefficient. This is a third-order tensor. Materials with high values of this tensor have a highly efficient response to the incoming field. The resulting polarisation can then be expressed as a Taylor-like expansion as a function of field:

Box 11.2   How to compute the dielectric tensors?

For insulators, the second-order derivative of the energy with respect to two electric fields yields the dielectric tensor. This computation is again done in several steps, just like the Born effective charges from Box 11.1. First, you need to compute the derivatives of the wave functions with respect to their wave vectors (rfelfd 1). Second, you use these derivatives (getddk 2) to compute the second-order derivatives (rfelfd 2 and rfdir 1 1 1). As in the previous case, it is advisable to use a grid of **k** points denser than in standard ground-state calculations.

#Structure			
acell	3*4.203 angstroms	#Actual Calculation	
rprim	0.0 0.5 0.5	ndtset	3
	0.5 0.0 0.5	#Ground-state calculation	
	0.5 0.5 0.0	prtden1	1
xred	0.0 0.0 0.0	prtwf1	1
	0.5 0.5 0.5	toldfe1	1.02-10
natom	2		
ntypat	2	#Compute du/dk	
typat	1 2	getwfk2	1
znucl	12 8 # Mg O	rfdir2	1 1 1
#Parameters		rfelfd2	1
ecut	24	tolwfr2	1.0d-18
pawecutdg	72		
ngkpt	8 8 8	#Compute d2E/d$\mathcal{E}$2	
nshiftk	4	getwfk3	1
shiftk	0.5 0.5 0.5	getddk3	2
	0.5 0.0 0.0	rfdir3	1 1 1
	0.0 0.5 0.0	tolvrs3	1.0d-10
	0.0 0.0 0.5	rfelfd3	2
kptopt	2		

Sometimes you can accelerate the computation of the wave functions at the ground-state level on a highly dense grid of **k** points by performing a first self-consistent calculation on a limited set, for example, $4 \times 4 \times 4$ grid for MgO, and then a non-self-consistent calculation on a dense grid, for example, $8 \times 8 \times 8$ grid for MgO.

$$P \approx P_0 + \alpha \mathcal{E} + \eta \mathcal{E}^2 + \gamma \mathcal{E}^3 \tag{11.9}$$

where P_0 is the spontaneous polarisation, $\mathcal{E}$ is the electric field, η is the dielectric tensor, and γ is the non-linear optical coefficient.

The vibrations of the atoms are described by quanta of motion called phonons. In the direct space, the phonons correspond to waves of atomic displacements. They are each characterised by a wavevector, which defines their orientation and periodicity, and a frequency, which corresponds to their energy. The phonons do not have an associated particle; they are called quasi-particles. They behave according to the Bose–Einstein statistics, which makes them bosons, in contrast to electrons, which are fermions.

The atoms are in a constant state of vibration. As they move during the vibrations around their equilibrium positions, they induce forces on the other neighbouring atoms, consequently modifying the internal energy of the structure. These energy changes need to be solved to characterise the phonons completely. In the following, we will solve these changes within the quasiharmonic approximation, where the energy development as a Taylor series is cut after the second-order terms (see Section 9.3).

12.1 Interatomic Force Constants

Figure 12.1 illustrates a model crystalline lattice whose atoms are interconnected by ideal elastic resorts. The total energy of such a vibrating crystalline system can be expressed as the ground-state electronic energy of the static system plus the contributions coming from all the pairs of atoms in the crystal that interact elastically. The interaction between each pair of atoms is characterised by its specific elastic constant, which is called the interatomic force constant by analogy to the elastic case. The entirety of the elastic-type interatomic force constants in the crystal form the interatomic force constants matrix, $\mathbf{C}$. Because this matrix describes the pair interactions between all the atoms in the crystal, $1 .. \mathbf{N_c}$, each that can be displaced along the three dimensions of the space, the size of the interatomic force constant matrix is $\mathbf{N_c} \times 3 \times \mathbf{N_c} \times 3$. Each entry of the interatomic force constants matrix, $C_{\alpha\beta}$, corresponds to the equivalent spatial component of an elastic constant of an interatomic bond between atoms α and β. Expressed as a function of energy, each element of this matrix corresponds to the change in total energy of the crystal when the

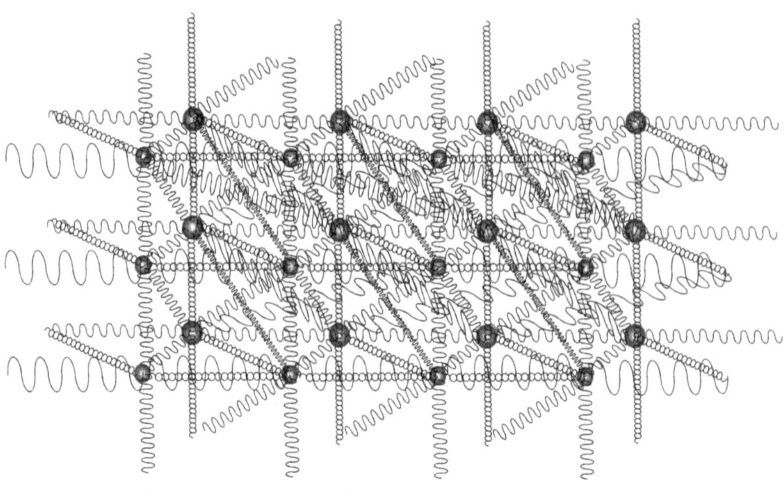

Fig. 12.1   The harmonic model of a crystal consists of all the atoms α, β, and so on, of all the cells a, b, and so on, being interconnected with each other via elastic resorts, each resort characterised by an elastic-like interatomic force constant $C_{\alpha,\beta}(a, b)$. Each force constant is a 3×3 tensor (Equation 12.2). For visibility, not all the elastic resorts are represented here.

two atoms α and β are displaced (see also Chapter 10). As the displacements can be done along any of the three directions of space, k, k', and k'', the elements of the interatomic force constants matrix are:

$$C_{\alpha,\beta}^{k,k'} = \frac{1}{2}\frac{\partial^2 E}{\partial\tau_\alpha^k \partial\tau_\beta^{k'}} \tag{12.1}$$

The notations can be worked a little more, as the different atoms α, and β may belong to different unit cells a, and b, respectively. This yields:

$$C_{\alpha,\beta}^{k,k'}(a, b) = \frac{1}{2}\frac{\partial^2 E}{\partial\tau_{\alpha,a}^k \partial\tau_{\beta,b}^{k'}} \tag{12.2}$$

With the expression for the interatomic force constants from Equation 12.2, Equation 9.8 can be rewritten in terms of forces. The force acting on the atom α in cell a is the sum of all the interatomic forces due to all atoms β from all the cells b of the entire lattice:

$$F_{\alpha,a}^k = \sum_{\beta,b}\sum_{k'} C_{\alpha,\beta}^{k,k'}(a, b)\tau_{\beta,a}^{k'} \tag{12.3}$$

But using fundamental mechanics formalism, the other form for writing the force acting on one atom α stems from the use of the Newtonian formulation with the mass, M_α, and the acceleration, a_α:

$$F_{\alpha,a}^k = M_\alpha a_\alpha^k = M_\alpha \frac{d\tau_{\alpha,a}^k(t)}{dt} \tag{12.4}$$

where t is the time, and the force is expressed with its components along the three directions of space. The acceleration is the second-order derivative of the coordinate with respect to time.

Now the two formulations for the force acting on the atom α, elastic from Equation 12.3 ($F = -kx$) and Newtonian from Equation 12.4 ($F = m\,a$), can be equated. The result is:

$$M_\alpha \frac{d\tau_{\alpha,a}^k(t)}{dt} = \sum_{\beta,b}\sum_{k'} C_{\alpha,\beta}^{k,k'}(a,b)\tau_{\beta,a}^{k'} \tag{12.5}$$

Equation 12.5 represents the primary tool for solving the vibrations of a lattice. It represents a square system of differential equations describing an ensemble of harmonic oscillators. The size of this system is given by the total number of atoms in the lattice and the number of directions of the space, $\mathbf{N_c} \times 3 \times \mathbf{N_c} \times 3$.

In the harmonic case (according to Section 9.3), which is treated here, the displacements of the atoms are periodic, and they are best described by a set of oscillatory movements, each with frequency ω_σ, around the equilibrium positions. Each of these movements σ is called a normal mode of vibration. Along each direction of the space k, the atomic displacements are:

$$\delta\tau_{\alpha,a}^k(t) = \sum_\sigma A_\sigma\, U_\sigma^\alpha(ak)e^{(i\omega_\sigma t)} + ct. \tag{12.6}$$

where $U_\sigma^\alpha(ak)$ represents the amplitude of a pattern of atomic displacements, and ω_σ is the frequency of each of the σ modes of vibration. Then replacing in Equation 12.5 the atomic displacement τ with the wave-like pattern from Equation 12.6 yields:

$$M_\alpha \omega_\sigma\, U_\sigma^{\alpha(ak)} = \sum_\beta C_{\alpha,\beta}^{k,k'}(a,b)\, U_\sigma^\beta(bk') \tag{12.7}$$

Solving this system is still impractical because the number of atoms in a crystal is huge. However, the use of translational symmetry in an ideal infinite crystal brings a first relief. This symmetry is also applied to the interatomic force constants matrix: Indeed, the actual cells a and b where the atoms α and β lie are not important *per se*, but the distance between the atoms is important. By convention, this is expressed as the difference $a - b$ between the cells. The interatomic force constants between the atoms α of the cell a and β of the cell b are the same as between the atoms α of the cell $a + n$ and β of the cell $b + n$:

$$C_{\alpha,\beta}(a,b) = C_{\alpha,\beta}(a+n,b+n) = C_{\alpha,\beta}(0,b-a) = C_{\alpha,\beta}(0,a-b) = C_{\alpha,\beta}(0,q*) \tag{12.8}$$

As the translational symmetry applies to the entire crystal, the set of translations $a - b$ defines the wavelength of a group of collective atomic displacements of normal modes of vibrations. They have the corresponding wavevector $\mathbf{q}^* = (a-b)^*$ in the reciprocal space. Consequently, the pair (a,b) can be replaced with the generalised $(0,b-a)$ vector, further simplified as $(0,q)$, by replacing all possible vectors $a - b$ with their corresponding q.

Then the wave of collective displacements can be expressed as a Bloch wave, in the same way as the electronic wavefunctions (Section 3.3):

$$U_\sigma^a(\alpha k) = e^{i\mathbf{q}R_a}\, U_{mq}(\alpha k) \tag{12.9}$$

(i) The first way to compute the entries of the matrix of interatomic force constants is to use the finite differences approach. For this, choose every pair of atoms and displace them simultaneously along two Cartesian axes (Box Figure 2):

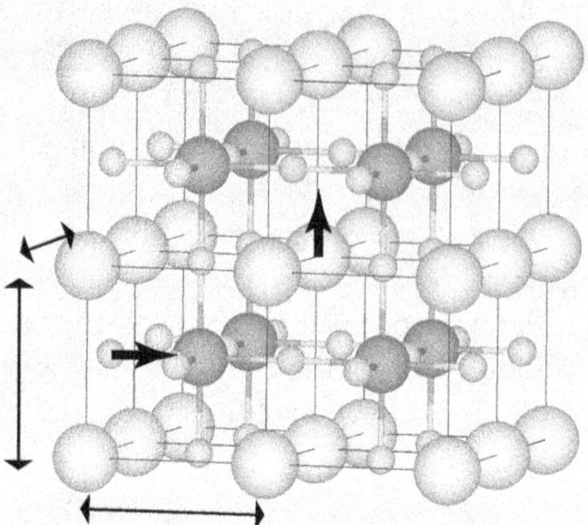

Box Figure 2. Building the interatomic force constant matrix involves computing the energy change during two atomic displacements.

Then monitor the change in energy; the second-order derivative of the change in energy with respect to the two perturbations (two displaced atoms) gives the force constant value. The displacements need to be performed in supercells whose sizes correspond to the wavevector **q** of the interatomic force constants.

(ii) The second way is to use the perturbation theory. For this, you need to specifically build the second-order energy derivatives with respect to atomic displacements. The code does it in several steps, in increasing order of derivatives of the energy and wavefunctions, before building the actual second-order terms. Specifically, for the computation of the interatomic force constants, you will need to specify:

rfphon	1	# flag for computing phonons
rfatpol	iatom jatom	# atoms from i to j are considered
		# in the energy derivatives
rfdir	1 1 1	# directions of the derivatives (k, k′, etc.)
nqpt	1	# number of q points (can be either 0 or 1)
qpt	0.25 0.25 0.5	# phonon wavevector

An efficient parallelisation can be obtained for large structures by computing each independent perturbation in a different dataset in a different run. For example, garnet has 80 atoms in the primitive unit cell and 7 atoms in its asymmetric part. Then run seven calculations, one for each atom in the asymmetric part of the garnet's unit cell, and each calculating one slice of the interatomic force constants matrix: rfatom 1 1, rfatom 2 2, ... , rfatom 7 7. In the end, save all the files and merge them later.

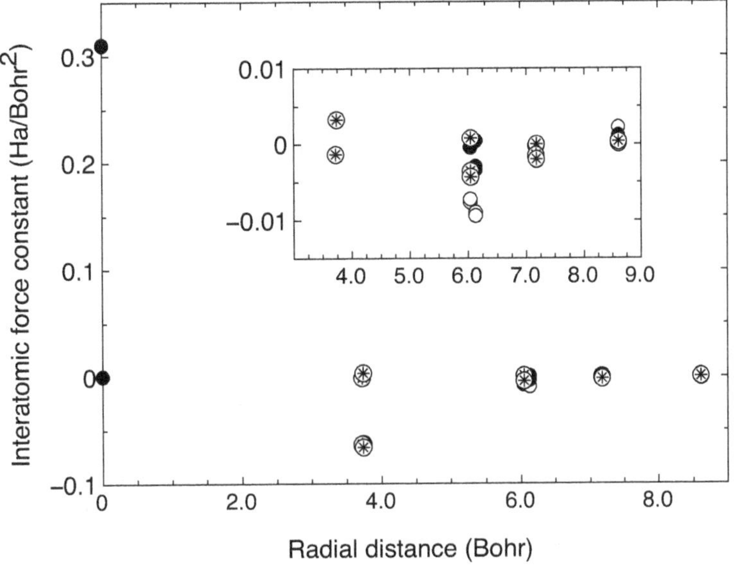

Fig. 12.2 The variation of the interatomic force constants with distance for the Zn and the O atoms in zincite, hexagonal ZnO. There is a strong decrease in the values of the force constants beyond the first coordination sphere. However, their values tend to zero only in the high-order coordination spheres. Note the difference in vertical scale on the inset.

where R represents the coordinate of the cell in the real space. In Equation 12.9, we have replaced the σ index of the vibrational mode with an index mq specifically dependent on the wavevector $\mathbf{q}$.

Thus, the normal modes of vibration described in Equation 12.9 are quantised, hence the index m, for each wavevector $\mathbf{q}$. They are called phonons, and they play a fundamental role in describing the lattice dynamics of solids. Their computation allows access to a wide variety of properties, going all the way to the free energy and the chemical potential.

12.2 The Dynamical Matrix

Equation 12.7 can now be rewritten in terms of the Bloch-like waves from Equation 12.9. At the same time, we can perform a Fourier transformation of the matrix of interatomic force constants (Box 12.2) and express it in the reciprocal space as a function of the wavevector of the phonons $\mathbf{q}$ (corresponding to the $a - b$ difference between the cells from above). The result is the dynamical matrix, which is fundamental in the determination of the dynamical properties of a lattice. Its expression is:

$$D_{\alpha k,\beta k'}(\mathbf{q}) = \sum_b C_{\alpha k,\beta k'}(0,b)e^{i\mathbf{q}R_b} \tag{12.10}$$

Box 12.2   **How to extract the interatomic force constants?**

Once the perturbation calculations are done (according to Box 12.1), the *anaddb* code can extract the values of various interatomic force constants (IFCs), using the following variables (note that the full input anaddb files are discussed at length in Chapter 13):

ifcflag	1	# flag for computing phonons
ifcana	1	# use the grid of **q** points to
		# interpolate the IFCs up to the required distance
ifcout	96	# number of IFCs to be printed
		# between the central atom and
		# its neighbours in decreasing distance
natifc	4	# number of atoms you ask IFCs for
atifc	1 2 3 4	# list of central atoms for the IFCs list

In one of the first detailed calculations about IFCs in silica [115], it was shown that even extended coordination spheres have a small but visible effect on the dynamics of the lattice. Such long-range force constants can be estimated from Fourier interpolation, as is discussed in the following pages. With the example earlier, the decrease of the IFCs can be seen as a function of distance in Figure 12.2.

It is important to note that the dynamical matrix is expressed in the reciprocal space, while the interatomic force constants matrix is represented in the direct space. With this formulation for the dynamical matrix, we can go back and rewrite Equation 12.7 to also obtain the displacement of the atoms in the reciprocal space, using the wavevector $\mathbf{q}$:

$$\sum_{\beta k'} D_{\alpha k,\beta k'}(\mathbf{q}) U_q(\beta k') = M_\alpha \omega_\sigma U_q(\alpha k) \tag{12.11}$$

Each dynamical matrix corresponds to a wavevector $\mathbf{q}$. It has dimension $\mathbf{N} \times 3 \times \mathbf{N} \times 3$, where $\mathbf{N}$ is the number of atoms in the *unit cell*, and the factor 3 comes from the three directions of the space along which the atoms are allowed to move during their vibrations. Note the difference to the initial interatomic force constants matrix, $\mathbf{C}$, which was a square matrix of dimensions $\mathbf{N_c} \times 3 \times \mathbf{N_c} \times 3$, with $\mathbf{N_c}$ the total number of atoms in the *crystal*.

Because the dynamical matrix is defined in a specific point $\mathbf{q}$ of the reciprocal space, this point becomes the wavevector of the phonons. This corresponds to the periodicity of the oscillatory movement of the atoms in the direct space, which is the inverse of the $\mathbf{q}$ point. Here we make the convention that all the atoms move *in-phase* within the same unit cell. With this convention, the general form for the displacement of the atoms in the direct, real space is given by:

$$\tau_{\alpha k} = U_{\alpha k} exp(-i\mathbf{q}\mathbf{r}) \tag{12.12}$$

The different energy derivatives are computed in several steps by the *abinit* code. After each step, they are written in a specific DDB file. Moreover, for large systems, you can divide the calculation of the derivatives into as many components as you wish (with a maximum of 3 directions × number of atoms).

Then every component is written on one line in the DDB files (Box Figure 3), which contains, in order, the direction and the type of the perturbation followed by the real and the imaginary values of the derivatives. For example, the energy derivatives with respect to the atomic displacements are listed as:

```
1    8    3    25    0.14167951238896D+00    0.12891965402986D-17
2    8    3    25    0.14242073771551D+00    0.10934069033585D-17
3    8    3    25    0.23255853085134D-02    0.16087765440968D-17
1    9    3    25   -0.12701502353682D+01    0.83778714166953D-19
2    9    3    25   -0.89609853041867D+00    0.30709662739883D-17
3    9    3    25   -0.10037294409847D+01    0.38622785666019D-17
1   10    3    25   -0.41538273539446D+00   -0.10074771772421D-17
2   10    3    25   -0.22337827351770D+00    0.17599244883618D-17
3   10    3    25    0.10110793340095D+00    0.50648478224019D-17
```

Box Figure 3. Each component of the dynamical matrix is written on one line in the DDB files.

At the end of all the calculations, the *mrgddb* utility merges all the DDB files into a single database, which contains all the calculated terms of Equation 9.12. The interatomic force constants are listed in the order of the wavevectors.

The *mrgddb* utility applies the symmetry on all the derivative matrices and reconstructs the missing terms in the entire matrix if the one in the reduced part is provided. This feature is useful when building the dynamical matrices for very large structures. Then you merge all the slices of the interatomic force constants and build one single file with all the perturbations. The *mrgddb* needs a list of file names, which you can put, as usual, in one file containing:

garnet.DDB

free form comment line

number of DDB files

tremolite.OU1_DDB

tremolite.OU3_DDB

tremolite.atom1.OU_DDB

tremolite.atom2.OU_DDB

...

tremolite.atom7.OU_DDB

In this example, the number of DDB files should be 9. Typically, the first DDB file contains the first-order derivatives of the energy with respect to atomic displacements, that is, the forces, and the second one the derivatives with respect to an electric field, that is, dielectric tensors.

This is the solution to Equation 12.11, and it represents a wave of atomic oscillatory movements. The amplitude of the wave is given by the prefactor U, and the wavelength and orientation of the wave are directed by the wavevector $\mathbf{q}$. In practice, the values of U are obtained by diagonalisation of the dynamical matrix.

12.3 Eigenvectors of the Dynamical Matrix

The eigenvectors of the dynamical matrix yield the atomic displacement patterns corresponding to the phonons with wavevector $\mathbf{q}$. They are matrices of dimension $\mathbf{N} \times 3$ whose entries represent the amplitudes of the atomic displacements during vibrations, that is, the $U_{mq}(\alpha k)$ quantities from Equation 12.9. The displacement patterns $U_{mq}(\alpha k)$ correspond to a given phonon m, with wavevector $\mathbf{q}$, and describe the displacements of all the atoms α of a unit cell, along the three directions of the space k (hence this $\mathbf{N} \times 3$ dimension).

The eigenvectors are complex numbers. Because of the time-reversal symmetry, the displacements of the atoms from a phonon in $\mathbf{q}$ combine with the displacements of the atoms from a phonon in $-\mathbf{q}$. This leads to cancelling the imaginary part of the displacement. Hence, the actual movement of the atoms corresponds only to the real part of the eigenvectors. The imaginary part is zero for the phonons with wavevector in Γ. The product between the eigenvectors and the phase factor yields the amount of displacement of the atoms in each cell in the crystal corresponding to that phonon. As stated earlier, the common convention is that all the atoms in the same unit cell have the same phase.

Figure 12.3 explains the phonons propagate in the lattice as a function of their wavevector. A phonon with a wavevector in Γ implies the same amplitude of displacements in each cell (Figure 12.3(a)). Using the example of a primitive cubic crystal, a phonon in $\mathbf{X} = \mathbf{1/2\ 0\ 0}$ implies a planewave alternance of the atomic displacements along the x axis (Figure 12.3(b)). In this case, every neighbouring cell vibrates in the opposite direction. A phonon in $\mathbf{M} = \mathbf{1/2\ 1/2\ 0}$ is the combination of two waves, one along the x axis and one along the y axis, both with a wavelength spanning two unit cells. The result is an alternance of displacements in columns: The atoms move in the same direction along columns parallel to the z axis (Figure 12.3(c)) and alternating directions along the x and y directions. According to this algorithm, for a phonon in $\mathbf{R} = \mathbf{1/2\ 1/2}$ $\mathbf{1/2}$, the atoms in all neighbouring cells vibrate in opposite directions (Figure 12.3d). A phonon in $\mathbf{T} = \mathbf{1/4\ 0\ 0}$ corresponds to an alternance of the amplitude that spans four unit cells. One possibility is to represent two cells with maximum amplitude and the other two with zero amplitude (Figure 12.3(e)). An alternative graphical representation has two consecutive cells with positive half amplitude and the other two with negative half amplitude (Figure 12.3(f)). Finally, a mode in a more general point of the Brillouin zone implies vibrations with different amplitudes in the other cells along the direction of the propagation of the phonon. The schemes

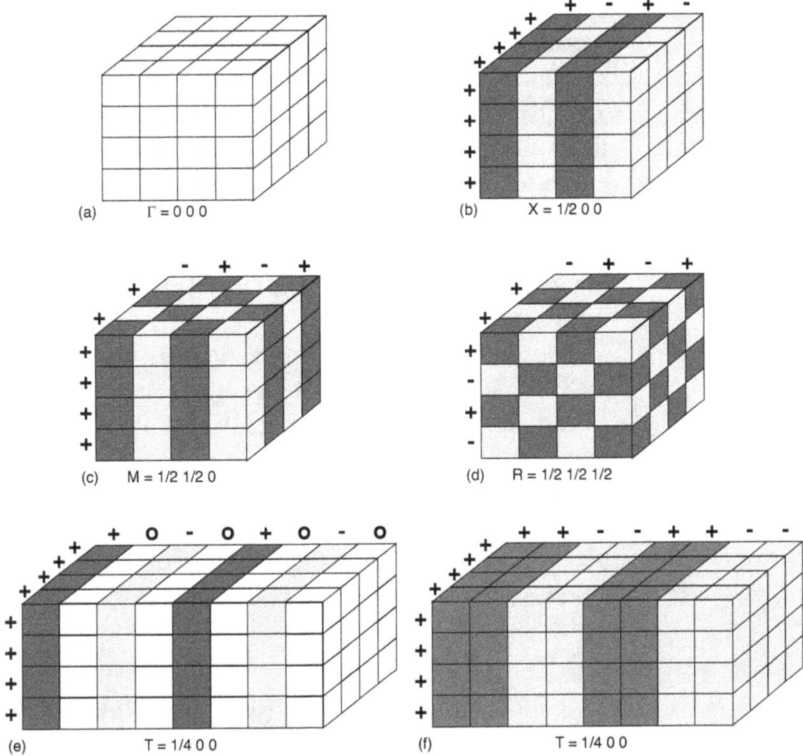

Fig. 12.3 The phonons are waves of displacements traversing the crystal lattice. The convention in representing atomic displacements is that the phase is preserved for all the atoms in one unit cell. In this diagram, the phase of the phonon wave varies in each cell depending on the phonon wavevector. In Γ (a) all cells have the same phase. The phonons in X = 1/2 0 0 (b), M = 1/2 1/2 0 (c), and R = 1/2 1/2 1/2 (d) exhibit alternance of displacements between neighbouring cells along the corresponding directions. The phonons in T = 1/4 0 0 (e and f) can be represented in various equivalent ways.

in Figure 12.3 correspond to the relative relations of phase in various cells across the crystal; as the phonons represent oscillatory movements, the atoms describe an entire back-and-forth alternant movement. This scheme shows how the relative phase of this movement changes at any moment between individual cells.

Because the lattice is in a constant state of vibration, the instantaneous position of any atom in the cell is a sum of the equilibrium position plus the sum of all possible displacements due to all possible phonons. Averaged over time, this apparent disorder corresponds to the thermal agitation of atoms. This can be graphically represented as the thermal ellipsoids.

12.4 Eigenvalues of the Dynamical Matrix

The eigenvalues of the dynamical matrix yield the square of the frequencies of the phonons with wavevector **q**. They correspond to the curvature of the energy plots from Figure 9.1, that is, they are linked to the second-order derivatives of the energy. If the eigenvalue is positive, then $\omega^2 > 0$, and ω is also positive. This is the case of the stable phonons: The atoms vibrate around their initial equilibrium positions. If the potential well opens up, the curvature decreases and the phonon frequency decreases. When ω^2 approaches 0, the energy curve flattens out, and the frequency ω approaches 0. This is the case of a soft mode. Finally, for the extreme last case $\omega^2 < 0$, the frequency ω is imaginary: Such a phonon cannot exist, and consequently, it represents an unstable mode, discussed in detail later on.

The lightest atoms in a structure are the ones that have the highest vibration amplitude and highest frequency of vibration. The highest vibrational frequencies of the fundamental modes are recorded in the H_2 molecule, at 3,441 cm^{-1}, while the OH group in hydrous minerals vibrates typically in the 3,000–3,500 cm^{-1} range at ambient conditions. For comparison, the vibrational modes in quartz, SiO_2, have frequencies below about 1,300 cm^{-1}. The amplitude of the atomic displacements during vibrations is on the order of a fraction of Å.

12.5 Vibrational Patterns

Access to the full dynamical matrix gives a unique insight into the nature of the vibrational properties of minerals. If, in experimental mineralogy, one measures the frequencies and makes deductions about the vibrational pattern by comparing spectra, in simulations, we directly compute the vibrational pattern at no additional cost.

For complex structures containing various molecular fragments, chemists distinguish intermolecular vibration modes, where such large fragments move as rigid bodies, and intramolecular vibrations that take place inside those fragments. For minerals and crystalline solids with different structural units easily identifiable, like anionic groups, the intermolecular vibrations are often called *lattice* modes. The intermolecular vibrations or lattice modes can be, for example, rigid translations (Figure 12.4(a)) or rigid rotations (Figure 12.4(b)), the latter being also called librations. The lattice modes are usually characterised by low frequencies because there is a relatively large mass that moves in phase during the vibration.

The intramolecular modes are the vibrations in which the atoms within the molecular fragments are displaced relative to each other. The actual amount of mass that is involved in the vibration is significantly smaller than in the case of the lattice modes. Consequently, the frequencies of the intramolecular modes are usually higher than those of the lattice modes.

For the intramolecular vibrations, there is a first distinction between *stretching* and *bending* modes (Figure 12.5). In the stretching modes,

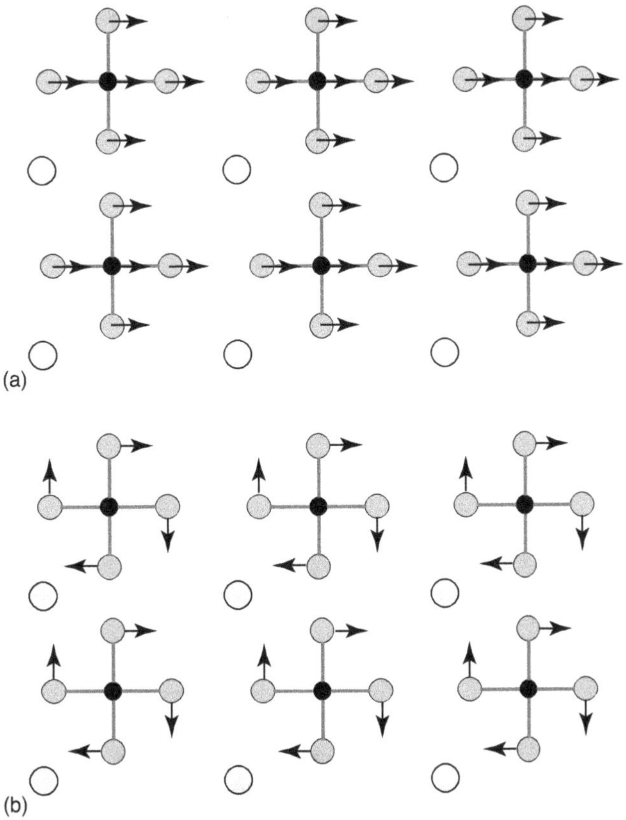

Fig. 12.4   Example of eigendisplacements of lattice modes. (a) rigid translation and (b) rigid rotation of one sublattice.

the distance between atoms changes during the vibrations: The atoms approach and distance to and from each other alternatively. When several atoms are involved in this movement, the mode is called symmetric if all the distances increase and decrease simultaneously. Otherwise, it is called asymmetric (first row in Figure 12.5).

In bending modes, the angles between atoms change during the vibrations, but the interatomic distances are kept constant. The bending can happen by preserving the planar geometry of the molecule (middle row example in Figure 12.5), or the atoms can vibrate outside of the planar geometry of the molecule (bottom row example in Figure 12.5).

Mineralogists loosely adapted the terminology of the intramolecular modes. There are stretching modes, breathing modes, bending modes, rotation modes, and so on. Such a description of the modes allows us to easily exchange information about the different phonon modes observed, for example, in Raman or infrared spectra, and to easily visualise them.

Figure 12.6 shows further examples and analysis of vibrations for a penta-atomic molecule formed of one central atom and four ligands arranged at the corners of a square. In two-dimensional (2D), there are 10 possible modes for a molecule with five atoms (each atom can move along two directions of the space). Each mode is characterised by its frequency

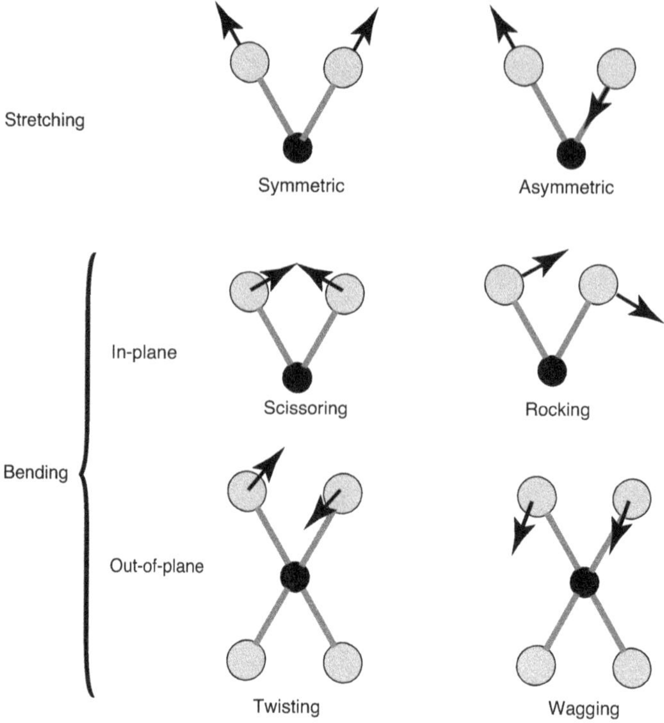

Fig. 12.5 General classification scheme for the intramolecular vibrational modes, which can also be applied to various atomic groups in crystals. Stretching modes involve changes in bond lengths, bending modes involve changes in interatomic angles.

and its atomic displacement pattern. For some modes, their vibrational patterns are related by symmetry (typically with the help of rotations). In this case, they also have the same frequency and are called degenerate.

The first two phonons, illustrated in Figure 12.6(a,b), show all the atoms moving in the same direction in space with the same amplitude. These correspond to a rigid shift of the entire molecule, which is equivalent to nothing moving, as the reference point can also move. The frequencies of these modes are consequently zero. Because there are two dimensions, a first such shift can be done along the x direction and a second along the y direction. These modes, degenerate, are called acoustic. The diagonalisation of the dynamical matrix of a molecule (or in the Γ point) always yields a set of such acoustic modes whose degeneracy equals the number of dimensions of the space.

The next phonon represented in Figure 12.6(c) shows the case of a rotational mode, or a rocking mode, where the central atom is fixed and each one of the four ligands is moving along one axis, such that in total their movement yields a rigid rotation of the coordinating polyhedron.

The two phonons from Figure 12.6(d,e) have the central atom fixed, while the four ligand atoms are translated along the horizontal or vertical axis. These modes are also doubly degenerate. As the central atom

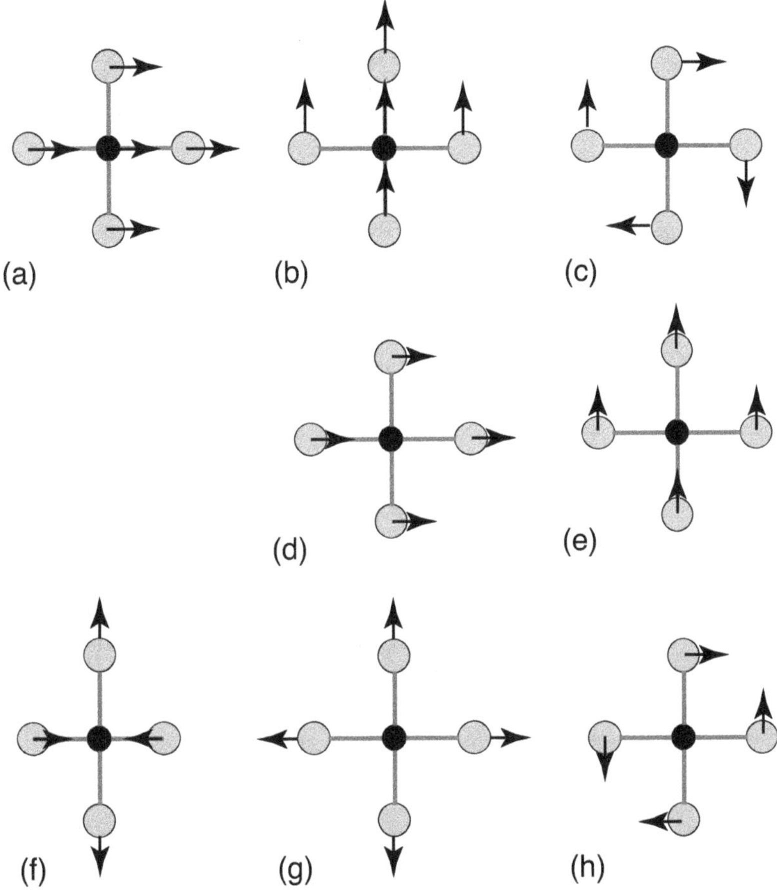

Fig. 12.6 Example of various phonon eigendisplacements for a model planar penta-atomic molecule with two atomic types. (a, b) rigid translation of all atoms; (c) rotational mode; (d) and (e) translational mode of a sublattice, double degenerate, polar; (d) anti-symmetric stretching, non-polar, non-degenerate; (e) breathing, or symmetric stretching, mode; (f) scissoring mode.

and the ligand atoms are of different types, they will carry a different electronic charge. The vibration leads to the formation of a local electric dipole moment: These modes are polar.

The next phonon type, represented in Figure 12.6(f), shows the central atom fixed while two opposite ligands move away from it and the other two opposite ligands move closer to it. There is no net charge forming in the cell because of the symmetry of the movement. Hence, this mode is non-polar. This mode is not degenerate, as there is no other equivalent movement of the atoms. This is an asymmetric stretching.

The phonon shown in Figure 12.6(g) shows the central atom fixed and all ligands moving away from it. This is a symmetric stretching, also called a breathing mode, because during the atomic movement, the volume of the group alternatively increases and decreases, just like breathing lungs. Most anionic groups in minerals, like the CO_3 in carbonates, the SiO_4 in silicates, the SO_4 in sulphates, and so on, have one breathing mode and several asymmetric stretching modes. In general, the breathing modes of the anionic groups in minerals yield peaks with large intensity in the Raman spectra. And as their frequency changes little with chemistry (because these vibrations involve mainly the ligands of the anionic groups), they can be successfully used to identify the class of a mineral.

For the phonon represented in Figure 12.6(h), the angles between every two ligands and the central atom change during the vibration, but the distances between the central atom and the ligands are constant. This is a bending mode, more precisely, a scissoring mode according to the classification from Figure 12.5.

In Figures 12.5 and 12.6, the movements of the atoms are represented by convention with arrows, which are unidirectional. Because these are all oscillatory movements, in reality, the atoms perform *back-and-forth* movements, with the amplitude $U_m(\alpha k)$. They go on one side up to $U_m(\alpha k)$, and then they reverse and go all the way to the other side of their position of equilibrium up by the same displacement $-U_m(\alpha k)$. The totality of the $U_m(\alpha k)$ displacements matrix represents the eigenvector of the phonon m. Up to now, the wavevector q was ignored from the notation, as discussed in detail in Section 12.6. Without a wavevector, the description of the atomic displacements corresponds to the case of isolated molecules. This can also be applied to glasses, as these structures do not have translational symmetry.

12.6 The Γ Point

In the next step, the eigendisplacements of the phonon modes can be placed in the context of the lattice. This corresponds to reintroducing the q wavevector in the description of the atomic displacements. With this, from now on, we will talk about vibrations in a crystal. We start with the centre of the Brillouin zone, the $q = 0$ wavevector, that is, the Γ point.

The dynamical matrix from the $\Gamma = 0$ point represents a particular case and is extremely important for mineralogists. The wavevector $\mathbf{q} = \Gamma = 0$ of the phonons corresponds in the direct space to an oscillatory movement that is identical in all unit cells. Indeed, its wavelength is infinite: $1/(\mathbf{q} = 0) = \infty$. This means that the amplitude of the atomic movements is the same in each unit cell and that the atoms move all in phase in the entire crystal.

Figure 12.7 shows the eigenvectors of several phonons with wavevector $\mathbf{q} = \Gamma$ for the case of a crystal in two dimensions. The molecule and modes discussed in Figure 12.6 are placed in the context of a regular crystalline lattice.

The first mode (Figure 12.7(a)) corresponds to a rigid translation of the molecule along one direction of the space. Because the wavevector

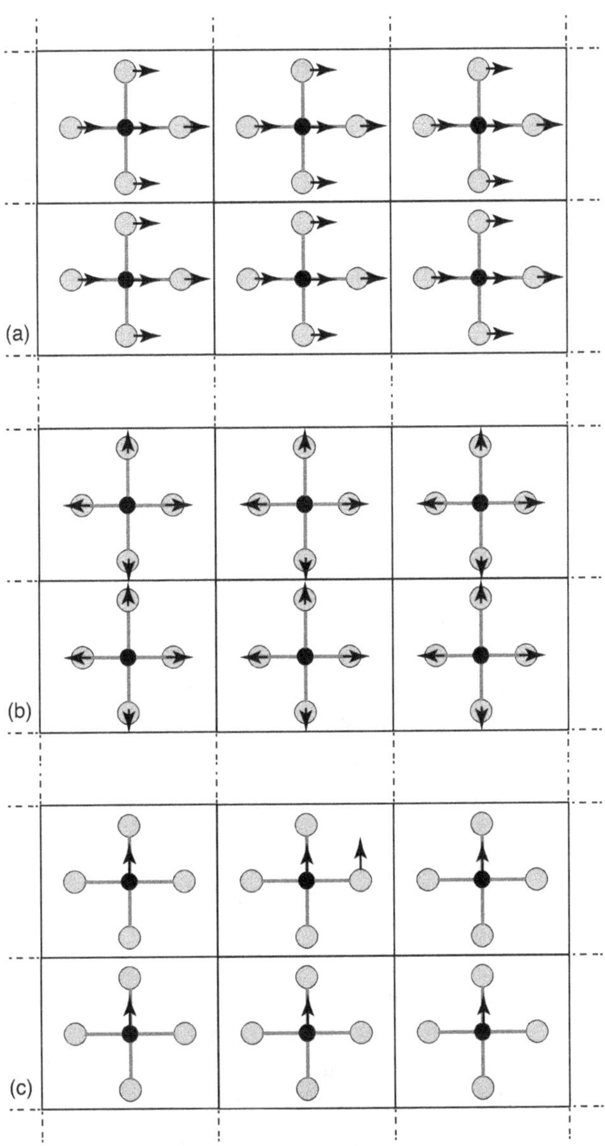

Fig. 12.7 Example of a few phonon eigendisplacements for a crystal containing one planar penta-atomic molecule with two atomic types, as in Figure 12.6. Note the translational symmetry of the lattice, as shown here by the repetition of the unit cells. Because the phonon is anchored in Γ, the atoms move in the same way in all the unit cells. (a) Rigid translation of all atoms. This is an acoustic mode. As the displacement is identical for all atoms, there is no net resultant movement – the frequency of this mode is 0. (b) Anti-symmetric or asymmetric stretching; (c) Rotational mode.

of the phonon is Γ, the same translation takes place in every cell of the lattice. Consequently, this corresponds to a change of origin, which is arbitrary. As in the previous case, this mode has a null frequency, as nothing is moving, and the origin of the cell is arbitrary. This mode is also double degenerate, as there are two directions of the space (one horizontal and one vertical). For a real three-dimensional crystal, there are three such modes.

These modes with $\omega = 0$ in Γ are called acoustic modes, as briefly mentioned earlier. Their name comes from the similarity to the transmission of acoustic, or elastic, waves. The elastic waves have wavelengths that are orders of magnitude larger than the size of a unit cell. For example, elastic waves originating from large earthquakes have typical wavelengths between a few hundred meters and a hundred kilometres. This represents more than 12 orders of magnitude larger than the unit cell. Even the wavelength of ultrasounds, at a centimetre scale, remains eight orders of magnitude larger than the unit cell. Hence, these waves correspond to an infinitesimal $\mathbf{q} = >0$ wavevector. When a longitudinal sound wave travels through a material, it applies a uniaxial compression along the direction of propagation that corresponds to a monotonous displacement of all atoms along that direction. This elastic distortion is the equivalent of a phonon mode with zero frequency at Γ.

The rest of the modes in Γ are called optical, not only to differentiate them from the acoustic modes but also because they can be *seen* using Raman or infrared spectrometers. The mode in Figure 12.7(b) is a breathing mode, that is, symmetric stretching, which is active in Raman. The mode in Figure 12.7(c) is polar, that is, asymmetric stretching, which is active in infrared. The Raman and the infrared spectroscopies, discussed at length in Chapter 14, became widely used in laboratories for exploring the vibrational properties of mineral structures. As the vibrational pattern contains information about the structure and the interatomic bonding, the spectra can be used for identifying minerals. These techniques are so powerful that even the last Mars rover embarked on two such Raman spectrometers on board to help identify minerals on the surface of Mars.

After its determination, the dynamical matrix is diagonalised to extract all the phonon information. For this, a post-processing utility is required in most cases. In the abinit package, *anaddb* is the processing utility that performs this analysis. The details of using this utility are shown in the Box 12.4 for the Γ point. There are several alternative options, like phonon[197], phonopy[244], and so on, which can be used with other *ab initio* codes.

12.7 Irreducible Representations (*irreps*)

The movement of the atoms is not chaotic but respects the rules of symmetry. This peculiar relation with the symmetry comes from the diagonalisation of the dynamical matrix. If we could observe the atoms in a crystal in real time, we would see that they are not static balls of electronic density, but they are in constant movement. Over time, this movement describes a rather chaotic trajectory (the thermal agitation)

Box 12.4   **How to compute the phonons in Γ?**

With the interatomic force constants (and thus also the dynamical matrix) computed in a previous step (see Box 12.1), you need to extract the eigenvalues and eigenvectors to assess the phonons. For this, we need to use one of the abinit post-processing tools: the *anaddb* code. In Abinit, the *anaddb* code is sequential only and runs at the command line. As with the main abinit code, you can create a file containing the names of the files, like:

```
ana.in            # input file with keywords
anaddb.out        # main output file
your_file.DDB     # DDB file
file1.eps         # plot phonon band structure
file2.gkk         # electron-phonon coupling
file3.ep          # electron-phonon output
fiel4.dkk         # derivatives for electron-phonon coupling
```

The basic ana.in file needs just a few keywords to perform the diagonalisation of the dynamical matrix in Γ:

```
                         # Flags
ifcflag    1             # Interatomic force constant flag
                         # grid of q-vectors
brav       1             # Bravais Lattice : 1-S.C., 2-F.C., 3-B.C., 4-Hex.)
ngqpt      1 1 1         # grid of q points
nqshft     1             # number of shifted grids
q1shft     0.0 0.0 0.0   # shifts of the grid
                         # define the q point(s) to compute the phonons
nph1l      1             # number of q points in 1st list of phonons
qph1l      0.0 0.0 0.0 1.0   # first list of phonons
                         # specify Γ point
                         # the 4th number is the scaling
                         # set it at 1.0 in the first list of phonons
nph1l      1             # number of q points
qph2l      1.0 0.0 0.0 0   # directions of non-analicity
           0.0 1.0 0.0 0   # the 4th number is the scaling
           0.0 0.0 1.0 0   # set it at 0.0 for the second list of phonons
           0.0 1.0 1.0 0
```

around equilibrium positions (the energetic minima). The diagonalisation of the dynamical matrix allows decomposing this chaotic trajectory in a linear combination of normal modes of vibrations. Even if it might not appear obvious at first sight, their existence is proven by the discrete character of the spectra recorded in vibrational spectroscopy, like Raman or infrared.

Table 12.1 Main symbols used to describe the phonon modes.

A	symmetric under principal rotation
B	antisymmetric under principal rotation
E	double degenerate (from German *entartet* = degenerate)
T	triple degenerate
g	symmetric under inversion (from German *gerade* = straight, even)
u	antisymmetric under inversion (from German *ungerade* = odd, uneven)
1, 2	index to differentiate between similar modes
′, ″	apostrophe to differentiate between different orientations

Furthermore, the relations between the atomic displacements of each phonon mode and each of the symmetry operations of the point group of that crystal can be checked. Some operations leave invariant the vibrational pattern, while others reverse it. An equivalent way of seeing this is counting the number of bonds that are left invariant by applying the symmetry operations on the vibrational pattern. Eventually, this behaviour defines a symmetry fingerprint for each phonon mode. Then, depending on the specific relation between the symmetry elements and the atomic displacements, using this symmetry fingerprint, particular labels can be defined, which can be associated with each vibrational mode. They are an important way of communication and exchange between spectroscopists, as the labels carry all the symmetry information with them. As both the atoms and the symmetries are represented as matrices, the symmetry fingerprints bear the name of *representations*.

In a crystal, several distinct phonon modes may behave in the same way under the action of the symmetry operations, so they have the same representation. The symmetry of the point group and the site symmetry of the atoms determine the possible type and the number of each type of phonon mode in a crystal. The sum of the total number of representations defines the irreducible representations of the phonons in the zone centre, in short, the *irreps*. The same concept can be applied not only to the Γ point but also to any other wavevector from the reciprocal space. In practice, it is common to use the irreps only in Γ.

In Γ, the different representations have specific symbols, their origin being in the German language (Table 12.1). Then they may bear apostrophes and number indices to distinguish between their different types [187]. From these, the u and the g indices are important markers for polar and non-polar modes, respectively.

For a given mineral structure, with the site symmetry known for each atom, the point group symmetry can be used to build a table with all the possible phonon representations. The Bilbao Crystallographic Server provides a tool for automatically determining the irreps [10]. Box 12.5 shows the procedure to obtain the irreps for the cubic Pm3m structure of $CaSiO_3$ with a perovskite structure. In this structure, the Ca, Si, and O atoms occupy the $1a$, $1b$, and $3d$ Wyckoff positions, respectively. With five atoms

We start with the cubic Pm3m structure (space group no. 221) of $CaSiO_3$ with a perovskite structure, where the Ca, Si, and O atoms lie respectively on the 0 0 0, 1/2 1/2 1/2, and the 1/2 0 0 positions. We load the SAM program (www.cryst.ehu.es/rep/sam.html) from the suite available at the Bilbao Crystallographic Server (www.cryst.ehu.es/). We choose space group no. 221 and pick the Wyckoff positions of the atoms (a). In the next step, we obtain all the information related to the representations (b), the spectroscopic properties of various modes, and we can choose from all the irreps, those corresponding to the set of atoms present in the structure, as shown in the table of mechanical representations (c) (Box Figure 4):

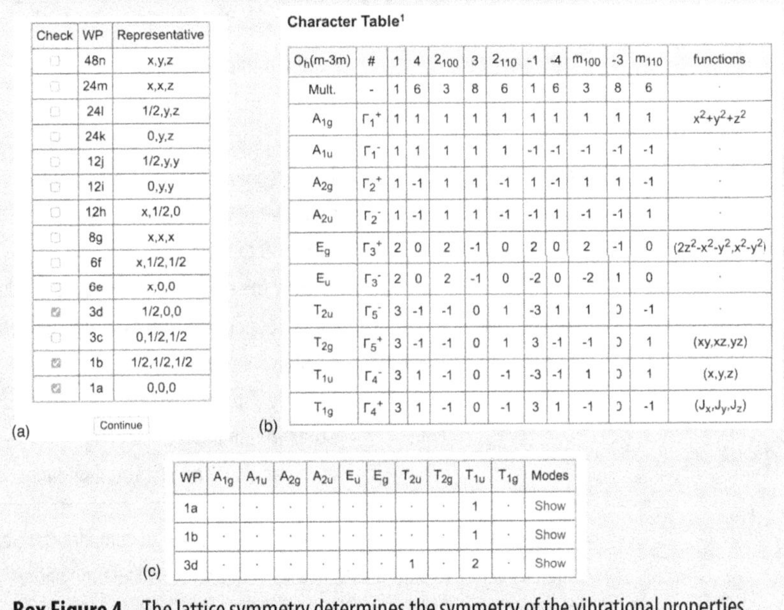

(a)

Check	WP	Representative
	48n	x,y,z
	24m	x,x,z
	24l	1/2,y,z
	24k	0,y,z
	12j	1/2,y,y
	12i	0,y,y
	12h	x,1/2,0
	8g	x,x,x
	6f	x,1/2,1/2
	6e	x,0,0
☑	3d	1/2,0,0
	3c	0,1/2,1/2
☑	1b	1/2,1/2,1/2
☑	1a	0,0,0

Continue

(b)

Character Table[1]

O_h(m-3m)	#	1	4	2_{100}	3	2_{110}	-1	-4	m_{100}	-3	m_{110}	functions
Mult.	-	1	6	3	8	6	1	6	3	8	6	
A_{1g}	Γ_1^+	1	1	1	1	1	1	1	1	1	1	$x^2+y^2+z^2$
A_{1u}	Γ_1^-	1	1	1	1	1	-1	-1	-1	-1	-1	
A_{2g}	Γ_2^+	1	-1	1	1	-1	1	-1	1	1	-1	
A_{2u}	Γ_2^-	1	-1	1	1	-1	-1	1	-1	-1	1	
E_g	Γ_3^+	2	0	2	-1	0	2	0	2	-1	0	$(2z^2-x^2-y^2,x^2-y^2)$
E_u	Γ_3^-	2	0	2	-1	0	-2	0	-2	1	0	
T_{2u}	Γ_5^-	3	-1	-1	0	1	-3	1	1	0	-1	
T_{2g}	Γ_5^+	3	-1	-1	0	1	3	-1	-1	0	1	(xy,xz,yz)
T_{1u}	Γ_4^-	3	1	-1	0	-1	-3	-1	1	0	1	(x,y,z)
T_{1g}	Γ_4^+	3	1	-1	0	-1	3	1	-1	0	-1	(J_x,J_y,J_z)

(c)

WP	A_{1g}	A_{1u}	A_{2g}	A_{2u}	E_u	E_g	T_{2u}	T_{2g}	T_{1u}	T_{1g}	Modes
1a	·	·	·	·	·	·	·	·	1	·	Show
1b	·	·	·	·	·	·	·	·	1	·	Show
3d	·	·	·	·	·	·	1	·	2	·	Show

Box Figure 4. The lattice symmetry determines the symmetry of the vibrational properties.

in the unit cell, there are 15 phonon modes. Box 12.5 shows that their movements are distributed between $4T_{1u}$ modes and 1 T_{2u} mode, each mode being polar, threefold degenerate.

12.8 Phonon Dispersion

The wavevector $\mathbf{q}$ of the phonons scans the entire Brillouin zone. The dependence of the energy of the vibration, that is, the frequency of the phonon, with the wavevector $\mathbf{q}$ yields the dispersion. The same general rules from electrons apply to the dispersion of phonons: The approach to the zone boundaries is done asymptotically, the bands are continuous

 How to calculate the phonon dispersion?

Starting from a regular grid of **q** points, the interatomic force constants can be obtained at any **q** point of the Brillouin zone using a Fourier interpolation of the interatomic forces [114].

The *anaddb* code builds the entire grid of **q** points (defined by the ngqpt keyword) using the symmetry and the wavevectors of the dynamical matrices from the _DDB file (computed only in the asymmetric part of the Brillouin zone). Then it interpolates the interatomic forces for various wavevectors and prints the result.

The few variables that need to be changed with respect to the previous example (Box 12.4) are:

```
                                    # Flags
ifcflag    1                        # Interatomic force constant flag
                                    # course grid of q-vectors
brav       1                        # Bravais Lattice : 1-S.C., 2-F.C., 3-B.C., 4-Hex.)
ngqpt      2 2 2                     # grid of q points
nqshft     1                        # number of shifted grids
q1shft     0.0 0.0 0.0              # shifts of the grid
                                    # define the q point(s) to compute the phonons
nph1l      1                        # number of q points in 1st list of phonons
qph1l      0.0 0.0 0.0 1.0          # first list of phonons
                                    # specify Γ point
                                    # the 4th number is the scaling
                                    # set it at 1.0 in the first list of phonons
```

A very reliable way of checking the convergence is to compare the phonon frequencies for a given wavevector that are obtained (i) from a direct calculation of the dynamical matrix in that wavevector and (ii) from the Fourier interpolation based on a grid of **q** points.

and derivable, there might exist avoided crossing, and so on. Usually, the vertical axis is the energy of the phonons. This can be expressed either in units of energy or, more and more commonly, in units of frequency. Because the Raman spectrometers, which are extensively used to analyse vibrational properties of minerals, give their output in units of cm^{-1}, most diagrams of phonon dispersion have their vertical axis also expressed in cm^{-1}. The horizontal axis in these diagrams is a path through the Brillouin zone. There is no formula for choosing the path, as the phonon bands in different minerals might present certain particularities along certain directions. The rule of thumb is to try a first extensive path to search for various anomalies or interesting features and then select, for example, at publication time, only the most relevant parts. If nothing else, a path touching the highest-symmetry points and segments of the Brillouin zone can be a good choice to represent the phonon bands.

```
List of bloks and their characteristics

Total energy                    - # elements :        1

1st derivatives                 - # elements :      129

2nd derivatives (non-stat.)  - # elements :    15876
qpt  0.00000000E+00  0.00000000E+00  0.00000000E+00    1.0

3rd derivatives                 - # elements :     3348
qpt  0.00000000E+00  0.00000000E+00  0.00000000E+00    1.0
     0.00000000E+00  0.00000000E+00  0.00000000E+00    1.0
     0.00000000E+00  0.00000000E+00  0.00000000E+00    1.0

2nd derivatives (non-stat.)  - # elements :    15129
qpt  5.00000000E-01  0.00000000E+00  0.00000000E+00    1.0

2nd derivatives (non-stat.)  - # elements :    15129
qpt  5.00000000E-01  5.00000000E-01  0.00000000E+00    1.0

2nd derivatives (non-stat.)  - # elements :    15129
qpt  0.00000000E+00  0.00000000E+00  5.00000000E-01    1.0

2nd derivatives (non-stat.)  - # elements :    15129
qpt  5.00000000E-01  0.00000000E+00  5.00000000E-01    1.0

2nd derivatives (non-stat.)  - # elements :    15129
qpt  5.00000000E-01  5.00000000E-01  5.00000000E-01    1.0
```

Fig. 12.8   The DDB files of abinit contain a summary of their content. In this example, a chain of calculations produced the dynamical matrices on a regular grid of **q** points. After all the calculations were finished, the dynamical matrices were merged into one large _DDB file (using the *mrgddb* utility). The resulting file contains all the dynamical information in one place, from which a few lines from the bulk and the last lines are shown here. Note that the *anaddb* utility can build the grid of **q** points using the symmetry and the asymmetric part of the reciprocal space.

The dispersion diagrams and the phonon displacement patterns are obtained by diagonalising the dynamical matrices in various points across the Brillouin zone, most of them general points void of any symmetry (Box 12.6). Of course, computing the dynamical matrices using DFPT on each point along an entire path through the Brillouin zone would require huge computational resources and be extremely lengthy. Fortunately, the Fourier integration of the interatomic forces provides a very convenient, accurate, and fast alternative. The procedure starts with determining the dynamical matrices on a regular grid of **q** points in the reciprocal space. The dynamical matrices are handled and merged using the *mrgddb* utility (Box 12.3, Figure 12.8).

The grid is similar to the grid of **k** points used in the case of the electrons. However, it is highly advisable not to shift it and to specifically include the Γ point. Including the zone-boundary points improves the fit for all the other general points in the Brillouin zone.

The short-range interatomic forces from the dynamical matrices in Γ describe the shortest interactions between the atoms and thus greatly influence the dispersion relations. The long-range forces are obtained from

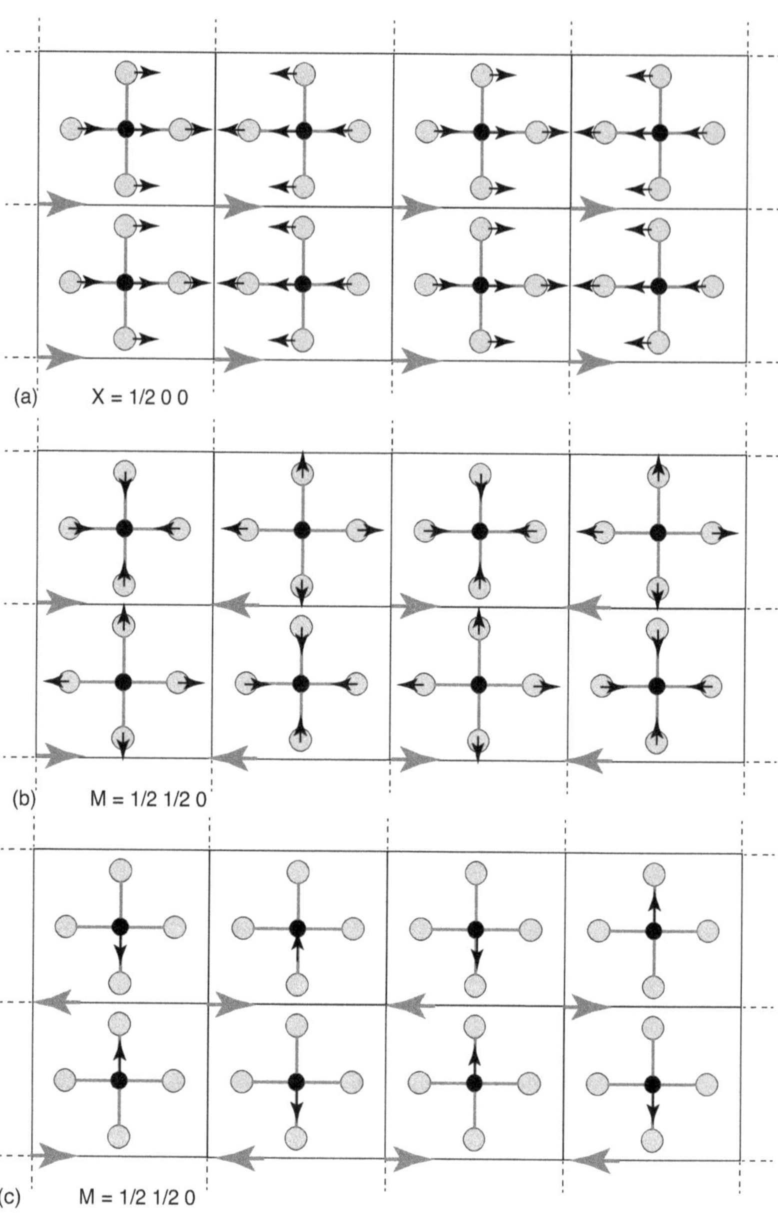

Fig. 12.9 Examples of atomic displacement patterns corresponding to several modes from Figure 12.7, anchored in various points of the Brillouin zone. (a) The acoustic mode from Figure 12.7(a), anchored in X = 1/2 0 0; (b) the acoustic mode anchored in M = 1/2 1/2; (c) the rotation mode from Figure 12.7(c), anchored in M = 1/2 1/2.

the dielectric functions, and their contribution determines the dispersion close to Γ. The intermediate-value forces for the next-neighbouring interactions are gathered from the dynamical matrices on various **q** points, like zone boundaries (first neighbours), **q** = 1/4 0 0, and so on, as shown in Figure 12.9.

12.9 Phonon Density of States

Just like for the electrons, the amount of phonon states as a function of energy gives the phonon density of state. The sum of the phonon states can be obtained either using the Gaussian or the tetrahedron method. The sum should be done over a very dense grid of $\mathbf{q}$ points. It is advisable to perform a convergence test on increasingly denser grids to ensure that details of the phonon band structure are correctly described.

Compared to electrons, it is much easier to extract the partial density of states for phonons. Because the eigendisplacements in every $\mathbf{q}$ point correspond to amplitudes of movements of *all* the atoms in the unit cell, it is straightforward to separate either individual atomic types or even individual atoms. For this, when the sum is performed over states, the different atomic contributions can be added to the different sums separately, which eventually yields the partial densities of state.

The procedure for computing the phonon density of states is straightforward (Box 12.7). Usually, the phonon density of states is plotted on the side of the diagram with the dispersion bands, with the vertical axis representing the energy. This is similar to the diagrams for electrons.

Several examples of diagrams with the phonon dispersion bands and the phonon densities of states are shown and discussed in detail in Chapter 13.

12.10 Instabilities

Figure 9.1 shows that the energy curve may be concave in certain situations. In this case, the curvature of the energy is negative, and the ω^2 is negative. There is no real solution, and ω becomes imaginary. Such a phonon cannot exist in reality but only in theory. The energy curve describes a double-well potential, with a local maximum at the former equilibrium positions and two local minima, one on each side. As the perturbation sets in, the atoms move away from their initial positions. Then the local minima act as attraction wells for the perturbed atoms. They cannot return to their initial positions, and the structure changes: We have a phase transition. If there is only one such unstable phonon mode, this defines a second-order phase transition. If there are several such unstable modes, then the transition needs to be first order [215].

In practice, one can never measure an unstable phonon mode. Long before the measurement is made, the atoms have already moved away from their initial equilibrium positions and start to vibrate around the local energy minima, which define new equilibrium positions. It is only in theory that these modes can be described.

Box 12.7   **How to calculate the phonon density of states?**

The procedure consists of employing Fourier interpolation of the interatomic forces to build the dynamical matrices on a very fine grid of **q** points, starting from a regular course grid of **q** points. This is done in the *anaddb* code. As always, it is advisable to generate several increasingly denser grids and check for convergence. The key variables are *prtdos* and *thmflag*. The code will employ either Gaussian interpolation (=1) or the tetrahedron method (=2) to interpolate between the states on the fine grid. In the first case, you can define a smearing parameter with the *dossmear* variable.

The *anaddb* input file needs to contain the following variables:

```
                          # Flags
ifcflag    1              # Interatomic force constant flag
thmflag    1              # Compute phonon density of states and the
                          # thermal properties including free energy from
                          # the quasiharmonic approximation


                          # The grid of q-vectors
brav       1              # Lattice: 1-simple cubic, 2-fcc, 3-bcc, 4-hexa)
ngqpt      2 2 2          # grid of q points
nqshft     1              # number of shifted grids
q1shft     0.0 0.0 0.0    # shifts of the grid
                          # define the q point(s) to compute the phonons
nph1l      1              # number of q points in 1st list of phonons
qph1l      0.0 0.0 0.0 1.0 # first list of phonons
                          # specify Γ point
                          # the 4th number is the scaling
                          # set it at 1.0 in the first list of phonons
```

To better understand how a second-order phase transition takes place, we can do another exercise, where we start with a given potential well and change its opening. This can be done by changing the distance between atoms. In ice X (Figure 12.10(a)), which is the high-density ionic phase of water ice, the oxygen atoms form a body-centred lattice, where the hydrogen atoms occupy the half-way positions between two oxygen atoms along the diagonal of the cube. At high pressures above 1.13 megabars, the displacement of the hydrogen atoms towards one of the two oxygens leads to an increase in energy (Figure 12.10(b)). The general trend is that under decompression, the distance between the oxygen atoms increases, and there is less and less electronic charge to constrain the movement of the hydrogen atoms. As a result, the hydrogen atoms can span large amplitude trajectories towards one or the other of the oxygen atoms while changing less and less the energy. The tipping point arises at about 113 GPa. Below this pressure, the potentials take a double-well shape. Any

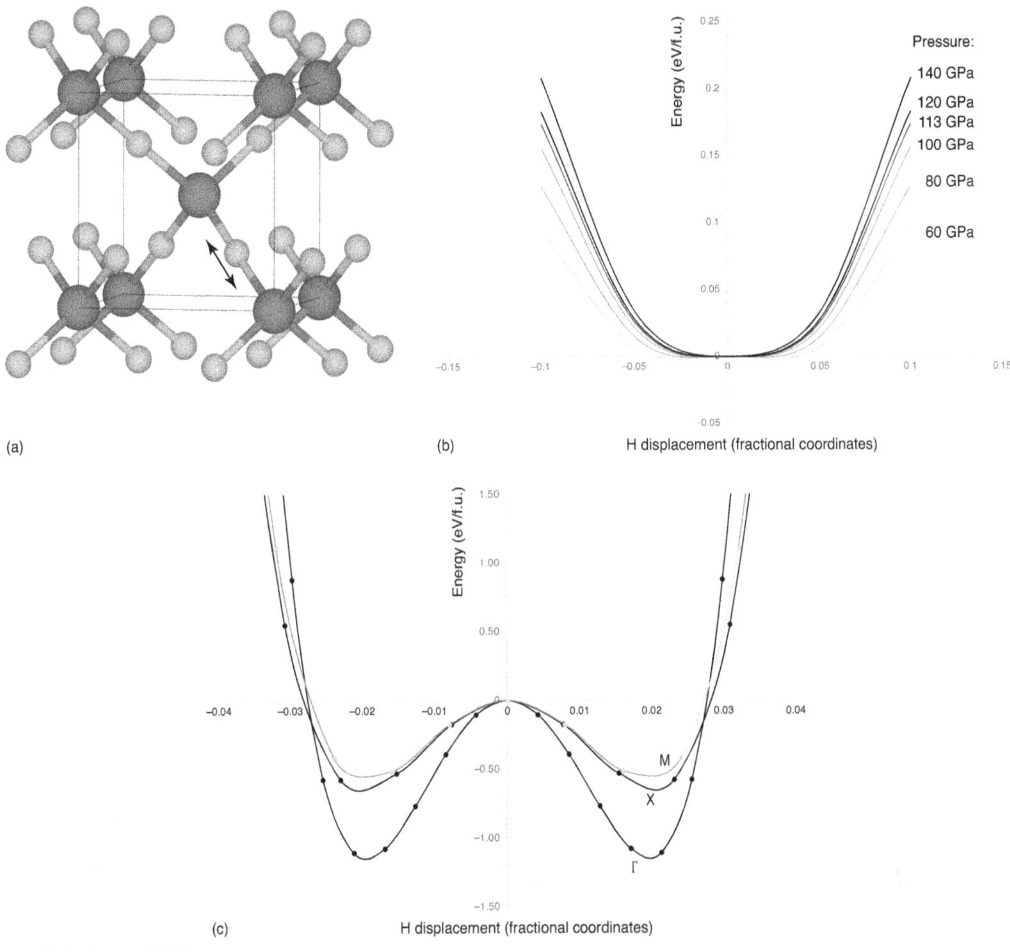

Fig. 12.10 Potential wells in water ice X, corresponding to the hydrogen displacements along the O - H - O diagonals of the Pn3m cube. (a) The crystal structure of the ionic high-dense ice X; (b) the potential in Γ opens up with decreasing pressure, that is, with decreasing distance between the O atoms. At low pressure, the electronic clouds of the O atoms are further apart from the median H atom, whose movement is less and less constrained. (c) The potential wells in three points of the reciprocal space (corresponding to the H displacements in various patterns that match the wavevector of that phonon) at 40 GPa. (b) The potential becomes a double-well potential below 113 GPa. At 40 GPa (c), there are several unstable phonon modes, and ice X becomes unstable upon decompression.

movement of the hydrogen atoms towards one of the two oxygens results in an irreversible displacement. The H atoms are locked in the minima of the potential wells. This leads to a phase transition. At 40 GPa, well outside the stability field of the ice X [59, 136, 135], deep double-well potentials develop, which are clear markers of the instability of ice X at these conditions.

## 13

Vibrational Spectroscopy

The vibrational properties are actual fingerprints of the dynamics of the atomic structure and the relative relations between structural units. The intramolecular modes discussed in Chapter 12 have specific normal modes, which can serve for quick and reliable mineral identification.

The vibrational pattern of atoms in a crystal can be sampled with the help of the interaction between the displacements of those atoms and an incoming electromagnetic field, like a laser. Different phenomena, like scattering or electromagnetic coupling, highlight different features of the vibrating lattice and thus shed light on different parts of the vibrational spectrum.

Besides the importance of vibrations for understanding the physical properties of crystals, the recent advances in spectroscopic apparatus made the investigation of the vibrational properties of minerals a routine endeavour nowadays. Miniaturisation and affordable prices offered most laboratories the possibility of acquiring and operating Raman spectrometers and, to a lesser extent, infrared spectrometers. Inelastic and Brillouin scattering require more sophisticated equipment. Still, in general, the techniques are made available today in various national and international facilities, like the European Synchrotron Research Facility (ESRF) in Grenoble, France, the Advanced Photon Source (APS) in Urbana-Champagne, Illinois, US, the Advanced Light Source (ALS) in Berkeley, California, US, or the Spring-8 in Sayo Town, Japan.

13.1 Infrared Spectroscopy

The principle of infrared spectrometry is the coupling between the incoming electromagnetic radiation and the dipole that forms during the atomic vibration [17, 22, 23, 50, 113, 114, 117, 118]. As a result, energy is transferred between the incoming laser and the structure via these dipoles. If no dipole is formed during a specific vibration, that has no signal in infrared. For centrosymmetric space groups, that is, those space groups that contain the centre of inversion as a symmetry element, this condition allows to distinguish between the infrared and the Raman active modes: The u modes are polar, so they are active only in infrared, the g modes are not polar, so they are active only in Raman. For non-centrosymmetric systems, there is no distinction between the g and the u nodes. They can all be active in both infrared and Raman. But in general, the modes that yield large dipoles are the most intense ones in infrared, with large splittings between the longitudinal component of an optical phonon (LO) and the transverse component of the same optical phonon (TO), that is defined as the LO-TO splitting. Each non-centrosymmetric phonon mode has its own LO-TO splitting. The modes that do not possess a large dipole are more intense in Raman.

The intensity of the infrared spectra is dependent on the direction of the wavevector, $\mathbf{q}$. The reflectivity intensity that is measured in an IR spectrometer can then be written in terms of the dielectric permittivity function as:

$$R_q(\omega) = \left| \frac{\sqrt{\epsilon_q(\omega)} - 1}{\sqrt{\epsilon_q(\omega)} + 1} \right|^2 \tag{13.1}$$

The dielectric function depends on the direction of the sampling, the frequency of the phonons, ω_m, and the frequency of the incoming radiation, ω:

$$\epsilon_q(\omega) = \sum_{\kappa\kappa'} q_\kappa \epsilon^0_{\kappa\kappa'}(\omega) q_{\kappa'} \tag{13.2}$$

When the incoming radiation has a frequency comparable to one of the vibrations of the mineral lattice, there is a net coupling and transfer of energy. This adds an ionic contribution to the dielectric tensor, which comes from the dipole generated during the vibrations. This contribution, named mode polarity, p_{mk}, is obtained as the product between the Born effective charges, Z^*, and the atomic displacements τ of the atoms α, corresponding to a particular phonon m at $q = \mathbf{0}$. The phonon has frequency ω_m, and the atomic displacements are projected along the direction k':

$$p_{m\kappa} = \sum_{\alpha k'} Z^{*\alpha}_{\kappa\kappa'} \tau_{mq=\mathbf{0}}(\alpha\kappa') \tag{13.3}$$

The mode polarity is also called mode-oscillator strength. Usually, the post-processing codes that treat the information from the DFPT print the mode polarities directly.

Thus, the contribution of the mode polarity to the dielectric response adds to the electronic contribution of the dielectric tensor. Together they define the dielectric tensor at the full response as a function of the electronic energy and the frequency of the incoming radiation, ω Box (13.1):

$$\epsilon_{\kappa k'}(\omega) = \epsilon^\infty_{\kappa k'} + \frac{4\pi}{\Omega_0} \sum_m \frac{p_{m\kappa} p_{m\kappa'}}{\omega^2_{mq=\mathbf{0}} - \omega^2} \tag{13.4}$$

Equation 13.4 also reveals a particularity. The $p_{m\kappa} p_{m\kappa'}$ factor is expressed vectorially as $\mathbf{p}_m\mathbf{q}$. Whenever the direction of the dipole resulting from the mode polarity is perpendicular to the direction of the polarisation of the incoming radiation, this term is zero, so there is no contribution of the phonon to the dielectric tensor. In this case, there is no absorption,

---

Box 13.1   **How to compute the infrared reflectivity spectrum?**

The energy derivatives with respect to electric fields yield the electronic contribution to the dielectric tensor, ϵ^∞. Then the *anaddb* utility combines the vibrational contribution from the iagonalisation of the dynamical matrix with the electronic contribution. The result is the frequency-dependent total dielectric function. The calculation requires only the second-order derivatives of the energy.

The input file of *anaddb* needs to contain the following keywords:

```
asr       2    & ensure the acoustic sum rule
               # that is, the sum of the Born effective charges
               # is zero along all the directions of the space
dieflag   2    # compute the frequency-dependent
               # dielectric tensor and
               # the infrared reflectivity
```

The dielectric function cannot be defined if there are unstable phonon modes. Also, remember that the dielectric function is defined only for insulators.

and the reflectivity reaches a maximum because of the lack of a coupling between the phonons and the photons. This is the case for the transverse modes, the TO. But when the direction of the dipole is aligned to the direction of the polarisation, the product $\mathbf{p}_m\mathbf{q}$ reaches a maximum. This is when the reflectivity drops to zero; all the incoming radiation is absorbed, and its energy is entirely transmitted to the phonon. Consequently, the frequency of the phonon increases. This is the case of the longitudinal modes, the LO.

The difference between the frequency of the LO and the TO components of the same mode is called LO-TO splitting. This is widely used in infrared spectroscopy to determine the alignment of the crystals and to identify particular phonon modes.

Equation 13.4 also shows that at very low frequencies, which are sufficiently small to allow nuclei to relax to their equilibrium positions under the applied field, the nuclei do not play a role anymore. This yields the static ϵ^0 dielectric permittivity, expressed in terms of the interatomic force constants matrix and the Born effective charges:

$$\epsilon^0_{\kappa\kappa'} = \epsilon^\infty_{\kappa\kappa'} + \frac{4\pi}{\Omega_0} \sum_{\alpha\beta} \sum_{ij} Z^{*\alpha}_{\kappa 1}[C(\mathbf{q}=0)]Z^{*\beta}_{\kappa'j} \tag{13.5}$$

Figure 13.1 shows the ideal dielectric spectrum of halite, NaCl, and quartz, SiO_2, as a function of frequency. The spectra start at small positive finite values. There is a coupling that occurs with the TO modes, which induces an instability. The reflectivity spectra start at a finite reflectivity value and reach 100% at each TO phonon mode. They remain at 100% until the laser frequency matches the frequency of the LO mode. The LO modes show minimum reflectivity, and the TO modes show maximum reflectivity. The absorption spectra are obtained as the complement

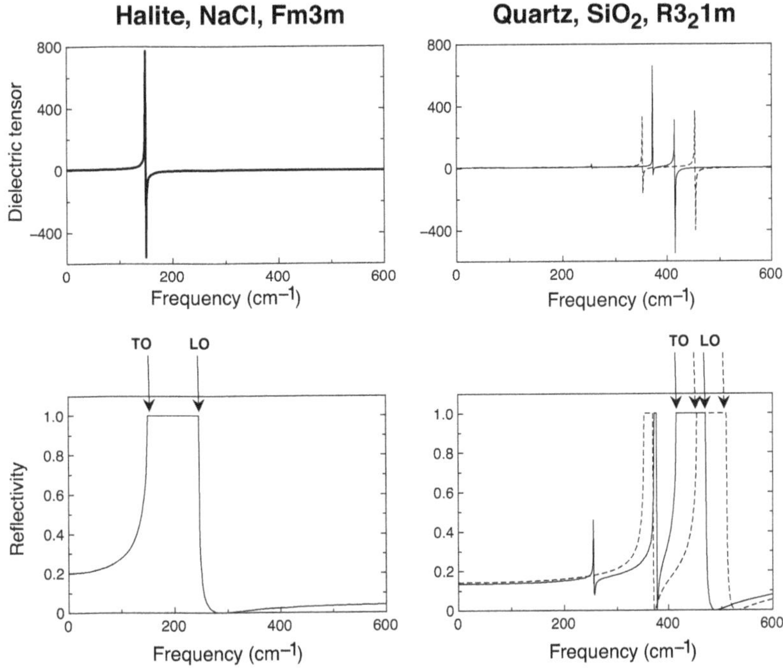

Fig. 13.1 Computed dielectric and reflectivity spectra of halite, NaCl, and quartz, SiO$_2$. The absence of any anharmonic effects makes the intensity of the spectra look significantly different from the experiment. Taking into account damping effects reduces the disparity.

of the reflectivity, 1-R. The major differences to experiment come from the idealised treatment of the reflectivity. The spectra have a damping factor stemming from the phonon–phonon interactions, which we neglect in the quasi-harmonic approximation. The damping adds an imaginary component to the dielectric function transforming it into a complex function, which we lack in the quasi-harmonic treatment. An arbitrary solution proposed for adjusting the calculated spectra of various hydrous minerals [18] was to consider a 4 cm^{-1} damping factor, which gave reasonable similarity of the computed spectra with the experimental ones for some hydrous minerals. In this case, the constant term of $4i$ is added to the frequency at the denominator, which thus becomes $\omega_m^2 - (\omega + 4i)^2$. This term transforms the ideal dielectric function into a complex function, smoothing the LO and the TO vibrational contributions.

13.2 Raman Spectroscopy

The scattering of the incoming electromagnetic radiation lies at the basis of the Raman effect [21, 133, 257]. The scattering occurs with an energy exchange between the phonon and the photon. As a result, in a Raman spectrum, we plot the change of energy of the photons, which is reflected in a frequency shift. The exchange can be made in the two directions. When the transfer occurs from the laser to the crystal, the effect is called Stokes, and when the transfer occurs in the opposite direction, from the crystal to the laser, it is called anti-Stokes. In the first case, as the phonon absorbs energy, it moves in an excited state, and the Stokes effect has a strong intensity. In the second case, the phonon loses energy as it de-excites, and the anti-Stokes effect has a weaker intensity.

The Raman spectrum looks like a collection of narrow and, usually, well-defined peaks of various intensities. The horizontal axis of the spectrum is expressed in frequency units, like cm^{-1} or THz, and corresponds to the energy shift of the radiation. The laser is typically reported by its colour, like green, ultraviolet (UV), and so on, or wavelength (expressed in nm).

The strongest peak occurs at 0 cm^{-1}. The large majority of the photons traverses the crystal unperturbed. In contrast, many others excite a phonon that immediately de-excites, so before exiting the sample, they gain whatever energy they lost upon entrance. The global result is a very intense peak, spreading up about 100–150 cm^{-1} wavenumbers. This is the Rayleigh peak.

The polarisability of the phonon mainly dictates the intensity of the rest of the Raman-active peaks. During the vibration, the movements of the atoms lead to a distortion of the electronic clouds, which in turn leads to changes in the dielectric tensor of the structure. This is what defines the Raman tensor. The changes happen on the time scale of the vibrations. Apart from the Raman tensor, there are pre- and post-factors that define the Raman cross-section. Then the Raman intensity of a phonon m is obtained using the following equation [21, 210]:

$$I = \frac{dS}{d\Omega} \frac{(\omega_0 - \omega_m)^4}{c^4} \|e_i \alpha_{i,o}^m e_o\|^2 \frac{\hbar(n_m + 1)}{2\omega_m} \qquad (13.6)$$

In this equation, the ω_0 and ω_m are the frequencies of the incoming laser and the phonon, respectively. c is the speed of light, and the e_i and e_o are vectors indicating the direction of the polarisation of incoming and outgoing photons. The phonon frequency results from the diagonalisation of the dynamical matrix from Chapter 12.

The n_m term is the Boson factor, which brings a first dependence of the intensity with the temperature, T (k_B is the Boltzmann constant):

$$n_m = 1/e^{\frac{\hbar\omega_m}{k_B T - 1}} \qquad (13.7)$$

The Raman tensor, α, needs to be computed for each phonon. It represents a measure of the polarisability of a given vibration, expressed as the change of the dielectric tensor induced by the vibrating atoms. It obeys strict rules of symmetry, which result from the intersection of the irreducible representation of the phonon with the symmetry of the dielectric tensor. Its general formula is:

$$\alpha_{ij}^{m} = \sqrt{\Omega_0} \sum_{\beta\kappa} \frac{\partial \chi_{ij}^{\infty}}{\partial \tau_{\beta\kappa}} U_{\beta\kappa}^{m} \tag{13.8}$$

where Ω is the volume of the unit cell, which acts as a normalisation factor, and U is the phonon eigendisplacement along direction κ for the atom α. If the dielectric tensor is written as the energy derivative, the Raman tensor becomes:

$$\alpha_{ij}^{m} = \sqrt{\Omega_0} \sum_{\beta\kappa} \frac{\partial}{\partial \tau_{\beta\kappa}} \frac{\partial^2 E}{\mathcal{E}_i \mathcal{E}_j} \tag{13.9}$$

Consequently, the Raman calculations necessitate third-order derivatives. These are available in DFPT implementations in several codes. Boxes 13.2 and 13.3 exemplify the calculation of the Raman tensors and the Raman spectrum using the abinit code. The energy derivatives are computed as in Equation 13.9. Usually, this step comes at the end of chained phonon calculations, where the lower-order terms are calculated in the previous steps. This calculation can be done only for insulators because the dielectric tensor derivative needs to have a finite value.

Special care needs to be paid to using grids of **k**-points that are dense. As always with derivatives, the higher the order of the derivative, the larger the numerical errors during the calculation. While increasing the kinetic energy cut-off increases the number of planewaves in the development of the electronic density, it does not necessarily bring more information to the system, as the large wavevectors correspond to highly oscillating planewaves, similar to background noise. However, dense grids of **k** points ensure a better sampling of the electronic density at small and medium planewave wavevectors, carry more information into the system, and thus help increase the numerical accuracy for the derivatives of the energy.

Once the third-order derivatives of the energy are computed, in the next step, the Raman tensors are calculated by combining the atomic displacements corresponding to the vibrational modes with the changes of the dielectric matrix under those atomic displacements. This coupling represents a post-processing step, which is realised using the *anaddb* utility. Box 13.3 illustrates this second step.

The Raman tensors are the critical ingredient for obtaining the intensity of the peaks, as in Equation 13.6. For single-crystal Raman spectra, the polarisation of the incoming laser and of the outgoing radiation needs to be defined. Then, as the crystal orientation can be changed, the different elements of the Raman tensors corresponding to each vibration can be measured. This is a valuable and interesting experiment in the laboratory with high-quality single crystals.

But in the field, when using, for example, a portable Raman spectrometer, or in the laboratory, when dealing with rocks or fine-crystalline powders, only an averaged spectrum can be obtained, combining over many or all possible orientations of the crystallites. Hence, the resulting powder

With the derivatives of the energy and wavefunctions computed in the previous steps (see Boxes 11.2 and 12.1), you can now add one more level of complexity by computing the derivatives of the energy with respect to three perturbations: two electric fields and one atomic displacement. The results are the ingredients of the Raman tensor, α.

Before you proceed with the third-order derivative, you need to be sure that at the end of the previous steps, that is, the calculation of the dynamical matrix, you *prepare* all the files in the proper formats by using the *prepanl 1* keyword (preparation of the non-linear calculations).

In the example below, the Raman ingredients are computed in the 5th step, and at this moment, your input file should contain:

```
getden5     1      # use ground-state electronic density from step 1
get1den5    4      # use density derivative from step 4
getwfk5     2      # use ground-state wavefunctions from step 2
get1wf5     4      # use wavefunctions derivatives from step 4
kptopt5     2      # use only time-reversal symmetry
optdriver5  5      # activate the non-linear response calculation


rf1elfd5    1      #first mixed perturbation
rf1phon5    1
rf1atpol5   1 36   # derivative to atomic displacements
rf1dir5     1 1 1  # along all three directions


rf2elfd5    1      #second mixed perturbation
rf2dir5     1 1 1


rf3elfd5    1      #third mixed perturbation
rf3dir5     1 1 1
```

Before starting your calculation, do not forget to use a dense grid of **k** points. At least in the *abinit* implementation [134, 258], in the simulations, you should not have any empty bands.

Raman spectrum is obtained by integrating the intensity of the peaks for all the crystal orientations. The powdered spectra can be obtained either with polarised or unpolarised lasers. The intensities of the two polarised components of the powder spectra, parallel and perpendicular, and the resulting total powder spectra can be computed as [210]:

$$I_{\parallel}^{powder} = C(10G_0 + 4G_2)$$

$$I_{\perp}^{powder} = C(5G_1 + 3G_2) \tag{13.10}$$

$$I_{total}^{powder} = I_{\parallel}^{powder} + I_{\perp}^{powder}$$

Box 13.3	How to calculate the Raman spectrum?

The DDB file with the third-order derivatives has, as expected, all the components listed in order:

1	6	2	38	3	38	$-0.10880583226733D + 01$	0.00000000000000D+00
2	6	2	38	3	38	0.63766012612963D+00	0.00000000000000D+00
3	6	2	38	3	38	0.15687920995309D+01	0.00000000000000D+00
1	7	2	38	3	38	0.10882496050130D+01	0.00000000000000D+00
2	7	2	38	3	38	0.63724291415030D+00	0.00000000000000D+00
3	7	2	38	3	38	$-0.15686283133929D + 01$	0.00000000000000D+00

where the indices correspond to the Cartesian direction of the derivation (atomic displacement, first field, second field) and the tensor's real and imaginary components. After you calculate the first-, the second-, and the third-order derivatives of the energy, you need to use the *mrgddb* utility once more to merge all these DDB files into a single one. The latter is then fed to the *anaddb* utility.

To specifically obtain the Raman tensors in an *anaddb*, you need to specify the supplementary keywords for the non-linear optical properties:

```
nlflag     1    # non-linear flag
                # use value of 1 for Raman
ramansr    1    # Raman acoustic sum rule
```

During the post-processing, the code first diagonalises the dynamical matrix to obtain the phonon frequencies and their corresponding atomic eigendisplacements. Then it constructs the Raman tenors for each mode by coupling the third-order derivatives in the amounts due by the mode eigendisplacements. In the end, the Raman tensors are 3×3 matrices, one for each mode:

;	-0.000000000	-0.000000314	-0.000000000
;	-0.000000314	0.000000000	0.000000372
;	-0.000000000	0.000000372	0.000000000

where C is a constant, and the three rotational invariants [211] are:

$$G^{(0)} = \sum_{i=x,y,z} \frac{(\alpha_{ii})^2}{/} 3$$

$$G^{(1)} = \sum_{i,j=x,y,z;i\neq j} \frac{(\alpha_{ij} - \alpha_{ji})^2}{/} 2 \tag{13.11}$$

$$G^{(2)} = \sum_{i,j=x,y,z;i\neq j} \frac{(\alpha_{ij} + \alpha_{ji})^2}{/} 2 + \frac{(\alpha_{ii} - \alpha_{jj})^2}{/} 3$$

The Raman scattering effect occurs when the wavelength of the incoming photons is the same as that of the vibrations, the condition of interference. The lasers have typical wavelengths in the hundreds to a thousand nanometres, which correspond to phonon wavevectors on the order of a

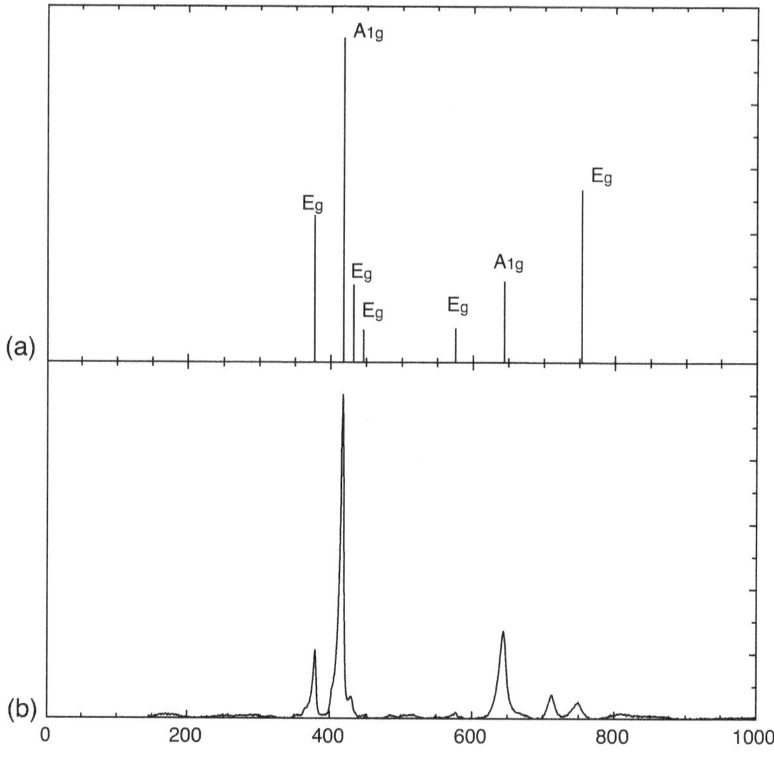

Fig. 13.2 Computed (a) (https://wurm.info) and experimental (b) (https://rruff.info) Raman spectra of corundum, Al_2O_3. The agreement between the two spectra is remarkable. Both peak positions and relative intensities are similar. Only the g modes are visible in Raman. The correct way to plot the theoretical spectra is to use delta functions on the position of the peaks with their heights proportional to the calculated intensity.

thousandth or less of the reciprocal space, placing the effect closer than 0.001 to the zone centre. Most modes are flat at such small wavenumbers; thus, the Raman spectrum reflects the modes in Γ. However, there might be cases for highly dispersive modes, where small differences in the peak position can be observed for different laser frequencies.

In general, the computed Raman spectra agree with the experimental measurements. The similarity is even more striking when comparing powders instead of single crystals. In most cases, the Raman spectra are dominated by the same peaks in both theory and experiment. Figure 13.2 shows the comparison between the theoretical Raman spectrum from WURM (https://wurm.info, Ref. [52]) and the measured spectrum from RRUFF (http://rruff.info). The position of the major peaks is almost identical in the two cases. Both the theoretical and the experimental spectra are dominated by the A_{1g} mode at 418 cm^{-1} computed and 416 cm^{-1} measured.

The agreement between computed and experimental spectra is better when comparing spectra obtained at the same density. Density functional theory (DFT) yields a volume shift relative to the experiment at any given pressure. The shift is larger at low pressures than at high pressures. As the phonon modes have a strong dependence on volume, any deviation from the experimental volume has dramatic consequences on the position of the phonon modes, which is directly reflected in the Raman peaks. For diamond, the computed Raman peak position can deviate by up to a dozen wavenumbers for theoretical structures, depending on the grid of $\mathbf{k}$ points. However, the spectrum computed at the same volume as the experimental one deviates by only 1 cm^{-1} [53].

In other cases, the differences in peak position persist even at the same density. Figure 13.3 shows the comparison between the computed and the measured Raman spectra of pyrope, $Mg_3Al_2Si_3O_{12}$. The upper panel shows the position of the computed Raman-active modes, irrespective of their intensity. A plethora of active modes are distributed throughout the frequency range spanned by the phonons in Γ. The middle panel shows the computed Raman spectrum with the intensity of the peaks. Bringing in the intensity filters out a large number of potential peaks. Finally, the lower panel shows the measured Raman spectrum of pyrope. The general patterns of the spectra are very similar. The computed peaks show a clear downward shift from the experimental values. Its origin is in the series of approximations and numerical limitations of the DFT. But so far, there is no analytical solution to correct this shift. Fortunately, the general features of the theoretical spectrum allow us to identify pyrope when looking at either the computed or the experimental spectra.

The Raman tensors can be calculated from finite differences as an alternative to perturbation theory. For this, the variation of the dielectric tensor needs to be computed as a function of atomic displacements. Then the Raman tenors are built from the phonon eigenvectors and the dielectric derivatives.

On the experimental side, Raman spectrometry developed recently as an affordable, easy-to-use, and non-destructive technique. With prices ranging anywhere between a few tens of thousands of euros for a simple portable Raman spectrometer to a few hundred thousand euros for a very sophisticated Raman scanner, this technique conquered the majority of mineralogy laboratories. Its usefulness is such that Perseverance, the last of the Mars rovers, embarked two Raman spectrometers on board to use them for mineral identification on the surface of the red planet.

Apart from the spectrometer itself, mineral identification requires reference spectra. Because this technique is so widespread and powerful today, several mineralogy laboratories put a lot of effort into providing collections of reference spectra. Most of them are freely available today, like the RRUFF (http://rruff.info) and the WURM (https://wurm.info) projects. Searching for a mineral spectrum in any of these databases is straightforward. However, identifying the mineral based on the spectrum is much more difficult. While some tools exist today for recognising spectra, they remain relatively primitive. Artificial intelligence techniques have a huge potential to come to the rescue of automatic mineral identification.

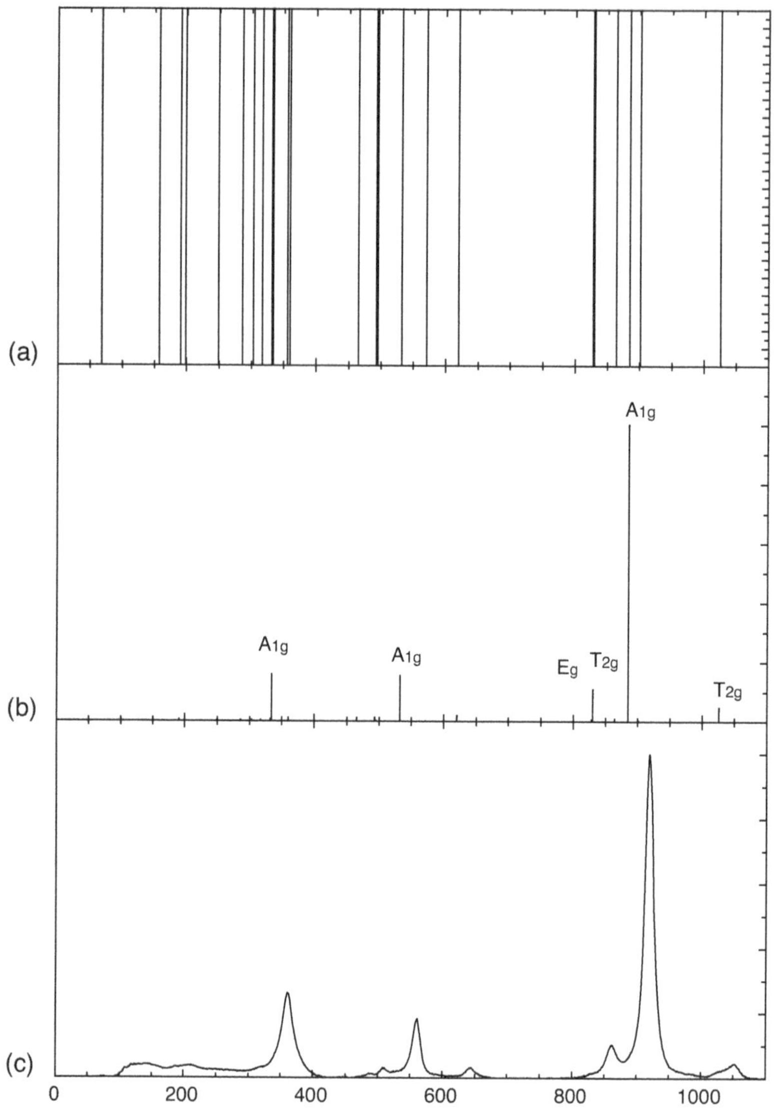

Fig. 13.3 The Raman spectra of pyrope garnet, $Mg_3Al_2Si_3O_{12}$. (a) The Raman-active modes are represented as vertical lines at their corresponding positions. (b) The computed Raman intensity for each mode leaves only a handful of peaks visible, while the others have relative intensity less than 1%. from the maximum peak. (c) The experimental spectrum presents all the same major computed peaks and thus confirms the simulations. However, there is a clear shift in the position of the computed peaks from the experimental ones. While the shift is more important at high frequencies, it is nevertheless not linear. The origin of the shift is multiple: differences in atomic position, lattice parameters, inherent approximations present at the LDA level, and so on.

13.3 Brillouin Spectroscopy

Another type of vibrational spectroscopy based on photon scattering by phonons is Brillouin spectroscopy [231]. If in Raman, the photons are scattered by the optical modes lying at high frequency, in the Brillouin spectroscopy, the scattering occurs on the acoustic modes. For this to work, the sampling of the phonons must be done very close to the Brillouin zone centre (Figure 13.4), hence the name of the spectroscopy. The typical

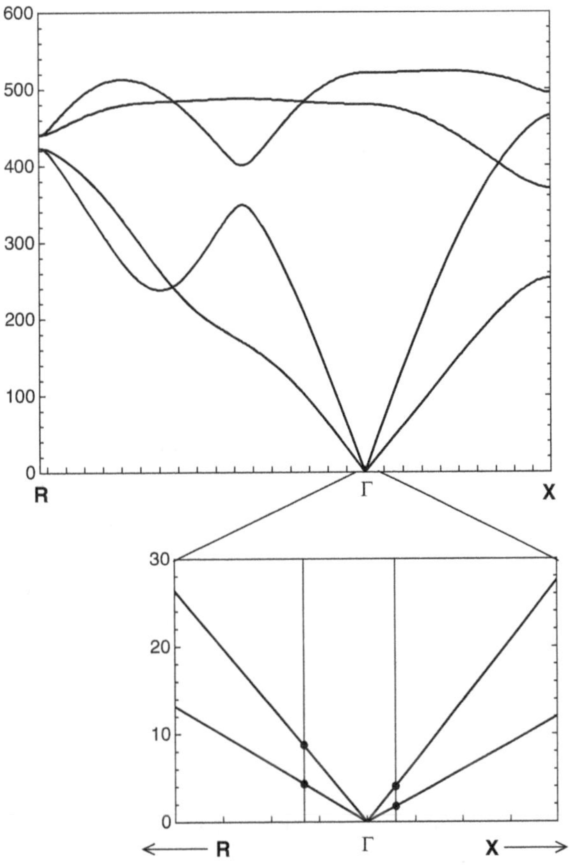

Fig. 13.4 Brillouin spectroscopy samples the dispersion of the acoustic modes close to Γ. In the upper panel, the dispersion of the phonon bands in the high-pressure bcc form of NaCl spans a range up to almost 600 cm^{-1}. Here we represent only the R - Γ - X bands. The lower panel shows a zoom around the Γ point where the modes would be sampled by Brillouin spectroscopy. For a laser of 1,064 nm, typical of Brillouin spectrometers, the scattering happens at distances in the reciprocal space corresponding to the vertical lines. The horizontal ticks represent 0.1 Å^{-1} in the upper diagram and 0.01 Å^{-1} in the lower diagram. The position of the Brillouin peaks corresponds to the intersection of the vertical lines with the dispersion of the acoustic modes; they are represented with thick black dots. Their position is different along the $\Gamma - R$ and $\Gamma - X$ directions because the reciprocal wavevectors of the structure are different along these two directions.

```
Speed of sound for this q and mode:
   in atomic units:      0.2154490607E-02
   in units km/s:                4.71336              TA
Partial Debye temperature for this q and mode:
   in atomic units:      0.1629338235E-02
   in SI units K  :            514.50371


Speed of sound for this q and mode:
   in atomic units:      0.2154490607E-02
   in units km/s:                4.71336              TA
Partial Debye temperature for this q and mode:
   in atomic units:      0.1629338235E-02
   in SI units K  :            514.50371


Speed of sound for this q and mode:
   in atomic units:      0.4499678752E-02
   in units km/s:                9.84391              LA
Partial Debye temperature for this q and mode:
   in atomic units:      0.3402891900E-02
   in SI units K  :           1074.54700
```

Fig. 13.5 The Brillouin spectroscopy samples the dispersion of the acoustic modes. The *anaddb* code writes the elastic analysis, as in the experimental Brillouin, corresponding to each phonon wavevector. For each $\mathbf{q} \to \Gamma$, there are two transverse acoustic modes (TA) with smaller frequencies, which in this case are degenerate because of the symmetry of the phonon branch, and one longitudinal acoustic mode (LA) with a higher frequency.

frequencies of the acoustic modes in this region of the reciprocal space lie in the range 10^{-2} cm^{-1} to about 10 cm^{-1}, and their sampling requires frequency resolution in the range of 10^{-3} cm^{-1} (Figure 13.5). Consequently, Brillouin spectrometry poses several technical problems inherent to sampling very low-frequency modes in the vicinity of the Γ point.

As was mentioned in the Raman section, when photons traverse a sample, their large majority only passes through the sample without any interaction, generating a bright Rayleigh peak. The Brillouin peaks of interest must be hunted out of that bright background. This is done inside the spectrometer, using Fabry–Pérot interferometers [231]. Just as in Raman, the maximum effect is obtained when the wavelength of the incoming laser is commensurate with the wavelength of the phonon.

The photoelastic tensor is the key ingredient that determines the intensity of the Brillouin peaks. The photoelastic tensor, p_{ijkl}, represents the change in dielectric tensor under strains. It is a fourth-rank tensor defined as:

$$\epsilon_1^\infty = \sum_{kl} p_{ijkl}\eta_{kl} \tag{13.12}$$

that can be written explicitly as the product of two matrices:

$$\left[\epsilon_1^\infty\,\epsilon_2^\infty\,\epsilon_3^\infty\,\epsilon_4^\infty\,\epsilon_5^\infty\,\epsilon_6^\infty\right] = \begin{bmatrix} p_{11} & p_{12} & p_{13} & p_{14} & p_{15} & p_{16} \\ p_{21} & p_{22} & p_{23} & p_{24} & p_{25} & p_{26} \\ p_{31} & p_{32} & p_{33} & p_{34} & p_{35} & p_{36} \\ p_{41} & p_{42} & p_{43} & p_{44} & p_{45} & p_{46} \\ p_{51} & p_{52} & p_{53} & p_{54} & p_{55} & p_{56} \\ p_{61} & p_{62} & p_{63} & p_{64} & p_{65} & p_{66} \end{bmatrix} \begin{bmatrix} \eta_1 \\ \eta_2 \\ \eta_3 \\ \eta_4 \\ \eta_5 \\ \eta_6 \end{bmatrix} \tag{13.13}$$

As with all tensors, the photoelastic tensor is symmetric, that is, $p_{14} = p_{14} = p_{1123} = p_{1231} = p_{2311} = p_{3112}$. The further application of symmetry dramatically reduces the number of elements. Nevertheless, direct computation of the photoelastic tensor from finite differences is a tedious process, and very few studies perform it. With the new implementations of the elastic response in DFPT, the photoelastic tensor can be obtained directly as energy derivatives (see Section 15.2).

Then the Brillouin intensity can be written as a function of the photoelastic tensor. In a simplified version, this is:

$$I_B \sim C(T,\omega)f(S)|e_i p e_j| \tag{13.14}$$

The prefactor C depends on temperature, T, and the frequency of the acoustic mode sampled close to Γ. The function f depends on the solid angle S sampled by the radiation in the crystal, and the last term is a cross-product between the polarising incoming and outgoing radiation and the photoelastic tensor.

The *anaddb* post-processing utility with the *thmflag 1* keyword directly computes the elastic wave velocities at any given **k** point, together with the equivalent Debye partial frequencies (Figure 13.5). These are listed at the end of each mode.

Mainly because of so many technical hurdles and despite its usefulness, Brillouin spectrometry remains a technique that is unfortunately rarely used in the laboratory today.

A Few Examples of Phonon Analysis

14.1 Halite, NaCl

Halite, NaCl, is the common kitchen salt. It has a simple and highly compact face-centred cubic structure, a prototype for many binary compounds. Wyckoff named it B1, as in the first binary structure, the NaCl-type structure. The Na and the O atoms lie on two interpenetrating fcc lattices, respectively, 1a (0 0 0) and 2b (1/2 1/2 1/2) Wyckoff positions in the Fm3m structure. Each Na and the Cl atom is sixfold coordinated, with the ligands bonded along the Cartesian axis. The primitive unit cell is a rhombohedron with an α angle of 60 degrees. This structure can also be seen as an alternation of three hexagonal layers perpendicular to the threefold diagonal axis, each displaced by one-third with respect to its neighbouring planes. According to polytype notation, halite has a 3C sequence. This sequence corresponds to a ..ABCABCABC.. packing, where each letter denotes one of the three possible anchor positions for the hexagonal planes.

With two atoms in the primitive unit cell, there are only six possible phonons. In Γ, the two phonon modes are triply degenerate as a result of the high symmetry of the lattice. For the space group 227 and two atomic positions (4a and 4b), the SAM application on the Bilbao Crystallographic Server (www.cryst.ehu.es/rep/sam.html) shows that the only two possible phonons are T_{1u} modes.

The first T_{1u} are the three acoustic modes with zero frequency. The second T_{1u} is a polar mode, corresponding to the opposite vibration of the two atomic types. As a result, a dipole forms, and the mode is active in infrared but not in Raman. The LO-TO splitting is 96 cm^{-1} at ambient pressure for halite. As pressure increases, the LO-TO splitting decreases as the atoms become less ionic.

Figure 14.1 shows the phonon dispersion bands for halite at ambient pressure conditions. The diagram follows a high-symmetry path through the Brillouin zone, which starts from the centre, and reaches, in order, the zone boundary along the x direction, then goes on the edges and middle of the faces of the Brillouin zone. Along the segments away from Γ, the phonons not only preserve the LO-TO splitting for the optical T_{1u} mode but also show a splitting for the acoustic modes, which this time is denoted with an A, standing for acoustic. Along the Γ-X segment, the degeneracy is 2+1, corresponding to two transverse modes (TA or TO)

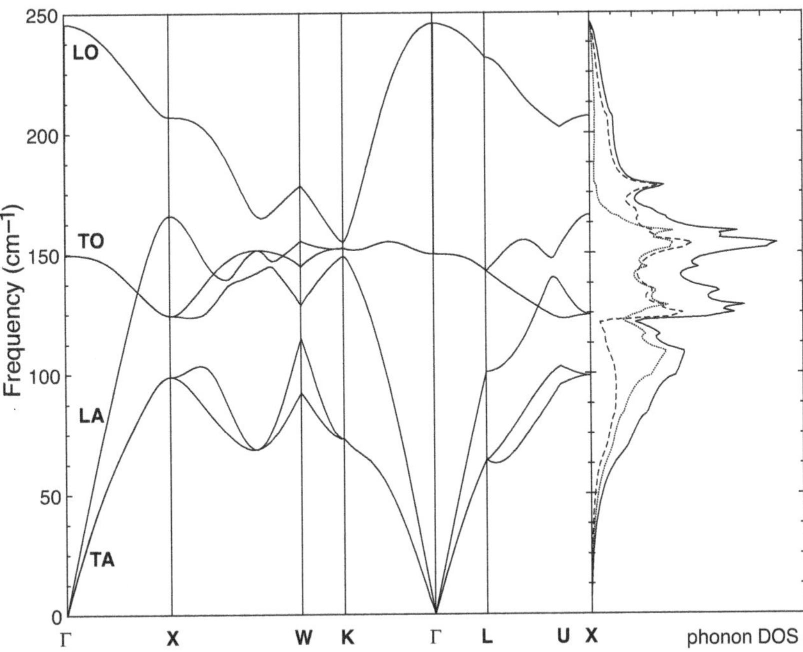

Fig. 14.1 Phonon band structure of halite, NaCl, computed at P = 0 GPa. All the phonon modes are positive, and the structure is dynamically stable. The first three modes, whose frequencies are 0 in Γ, are the acoustic modes, the remaining bands being the optical modes. All the modes are highly dispersive. The dispersion diagram shown here follows a high-symmetry path through the Brillouin zone: X = 1/2 0 0; W = 1/2 1/4 0; K = 3/8 3/8 0; L = 1/2 1/2 1/2; U = 1/2 1/8 1/8. The right panel shows the total phonon DOS (solid line) and its projection on the Na (dashed line) and Cl atoms (dotted line).

and one longitudinal mode (LA or LO). The same symmetry is retrieved along the L-Γ segment. Away from the high-symmetry segment, both the acoustic and the optical modes become fully degenerate. As a result, all the individual six phonon branches can be distinguished along the X-W, W-K, L-U, and U-X segments.

Contrary to the electrons, the phonons group much less frequently in packets of bands separated by energy gaps. Their frequencies depend on the structure topology and the masses of the varying vibrating atoms. When groups of bands are intertwined and have similar frequencies, they usually have similar vibrational patterns. For example, translation, bending, or stretching modes of similar anionic groups tend to have similar energy. The symmetry of the structure and the multiplicity of the groups rule the degeneracy of the phonon bands.

Figure 14.2 shows the phonon band structure for the B1 structure of NaCl computed at 100 GPa. The phonons are more dispersive than at low pressure, their spectrum going up to 600 cm^{-1}. The high-pressure structure is denser, and the atoms lie closer to each other. During their oscillatory movement, the nuclei encounter larger electronic density opposing their displacement. As a result, the amplitude of the movement decreases, but the frequency, and hence the phonon energy, increases. Usually, most of the vibrational modes see their frequency increase under pressure.

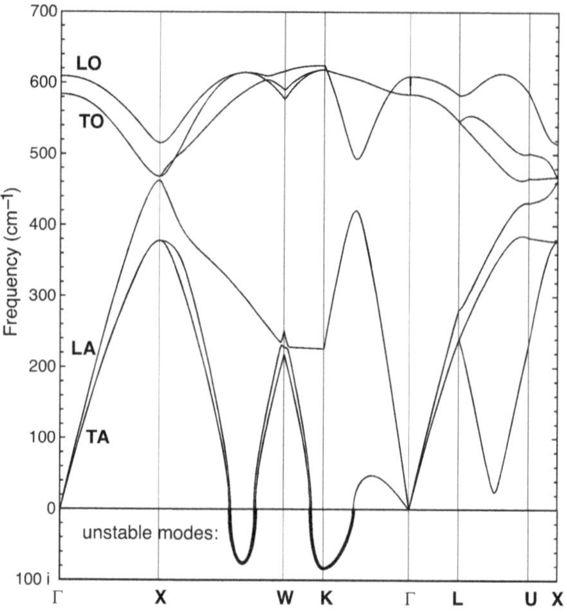

Fig. 14.2 Phonon band structure of halite, NaCl, computed at P = 100 GPa. The path through the Brillouin zone is the same as in Figure 14.1. Several modes along the X-W and around K have imaginary frequencies. These are unstable modes and, by convention, are represented on the phonon band diagram with frequencies below zero.

But the most interesting feature of the diagram is the presence of unstable phonon modes. The TA modes along Γ-K, K-W, and X-W become imaginary. This implies mechanical instability that develops under compression. Indeed, the stable high-pressure structure of NaCl is the B2 structure.

Figure 14.3 shows the variation of the T1u optical mode of halite as a function of pressure in the 0–100 GPa range. The frequency of the TO mode increases from 150 up to 584 cm^{-1} and the frequency of the LO mode from 245 up to 610 cm^{-1}. This pressure variation determines the microscopic formulation of the Grüneisen parameters, which are defined as the derivative of the phonon frequency with respect to volume:

$$\gamma_m = -\frac{V}{\omega_m}\frac{\partial \omega_m}{\partial P} \qquad (14.1)$$

where m denotes the phonon mode with frequency ω_m and V is the volume. The Grüneisen parameters are dimensionless. They can also be expressed in various equivalent ways involving thermal dilatation, bulk moduli, or elastic wave velocities.

At high pressure, as predicted by the elastic instability of the B1 phase, halite undergoes a structural phase transition towards another high-symmetry cubic structure, B2. This B2 structure, called the CsCl-type, has Pm3m symmetry. Each atomic type forms a simple cubic lattice. The two sublattices are interpenetrating: The Na atoms lie on the corners of a cube, on the 1a (0 0 0) Wyckoff positions, and the Cl atoms lie in

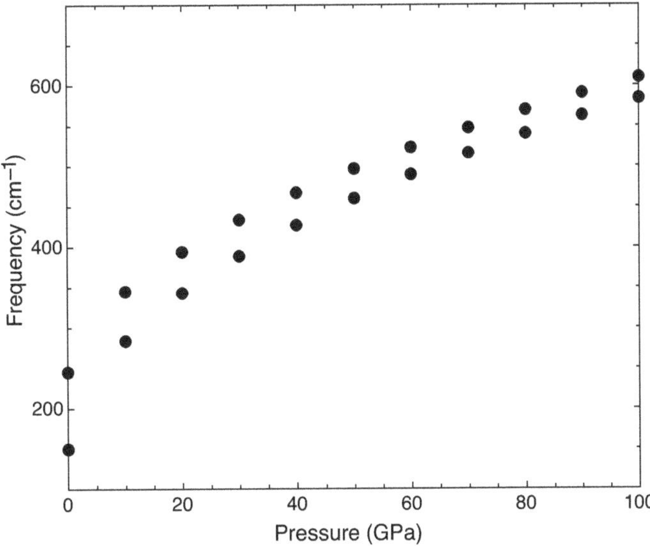

Fig. 14.3 Variation of the TO and LO modes of NaCl as a function of volume. With compression, the frequencies increase, and the LO-TO splitting decreases. The derivative defines the microscopic expression of the Grüneisen parameter.

the middle of the cube, on the 1b (1/2 1/2 1/2). Each atom is eightfold coordinated. With two atoms in the unit cell, the B2 structure also has six phonons.

Figure 14.4 shows the dispersion diagram of the phonon bands of the B2 structure of NaCl at 100 GPa. Because of the symmetry, there are two triply degenerate phonons in Γ, both with T_{1u} character. The vibrational spectrum of the B2 structure reaches up to almost 600 cm^{-1} frequencies. The modes are highly dispersive along the Γ-X and R-Γ directions and less so otherwise.

14.2 Sphalerite, ZnS

Sphalerite, ZnS, has the B4-type structure, with the Zn and S atoms in the respective 4a and 4c of the F-43m structure. There are six phonon modes, whose dispersion is shown in Figure 14.5. The structure is dynamically stable.

There are two sets of bands. The low-frequency bands are dominated by the Zn atoms, which are heavier, and the high symmetry bands by the S atoms, which are lighter. Box 14.1 shows the critical keywords of the abinit file used to generate the dispersion diagram.

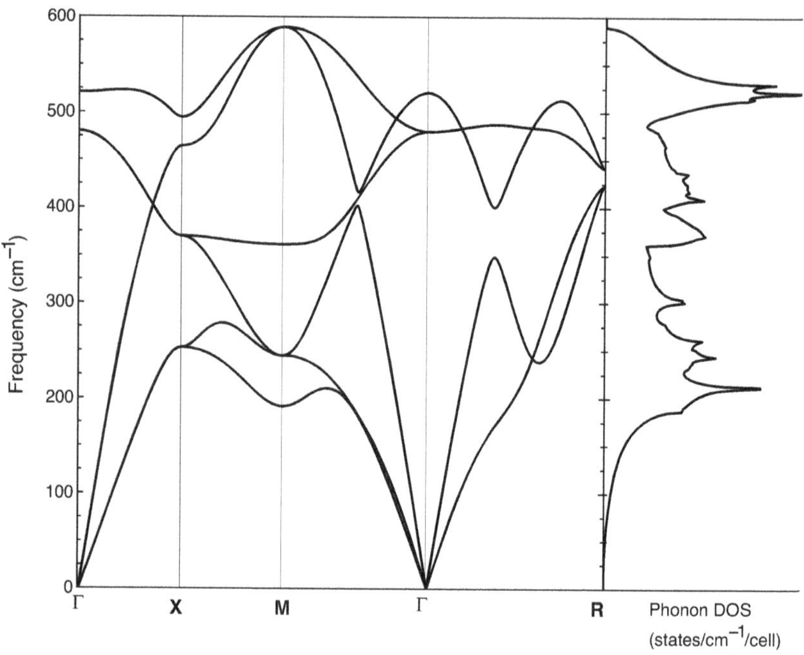

Fig. 14.4 Phonon band structure of the B2, Cs-Cl-type, structure of NaCl, computed at P = 100 GPa. All the phonon modes are positive, and the structure is dynamically stable, with a large spread in frequency. The phonon modes are dispersive along the Γ-X and Γ-R directions.

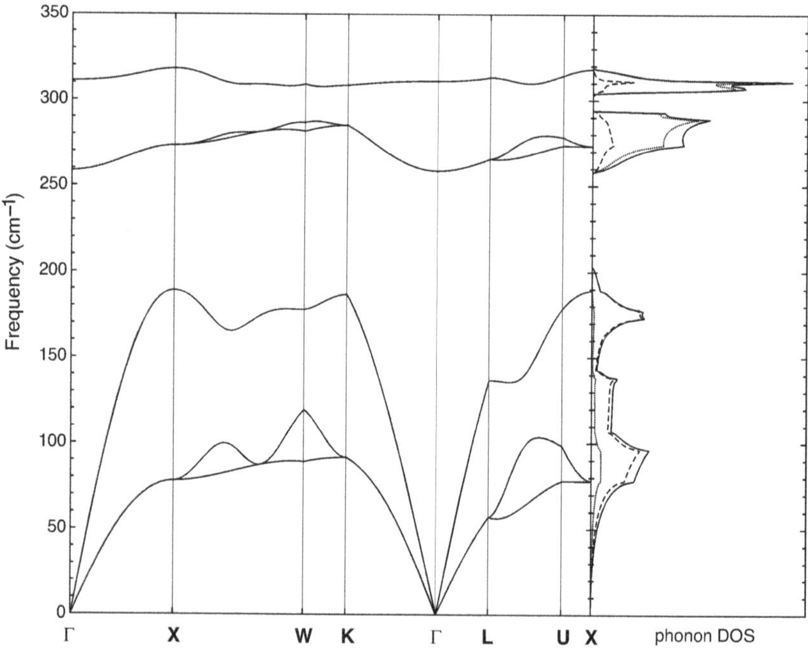

Fig. 14.5 Phonon band structure of sphalerite, ZnS. All the phonon modes are positive, and the structure is dynamically stable. Zn dominates the lower frequency bands (shown with a dashed line in the phonon DOS) and S the higher frequency bands (dotted line).

Box 14.1   **How to calculate the phonon dispersion of sphalerite, ZnS?**

First, compute the ground-state wavefunctions, followed by the dielectric response, and then the dynamical matrices in Γ. Use a dense grid of **k** points, like $8 \times 8 \times 8$, shifted four times (best for fcc lattices). Your input file should contain:

```
ndtset 3
prtden1 1 prtwf1 1 kptopt1 2 toldfe1 1.0d-10
getwfk2 1 rfdir2 1 1 1 rfelfd2 2 tolwfr2 1.0d-18 kptopt2 2
prtden3 1 getwfk3 1 getddk3 2 rfdir3 1 1 1 toldfe3 1.0d-10 rfelfd3 3
rfphon3 1 rfatpol3 1 2 kptopt3 2 nqpt3 1 qpt3 0.0 0.0 0.0
```

Then, you calculate the dynamical matrices on a dense grid of **q**-points. The grid of **q** points needs to be included in the grid of **k** points, that is, $\mathbf{q} + \mathbf{k} = \mathbf{q}'$. Use the previously computed ground-state wavefunctions and density.

```
ndtset 7
jdtset 11 12 13 14 15 16 17
qpt11 5.00000000E-01 0.00000000E+00 0.00000000E+00
qpt12 0.00000000E+00 5.00000000E-01 0.00000000E+00
qpt13 5.00000000E-01 5.00000000E-01 0.00000000E+00
qpt14 0.00000000E+00 0.00000000E+00 5.00000000E-01
qpt15 5.00000000E-01 0.00000000E+00 5.00000000E-01
qpt16 0.00000000E+00 5.00000000E-01 5.00000000E-01
qpt17 5.00000000E-01 5.00000000E-01 5.00000000E-01
kptopt 2 prtden 0 prtwf 0 tolvrs 1.0d-10
rfdir 1 1 1 rfphon 1 rfatpol 1 2 getwfk 1 nqpt 1
```

Once the simulation is finished, merge the dynamical matrices and use *anaddb* to interpolate the interatomic force constant matrices to obtain the phonons at any point in the Brillouin zone.

Compute the phonon dos:

nfreq	600	# number of frequency bins
frmin	0	# minimum frequency
frmax	0.002733801168727	# maximum frequency (in Ha)
prtdos	2	# phonon DOS with tetrahedron method

For the phonon dispersion along the Brillouin zone, use the following:

prtphbands	2	# print phonon bands in *gnuplot* format.
ndivsm	100	# number of points along the smallest segment
nqpath	8	# number of segments ($+1$)
qpath		# the phonon path in the Brillouin zone
0.0 0.0 0.0	0.5 0.0 0.0	2/3 1/3 0.0
0.0 0.0 0.0	0.0 0.0 0.5	0.5 0.0 0.5
2/3 1/3 1/2	0.0 0.0 1/2	

14.3   Zincite, ZnO, and Wurtzite, ZnS

Both zincite, ZnO, and wurtzite, ZnS, are isostructural, with the $P6_3m1$ space group. The Zn atoms are on $2a$, and the S or O $2b$ Wyckoff position. There are 12 phonon modes, which in Γ are decomposed as $2A_1 + 2E_1$.

The phonon band dispersion for the two minerals is represented in Figure 14.6. Because of the structural similarities, the two diagrams have the same topology. The phonon branches show larger dispersion perpendicular to the trigonal basis, such as the acoustic modes along the Γ-A or Γ-K directions, and less along the directions parallel to the trigonal planes, like along M-K or A-L-H-A segments.

The low-frequency phonon bands are dominated by the lattice modes. The entire trigonal basis planes or large fragments of them vibrate against each other. The E_1 mode in Γ corresponds to the alternating movement of the basal planes parallel to each other and perpendicular to the vertical sixfold axis (Figure 14.7). A similar mode in A, at 73 cm^{-1}, shows a double alternance of the movement of the planes because of the phonon wavevector.

The high-frequency bands correspond to *intramolecular* modes, which are *in-plane* vibrations for these structures. The Zn atoms vibrate along the sixfold against the basal plane, as in the E_1 mode at 249 cm^{-1} in Γ (Figure 14.7). At even higher frequencies lies a stretching-like mode of the Zn-O bonds within the basal planes. Because of the interconnectivity between the atoms, the movement is slightly eccentric from the Zn-O bond distances. Such vibration is the 527 cm^{-1} mode in L, in which the double alternance of the Zn-O stretching due to the zone-boundary wavevector can be observed.

14.4   Jadeite, NaAlSi$_2$O$_6$

Jadeite, NaAlSi$_2$O$_6$, is a high-pressure pyroxene in various metamorphic and sub-crustal rocks. It has a monoclinic structure, with 20 atoms in the primitive unit cell.

With such a large number of atoms and large cells, the phonon band dispersion could be obtained by interpolating the interatomic force constants only from the Γ point dynamical matrix. However, the interatomic forces might not be fully converged. A further calculation in another **q** point can help with converging the forces, or at least with checking the convergence.

The dispersion diagram for the 60 phonon modes is shown in Figure 14.8. The low symmetry reduces the degeneracy of the bands, and the large number of atoms is responsible for the existence of so many bands.

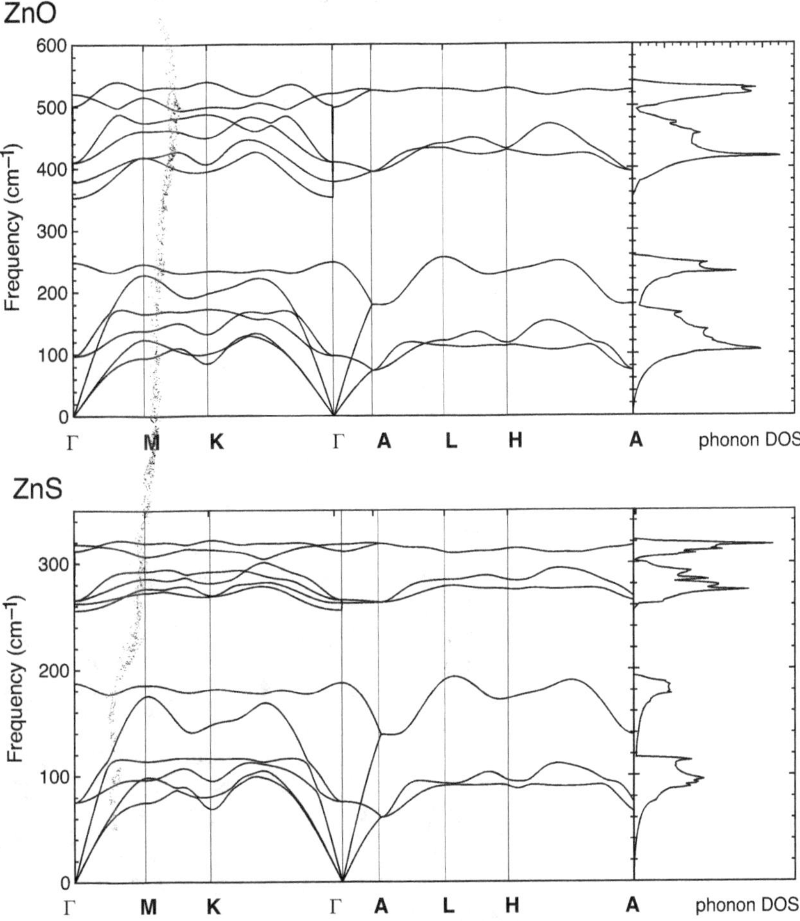

Fig. 14.6 Phonon band structure of zincite, ZnO, and wurtzite, ZnS. As the two minerals are isostructural, their phonon band dispersions have similar features. The vertical shift is due to the different masses of O and S. M = 1/2 0 0; K = 2/3 1/3 0; A = 0 0 1/2; L = 1/2 0 1/2; H = 2/3 1/3 1/2.

Figure 14.9 illustrates the corresponding phonon density of states (DOS). The DOS reveals the presence of several groups of bands separated by smaller or larger frequency gaps. They are more visible in the DOS than in the dispersion diagram.

The same figure also shows the projected DOS on the four atomic types. Vibrations of Si and O dominate the two highest sets of bands. Al dominates the intermediate region, from about 200 up to about 700 cm^{-1}, and Na dominates the low-frequency peaks.

The right side of the diagram in Figure 14.9 shows the computed Raman spectrum with intensities. There is a clear difference between the phonon DOS and the Raman spectrum, showing that we cannot always rely on Raman to obtain a representative sample of the vibrational properties of a mineral. Adding infrared spectra is a first improvement, as many

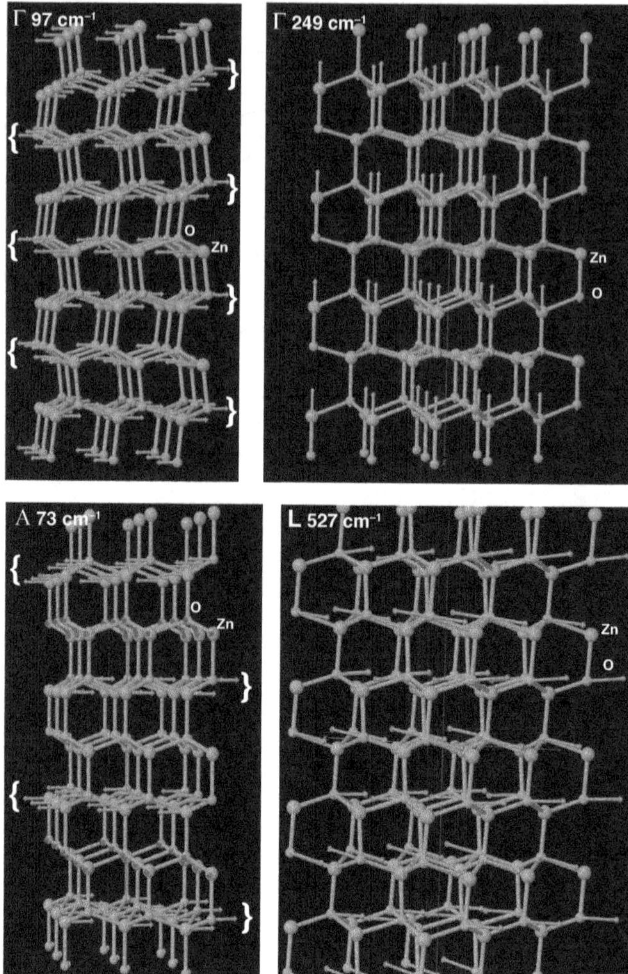

Fig. 14.7 A few examples of vibrational patterns of the phonon modes in zincite, ZnO. The E_1 modes at $97 \mathrm{cm}^{-1}$ in Γ and its equivalent at $73~\mathrm{cm}^{-4}$ in A are lattice modes. The E_1 mode at $249~\mathrm{cm}^{-1}$ in Γ and the 527 cm—1 mode in L are in- and out-of-plane vibrations, the equivalent of the *intramolecular* modes from chemistry; as such, they typically have higher frequencies.

infrared bands would be found in jadeite at high frequency, dominated by the various Si-O antisymmetric stretching modes, which for this structure are polar.

For measuring the phonon DOS, you can also use inelastic X-ray scattering or neutron inelastic scattering spectroscopies. In both techniques, the incoming radiation, X-rays in the first case and neutrons in the second case, interacts with the nuclei and is scattered by the nuclei movement during their vibrations. Thus, the inelastic scattered radiation samples the vibrational spectrum. The energy transfer is more effective at particular frequencies and modes of vibrations.

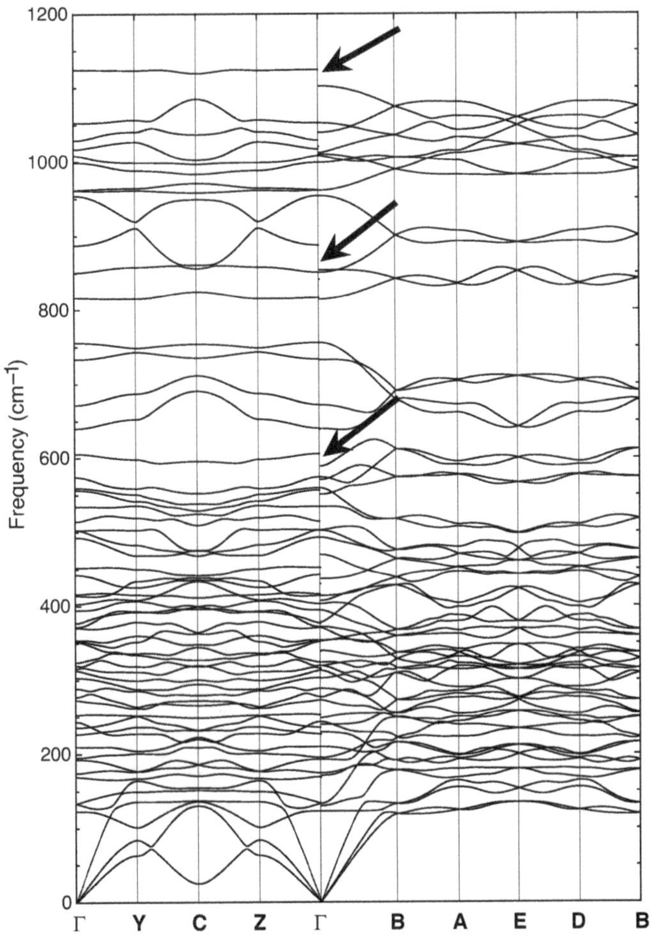

Fig. 14.8   Phonon band structure for jadeite, NaAlSi$_2$O$_6$, which is a major rock-forming pyroxene. The structure is monoclinic, with 20 atoms in the reduced unit cell. The arrows indicate a few large LO-TO splittings, easily observable.

14.5 Pyrite, FeS$_2$

Pyrite, FeS$_2$, is a common sulphide mineral with a cubic structure and a small electronic gap (Figure 4.20). Because of the small electronic gap, the dielectric tensors are very large. The electronic dielectric tensor has a value of $\epsilon^\infty = 23.05$, and the full dielectric tensor at 0 frequency is $\epsilon^0 = 30.72$.

Figure 14.10 shows the phonon dispersion bands for pyrite, its corresponding DOS, and the integral of the phonon DOS over energy. The phonon band structure is highly dispersive. Both Fe and S contribute to most phonon bands, except the last set. The highest-frequency phonon modes are dominated by S, with modes like the S$_2$ stretchings.

Because of the large mass of the atoms, the frequencies lie below 500 cm^{-1}.

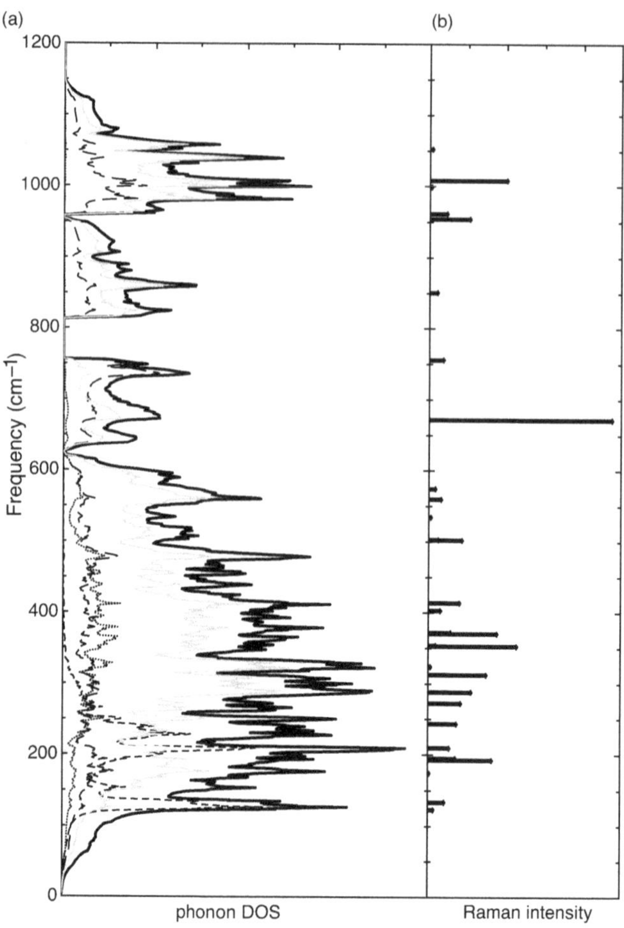

(a) Phonon DOS for jadeite. The solid black line shows the total DOS. The partial DOS contributions from Na, Al, Si, and O are represented with short-dashed, dotted, long-dashed, and solid light grey lines, respectively. (b) The theoretical Raman spectrum of jadeite, with peak position and intensity, samples only a small subspace of the entire vibrational spectrum of jadeite.

14.6 Lower-Mantle Perovskites: CaSiO$_3$ and Bridgmanite, MgSiO$_3$

The CaSiO$_3$ perovskite is the third mineral of the Earth's lower mantle by volume. At mantle pressure and temperature conditions, its structure is Pm3m cubic, similar to the mineral perovskite, CaTiO$_3$. In the ideal cubic perovskite structure, the Ca, Si, and O atoms lie respectively on the 1a, 1b, and 3c Wyckoff positions. With five atoms in the primitive unit cell, there are 15 phonon modes in all, with an irrep decomposition as $4T_{1u} + T_{2u}$ (see explanation in Box 12.5).

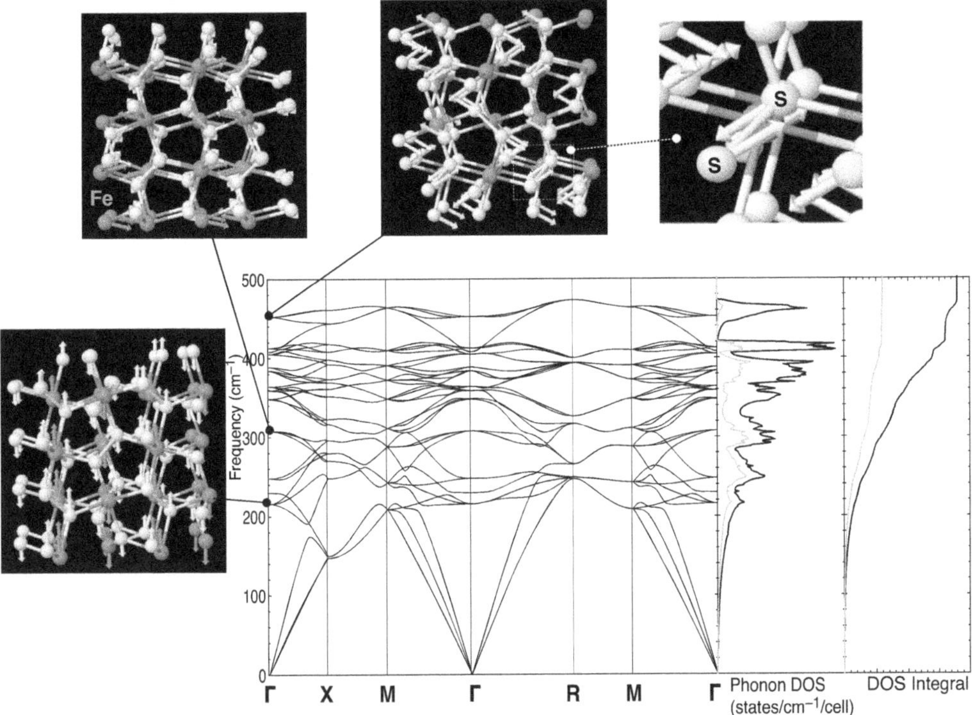

Fig. 14.10 Phonon dispersion for pyrite, FeS_2, the phonon density of states, and its integral over energy. The grey fine dotted line is the integral of the Fe participation to the phonons; the remaining contribution is due to S. The structural insets show the vibrational pattern for a few modes. The first mode in Γ resembles a lattice mode. Sulphur participates in the highest-frequency phonon bands, like in the S-S stretching mode at 452 cm^{-1} shown in the upper insets.

Figure 14.11 shows the phonon band dispersion of the ideal cubic perovskite at two pressures and 0 K temperature. Cubic CaSiO₃ perovskite shows the presence of unstable phonon modes at all pressures (and low temperatures). These modes have imaginary frequencies, which by convention, are represented on the negative side of the energy/frequency axis. The unstable modes are found in M and R high-symmetry points of the Brillouin zone and along the entire M-R branches.

The analysis of the eigendisplacements shows that in both M and R, the unstable phonon mode corresponds to rigid rotations of the SiO₆ octahedra (Figure 14.12). While the phonon is the same in both M and R, the difference in wavevector M or R changes the onset of the instability in the structure. For the simple cubic Brillouin zone, M is the middle of the edges of the cube, the (1/2 1/2 0) family of points. Note that (1/2 1/2 0) is symmetry equivalent to (−1/2 −1/2 0), (1/2 0 1/2), (0 −1/2 1/2), and so on. R is the corner of the cube, the (1/2 1/2 1/2) family of points, including (−1/2 −1/2 1/2), (1/2 −1/2 1/2), (1/2 −1/2 1/2), and so on.

Consequently, the unstable phonon that rotates the octahedra when anchored in the M (1/2 1/2 0) point corresponds in the direct space to

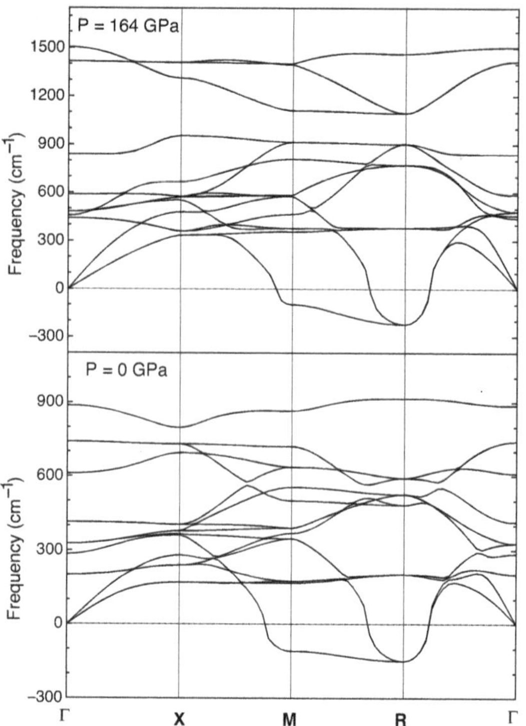

Fig. 14.11 The phonon band dispersion of the cubic Pm3m structure of the CaSiO$_3$ perovskite at 164 GPa and at 0 GPa. There are unstable phonon bands in M and R, which make the structure dynamically unstable at all pressures. Because of the presence of instabilities at low pressures, the CaSiO$_3$ perovskite is unquenchable as a cubic phase at ambient conditions. The M-R branch at low pressure is almost flat, which is responsible for the amorphisation of CaSiO$_3$ in all recovered samples at ambient conditions.

columns of octahedra parallel to the z direction, along which all octahedra rotate in phase. Because M is in the zone boundary, the periodicity of the rotation spans two unit cells, so the octahedra rotate in alternating directions along the x and y directions (Figure 14.12).

The same unstable phonon, when anchored in the R (1/2 1/2 1/2) point, corresponds in the direct space to columns of octahedra parallel to the z direction, but along which all octahedra rotate in anti-phase. Combined with the boundary condition of the two other directions, the eigendisplacements correspond to each octahedron rotating in anti-phase with respect to all its six neighbouring octahedra (Figure 14.12).

The M-R branches show weakly dispersive unstable phonons at low pressures. The low dispersion is a signature of order–disorder instability. The rotations of the octahedra can be frozen at any point along any of the M-R branches with the same energetic gain (their frequency is almost the same). When such freezing-in occurs simultaneously in multiple **q** points, this leads to various degrees of octahedral rotations in different parts of the same crystal. The resulting strain fields that are induced lead to the breakup of the structure and its amorphisation.

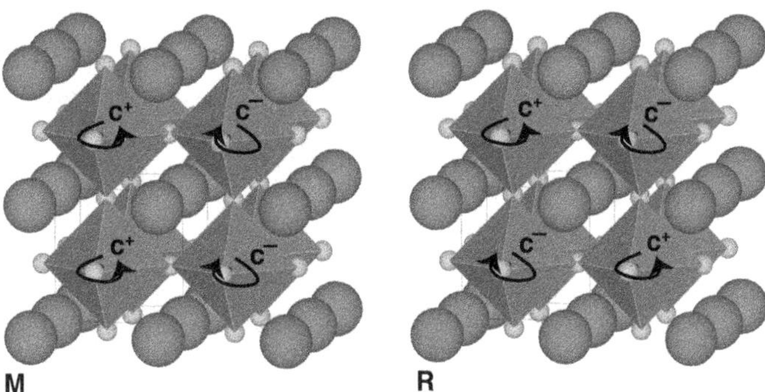

Fig. 14.12 The unstable phonon modes in CaSiO$_3$ correspond to rotations of the octahedra. As all SiO$_6$ octahedra share each of the six oxygen atoms with neighbouring octahedra, the clockwise rotation of any unit implies a counterclockwise rotation of its in-plane neighbours. The instability corresponds to in-phase rotations of the columns of octahedra, as in the left panel, or as anti-phase rotations of the octahedra along the same columns, as in the right panel. The left panel corresponds to the instability in M and is named a^+, and the right panel to the instability in R and is named a^-.

We initially consider only the instability in the M and the R points. The instability can fit in a supercell whose maximum extent is a $2 \times 2 \times 2$ multiple of the initial simple cubic cell. Because the only difference between the two families of rotations is the sense of the rotation, we can then define a column of octahedra that rotate in the same direction as a^+ and the column of octahedra that rotate in opposite directions as a^- [109]. Then we make use of the multiplicity of the M and R points. And finally, we allow combinations of phonon modes frozen in the structure in up to three M and/or R points, each along one of the Cartesian directions. In all, there are 26 possible structures, where the rotations can have similar or different amplitudes:

- no unstable mode   (parent cubic Pm3m phase)   1 structure
- one frozen phonon mode   M or R, (+) or (−)   2 structures
- two frozen phonon modes   M and/or R   6 structures
- three frozen phonon modes   M and/or R   14 structures

After removing all the symmetry-equivalent combinations of rotations, using the Landau theory of the order parameter to describe phase transitions, we are left with only a limited number of possible ways of freezing the instability in the structure [79]. They are listed in Table 14.1.

The relaxation of the CaSiO$_3$ structure according to the different distortion patterns [234, 58, 57] yields the I4/mcm structure to be the most stable one at all pressures and low temperatures.

The parent structure of MgSiO$_3$ perovskite is the same ideal cubic structure as for the CaSiO$_3$ perovskite. The cubic structure is once more unstable. Out of the different possible supercells and combinations of wavevectors for the instability, M + M + R prevails, where each of the three points is along a different direction. This determines the Pbnm space group for the MgSiO$_3$ perovskite, with 20 atoms in the Pbnm space group.

Table 14.1 Structural distortions resulting from rigid octahedral rotation in $2 \times 2 \times 2$ supercells of cubic perovskites.

Space group	Octahedral rotation	Unstable phonon
$Pm\bar{3}m$	$a^0\, a^0\, a^0$	
$I4/mcm$	$a^0\, a^0\, c^-$	R
$Imma$	$a^0\, b^-\, b^-$	R + R
$R\bar{3}c$	$a^-\, a^-\, a^-$	R + R + R
$P4/mbm$	$a^0\, a^0\, c^+$	M
$I4/mmm$	$a^0\, b^+\, b^+$	M + M
$Im\bar{3}m$	$a^+\, a^+\, a^+$	M + M + M
$Cmcm$	$a^0\, b^-\, c^+$	R + M
$P4_2/nmc$	$a^+\, a^+\, c^-$	M + M + R
$Pnma$	$a^-\, b^+\, a^-$	R + R + M

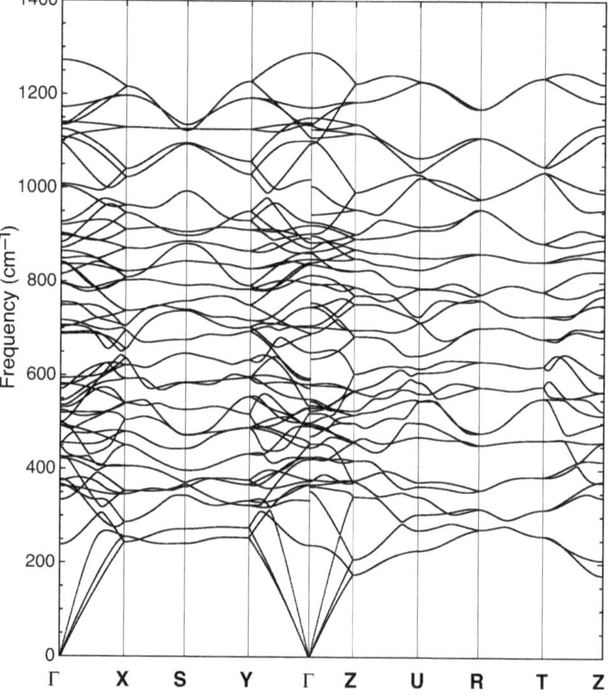

Fig. 14.13 Phonon band dispersion for bridgmanite, $MgSiO_3$, computed at P = 120 GPa. All the phonon modes are positive, making the distorted perovskite structure with Pbnm symmetry stable. The phonon branches are highly dispersive from the Brillouin zone centre towards the zone boundary and much flatter along the edges of the zone boundary. LO-TO splittings are visible in the Γ point.

Orthorhombic bridgmanite is the major mineral of the Earth's lower mantle, and as such, it is the most abundant mineral on Earth by volume.

In bridgmanite, the phonon modes in Γ decompose as $7A_g + 5B_{1g} + 7B_{2g} + 5B_{3g}$ for the Raman-active modes and $8A_u + 10B_{1u} + 8B_{2u} +$

10B$_{3u}$ for the infrared-active modes. The Raman-active modes are clustered in three main groups whose frequencies increase with increasing pressure [63].

The phonon band dispersion is represented in Figure 14.13. As a result of the octahedral tilts and cation distortions, all the phonons are positive, and the structure is dynamically stable.

15

Advanced Topics in Density Functional Perturbation Theory

In Chapters 9–12, we focused on the computation of the phonons, their dispersion, and their corresponding atomic eigendisplacements. There is even more information that can be extracted from the analysis of the phonons. The most important and straightforward is obtaining the thermodynamic parameters as a function of temperature for a given volume. This exercise stands at the base of the construction of phase diagrams.

15.1 Thermodynamics

Apart from the electrons, the movement of the atoms brings another important component to the free energy. They influence the entropy and the internal energy via a vibrational term. In the following, we remain in the quasi-harmonic approximation. This signifies that the vibrations do not interact with each other.

The phonon contribution to the energy is the sum of the energy of all the oscillators. The two are related via temperature and the Boltzmann constant: $F = -k_B T \ln Z$. In this equation, Z represents the partition function, which is the measure of the statistical distribution of the vibrations as a function of energy.

However, as the simulations provide the phonon density of states instead of a partition function, it is more convenient to express the energy variation in terms of this phonon density of states, $g(\omega)$, where ω is the phonon frequency. Four formulas relate the vibrational contribution to four macroscopic parameters:

* The heat capacity:

$$C_V = 3Nk_B \int_0^{\omega_L} \left(\frac{x}{e^{x/2} - e^{-x/2}} \right)^2 g(\omega) d\omega \qquad (15.1)$$

* The entropy:

$$S = 3Nk_B \int_0^{\omega_L} \left[\frac{xe^x}{e^x - 1} - \ln\left(e^x - 1\right) \right] g(\omega) d\omega. \qquad (15.2)$$

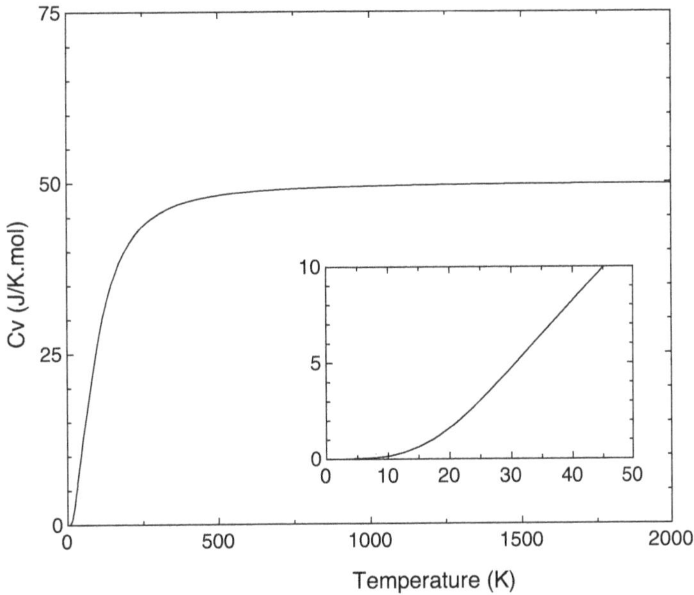

Fig. 15.1 Temperature-dependence of the heat capacity, Cv, of sphalerite, ZnS. Remark the T³ dependence at low temperatures, as shown in the inset. The values are obtained using the quasi-harmonic approximation and Equation 15.4.

* The internal energy:

$$\Delta E = 3N \int_0^{\omega_L} \left(\frac{e^x + 1}{e^x - 1}\right) \frac{\hbar\omega}{2} g(\omega) d\omega \tag{15.3}$$

* The heat capacity:

$$\Delta F = 3Nk_BT \int_0^{\omega_L} \{\ln\,(e^x - 1) - x/2\}\, g(\omega) d\omega \tag{15.4}$$

In Equations 15.1–15.4, N is the number of atoms in the unit cell, k_B is the Boltzmann constant, and T is the temperature.

Figure 15.1 shows the computed heat capacity, and Figure 15.2 displays the computed entropy of sphalerite, ZnS. The two thermodynamic functions are obtained within the quasi-harmonic approximations and Equations 15.1–15.4. While low- and moderate temperatures show an excellent agreement to experiment, there are deviations at high temperatures due to anharmonic effects.

The phase diagrams for pure phases are determined by the Gibbs free energy of the different polymorphs. The vibrational contributions to the entropy and the internal energy need to be added to the enthalpy. They are calculated, for example, using the *anaddb* utility of abinit (Box 15.1):

$$G = E_0 + E(T) + PV - S(T)T \tag{15.5}$$

where the E_0 is the ground-state electronic energy computed in the static density functional theory (DFT) simulations. There are other possible

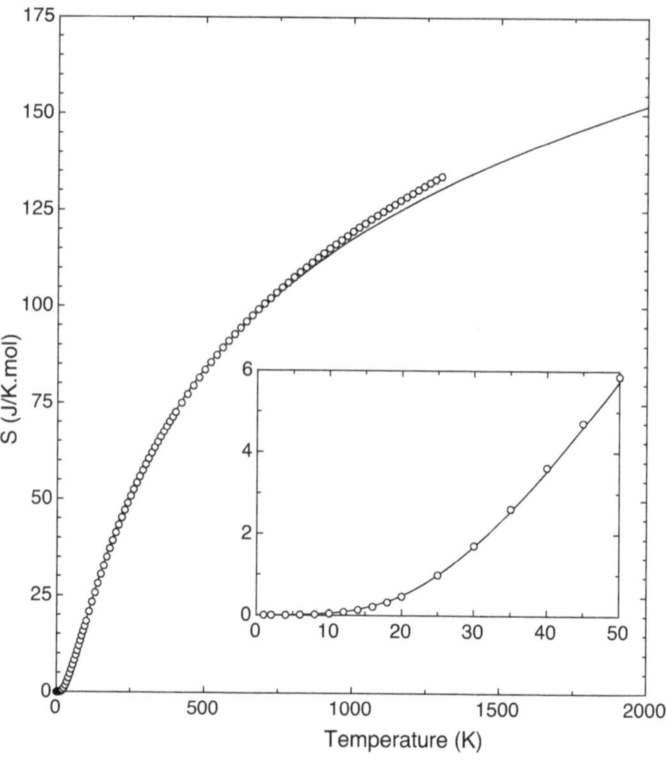

Fig. 15.2 Temperature-dependence of the entropy, S, of sphalerite, ZnS. The solid line shows the computed values within the quasi-harmonic approximation and Equation 15.2. The empty circles are measured values [124]. Remark the almost perfect agreement between the two at low- and moderate temperatures and the beginning of the theoretical deviation from experimental values at high temperature. The deviation is mainly due to the limitations of the quasi-harmonic approximation, which fails at temperatures higher than about 2/3 of the melting temperature, where anharmonic effects and phonon–phonon interactions become important.

Box 15.1　　**How to calculate the thermodynamic parameters?**

The quantity central to calculating the thermodynamical properties is the phonon density of states, which is obtained by interpolating over a dense grid of **q** points. The denser the grid, the better the quality of the interpolated grid. In the input file of *anaddb*, you need the following input variables:

```
           # Flags
ifcflag  1   # interpolate interatomic force constants
thmflag  1   # compute thermodynamical properties
ifcana   1   # analyse interatomic force constants
asr      1   # impose acoustic sum rule
```

```
chneut      1              # impose (Born) charge neutrality
prtdos      2              # print phonon DOS

                           # Thermal information
nchan       1500           # number of frequencies
nwchan      20             #
thmtol      20             # convergence criterion
ntemper     120            # no. of temperatures
temperinc   25             # thermal increment
tempermin   0              # minimum temperature

                           # Interpolation grid
ng2qpt      128 128 128    # sample the Brillouin zone
ngrids      4              # number of grids of increasing size
q2shft      0.5 0.5 0.5    # shift of the grids

                           # Computed grid
brav        1              # Bravais lattice type (simple cubic)
ngqpt       4 4 4          # grid of computed dynamical matrices
nqshft      1              # shift of the grid
q1shft      0.0 0.0 0.0    # includes Γ
```

entropic contributions for various crystals: the electronic entropy for metallic systems, the magnetic entropy for magnetic systems, and the mixing entropy with all its configurational terms for the solid solutions. They all need to be considered when building phase diagrams, as they all affect the free energy.

15.2 Strain Perturbations and a Modern View of the Elastic Constants

Apart from the electric field and the atomic displacements, another common perturbation is the strain. The elastic constants form a fourth-rank tensor, which relates the applied stress to the resulting crystal strain in the limit of small stresses. It can be calculated from finite differences (Chapter 8).

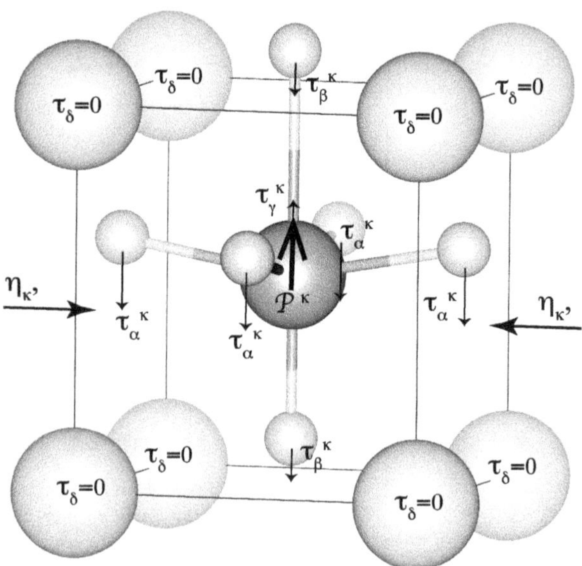

Fig. 15.3 The effects of strain on the crystal structure are exemplified on the ferroelectric mineral macedonite, $PbTiO_3$. A strain perturbation ($\eta_{\kappa'}$) applied to the crystal displaces the atoms (τ), and consequently induces changes in the interatomic force constants and polarisation ($\mathcal{P}$) (unless these are fixed by symmetry, like for the δ atom, which is placed on the 0 0 0). In this case, obtaining the elastic constants tensor from perturbation theory requires solving the responses to atomic displacements and electric fields, as represented in Equation 15.13.

An alternative way for its calculation is using perturbation theory. The stress on a crystal is defined as the derivative of the internal energy as a function of volume: $\sigma = dE/dV$. This suggests that the following relation could be used to obtain the elastic constants directly as a double derivative of the energy:

$$C_{ijkl} \sim \frac{d\sigma_{ij}}{d\eta_{kl}} \sim \frac{d(dE/d\eta_{ij})}{d\eta_{kl}} \sim \frac{d^2 E}{d\eta_{ij}d\eta_{kl}} \tag{15.6}$$

In this equation, the strain η is treated as a perturbation. The strain affects the underlying crystal structure and may induce polarisation and atomic displacements, $\eta = \eta(\mathcal{P}, \tau)$ (Figure 15.3). Consequently, to solve Equation 15.6, the electric and displacement response needs to be taken explicitly into account.

When the mechanical properties were computed from finite differences, that is, calculating the stresses induced by deformation, the crystal structure was relaxed under symmetry constraints, so the atom can accommodate both the electric and the displacement response. In DFPT, these responses have to be calculated and incorporated explicitly to solve Equation 15.6. This requires partial derivatives of the energy with respect to all other perturbations.

Some, like the electric susceptibility or the Born effective charges, were already discussed earlier. Others are new and specific to the elastic field. The six coupling tensors are calculated as:

* the clamped-ion (or frozen = unrelaxed structure) elastic tensor:

$$\bar{C}_{ijkl} = \frac{\partial^2 E}{\partial \varepsilon_{ij} \partial \varepsilon_{kl}}\bigg|_{\tau,\mathcal{E}} \tag{15.7}$$

* the clamped-ion dielectric susceptibility tensor:

$$\bar{\chi}_{jj'} = -\frac{\partial^2 E}{\partial \mathcal{E}_j \partial \mathcal{E}_{j'}}\bigg|_{\tau,\varepsilon} \tag{15.8}$$

* the clamped-ion piezoelectric tensor:

$$\bar{e}_{j\alpha\beta} = -\frac{\partial^2 E}{\partial \mathcal{E}_j \partial \varepsilon_{\alpha\beta}}\bigg|_{\tau} \tag{15.9}$$

* the Born effective charge tensor:

$$Z_{\alpha k j} = -\Omega \frac{\partial^2 E}{\partial \tau_{\alpha k} \partial \mathcal{E}_j}\bigg|_{\varepsilon} \tag{15.10}$$

* the clamped-ion force-strain coupling (or internal strain) tensor:

$$\Lambda_{\kappa k\alpha\beta} = -\Omega \frac{\partial^2 E}{\partial \tau_{\kappa k} \partial \varepsilon_{\alpha\beta}}\bigg|_{\mathcal{E}} \tag{15.11}$$

* the force constant tensor:

$$K_{\alpha\kappa\beta\kappa'} = \Omega \frac{\partial^2 E}{\partial \tau_{\alpha\kappa} \partial \tau_{\beta\kappa'}}\bigg|_{\mathcal{E},\varepsilon} \tag{15.12}$$

The determination of all strain-independent derivatives was done in previous steps. The remaining strain derivatives need to be calculated in an additional step. Box 15.2 shows how to compute the elastic constant tensor in DFPT.

With these six tensors, we can build a *big* Hessian matrix that relates the different perturbations between them. Its simplified form is:

$$
\begin{array}{cccc}
 & Displacement & Strain & Electric\ field \\
\begin{array}{c} Displacement \\ Strain \\ Electric\ field \end{array} &
\left(\begin{array}{ccc}
K & -\Lambda & -Z \\
-\Lambda & C & \bar{e} \\
-Z & \bar{e} & \bar{\chi}
\end{array} \right)
\end{array} \tag{15.13}
$$

Equation 15.13 holds all the necessary information for computing all the strain-related perturbations [23]. Its solution involves complicated combinations of derivatives. It was first solved for norm-conserving s [128] and later for PAW [177]. Its implementation is possible only because all the physical quantities are expressed (in *abinit*) using reduced coordinates and the metrics of the space. From the beginning, we have seen that a crystal structure is a vectorial space, with a base (= unit cell) and the atoms populating this base. With this formalism, the strain derivatives of the energy are done only relative to the basis of the crystallographic vectorial space. Thus, the norm of the space is preserved, and the actual derivation equations can be written fully analytically.

Taking into account the coupling between the different perturbations and the ion relaxations, we obtain the full elastic constants tensor as:

$$C_{ijkl} = \bar{C}_{ijkl} - \Omega^{-1} \Lambda_{\kappa\kappa'\alpha\beta} (K^{-1})_{i\kappa j\kappa'} \Lambda_{\kappa\kappa' kl} \tag{15.14}$$

Box 15.2   **How to calculate the elastic constants matrix?**

The elastic response is performed just as with all the other responses. While the calculations of the responses to various perturbations can be mixed, it is wiser to keep them separate. This helps to keep things in perfect order and reuse various terms and derivatives for later. The key variable is *rfstrs*, which has to be non-zero. The same input is usable for both norm-conserving *s* and PAW. The only difference between the calculation with norm-conserving *s* and within the PAW formalism is in the use of *s*, and of the variable *pawecutdg* [177].

```
kptopt4    2          # use time-reversal symmetry for k-points
prtden4    0          # do not print electronic density
prtwf4     0          # do not print wavefunctions
tolvrs4    1.0d-10    # use strong convergence criteria
rfdir4     1 1 1      # perturbations along all directions X Y Z
rfstrs4    3          # strain perturbation
```

Note that this is a PAW calculation, so add the following:

```
pawecutdg    64    # energy cutoff for inside PAW spheres
pawovlp      10    # allowed overlap of PAW spheres
                   # do not set it too large!
```

Once the calculation is done, as usual, merge the different DDB files, including the one with the elastic response. The last step is done in the *anaddb* post-processing tool. Specify the following in the input file:

```
elaflag     3    # compute full elastic tensor
instrflag   1    # compute full internal strain tensor
piezoflag   2    # compute full piezoelectric tensor
```

The last step for computing the coupling is done in the post-processing stage in the *anaddb* utility (Box 15.2). Table 15.1 compares the elastic constants of bridgmanite, $MgSiO_3$, computed at 0 GPa, using finite differences and linear response. The results are very similar between the two techniques. Further improvements can be achieved by increasing the density of the grid of **k** points and decreasing the amplitude of the finite strain in the finite differences calculations.

But the linear response has a huge advantage over the finite differences method. It provides all the strain-related tensors at the end of the same calculation (Box 15.2). When all the couplings are correctly accounted for, there are two more tensors to obtain, which have tremendous use in practical life.

The piezoelectric tensor characterises the electric field induced by strain. It appears only in non-polar crystals. Its full form is:

$$e_{j\alpha\beta} = \bar{e}_{j\alpha\beta} + \Omega^{-1} Z_{\kappa k j} (K^{-1})_{\kappa k \kappa' k'} \Lambda_{\kappa' k' \alpha\beta} \qquad (15.15)$$

Table 15.1 Elastic constants (in GPa) of bridgmanite, computed at several pressures from finite differences (FD) and elastic linear response (LR). All calculations were realized using norm-conserving (NC) pseudopotentials in LDA.

Method	C11	C22	C33	C12	C13	C23	C44	C55	C66	K	G
FD	491	571	480	131	131	148	162	177	205	263	184
LR	488	567	477	128	140	145	160	176	203	260	183

The full dielectric constant tensor contains all the corrections after all the contributions have been accounted for:

$$\chi_{jj'} = \bar{\chi}_{jj'} + \Omega^{-1} Z_{\kappa kj}(K^{-1})_{\kappa k \kappa' k'} Z_{\kappa' k' j'} \tag{15.16}$$

All the resulting relaxed-ion tensors are listed in the output of the *anaddb* file (see Figure 15.4).

15.3 Mass-Dependent Isotope Fractionation

The dynamical matrices explicitly contain atomic masses. As such, a straightforward application of the dynamical theory of solids is to predict the isotope fractionation. The idea is based on the formalism developed in the 1940s by Bigeleisen and Meyer [31], who used the partition function to explain how vibrations contribute to the partitioning of isotopes. Their final formulas are used in almost every paper published today about isotope fractionation, and the different authors of these papers usually even preserved the same notation.

The equations are valid only for mass-dependent fractionation because the atomic masses do not intervene directly in the DFT equations.

The underlying idea is the coupled mineral and isotope exchange reaction between two minerals, A and B, which have isotope concentrations, X and X'. The equilibrium constant of the isotope exchange reaction is:

$$AX + BX' \Longleftrightarrow AX' + BX \tag{15.17}$$

The equilibrium constant can then be expressed in terms of the partition function ratios of the chemical species involved in the isotopic exchange:

$$\alpha_{A/B} = \frac{(Z'/Z)_A}{(Z'/Z)_B} = \frac{(s'/s)_A}{(s'/s)_A} \cdot \frac{f_A}{f_B} \tag{15.18}$$

where prime indicates the heavy isotope (the missing prime is the light isotope), Z is the partition function, and f is the partition function ratio. The products $(s'/s)\,f$ are called the reduced partition function ratios (RPFRs) of the mineral. Sometimes, the RFPRs are named the β func-

```
Elastic Tensor (clamped ion) (unit:10^2GP):

    2.6969646    0.2776236    0.7072289    0.0000748    0.0001258   -0.0000306
    0.2776826    2.6959358    0.7072252   -0.0001773   -0.0000133   -0.0000309
    0.7054905    0.7057973    8.3031132    0.0000801   -0.0000617    0.0001213
   -0.0000007    0.0000002   -0.0000087   -0.7976813    0.0000031   -0.0000009
   -0.0000032   -0.0000007   -0.0000024    0.0000031   -0.7977228   -0.0000005
    0.0000135    0.0000132    0.0000021   -0.0000009   -0.0000005    0.2831093

Elastic Tensor (relaxed ion) (unit:10^2GP):
 (at fixed electric field boundary condition)

    2.4505451    0.1098596    0.8439747    0.0000646    0.0001653   -0.0000952
    0.1099186    2.4491607    0.8441400   -0.0002364    0.0000035   -0.0000955
    0.8422363    0.8427120    7.1921379    0.0001211   -0.0000853    0.0001701
   -0.0000109   -0.0000588    0.0000324   -0.3095999    0.0000068   -0.0000009
    0.0000364    0.0000161   -0.0000260    0.0000068   -0.3094867   -0.0000004
   -0.0000511   -0.0000515    0.0000509   -0.0000009   -0.0000004    0.2831093
```

inverse

```
Compliance Tensor (clamped ion) (unit: 10^-2GP^-1):

    0.3818371   -0.0315090   -0.0298397    0.0000398    0.0000630    0.0000506
   -0.0315214    0.3819899   -0.0298515   -0.0000909   -0.0000090    0.0000511
   -0.0297641   -0.0297934    0.1255096    0.0000164   -0.0000139   -0.0000602
   -0.0000000    0.0000005   -0.0000013   -1.2536335   -0.0000048   -0.0000038
   -0.0000014   -0.0000001   -0.0000002   -0.0000048   -1.2535683   -0.0000021
   -0.0000165   -0.0000160    0.0000019   -0.0000038   -0.0000021    3.5322049

Compliance Tensor (relaxed ion)  (unit: 10^-2GP^-1):
 (at fixed electric field boundary condition)

    0.4252316   -0.0019848   -0.0496666    0.0000708    0.0002408    0.0001721
   -0.0020021    0.4254957   -0.0497054   -0.0003448    0.0000174    0.0001728
   -0.0495622   -0.0496234    0.1506810    0.0000865   -0.0000686   -0.0001239
   -0.0000197   -0.0000860    0.0000269   -3.2299750   -0.0000709   -0.0000101
    0.0000540    0.0000260   -0.0000211   -0.0000708   -3.2311570   -0.0000046
    0.0000853    0.0000859   -0.0000451   -0.0000101   -0.0000046    3.5322052
```

```
Proper piezoelectric constants (relaxed ion) (unit:c/m^2)

   -0.00000407   -0.00007480    2.11194189
    0.00000962   -0.00006315    2.11156714
    0.00000163    0.00010000   -1.65080816
    0.00006377    1.41953604   -0.00006319
    1.41865438    0.00001846   -0.00031061
   -0.00000314    0.00000110    0.00062969
```

Fig. 15.4 Elastic tensorial properties were computed for macedonite, $PbTiO_3$. The elastic constants tensor is computed in DFPT in the *clamped* version – meaning the atoms are kept fixed. Then the interatomic force constants tensor and the dielectric tensor are combined with the elastic one to obtain the fully *relaxed* version of the elastic tensor. Macedonite is a non-polar insulator, and consequently, it is piezoelectric. The calculations of the dielectric tensors and the elastic constants tensor allow for the automatic determination of the piezoelectric tensors. Various key features of the output are outlined (units, *clamped* vs. *relaxed*, electric field condition). The compliance tensor is the inverse of the elastic tensor and vice versa.

tions. For minerals, the RFPRs can be expressed as a function of the quasi-harmonic phonon frequencies:

$$\beta = (s'/s) \cdot f = \prod_i \frac{u'_i exp(-u'_i/2)/[1 - exp(-u'_i)]}{u_i exp(-u_i/2)[1 - exp(-u_i)]} \tag{15.19}$$

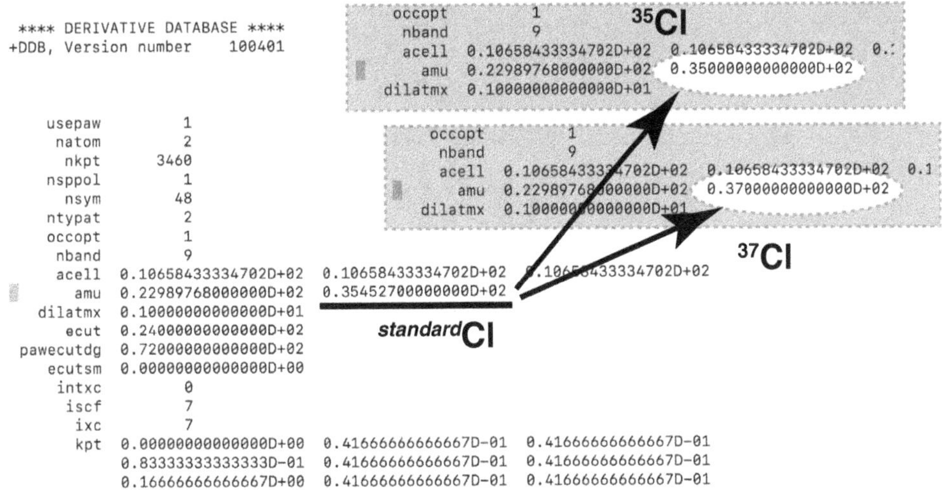

Fig. 15.5 Because of the Born–Oppenheimer approximation and the classical treatment of the nuclei, that is, non-quantum, the interatomic force constants are invariant under a change of isotopes. Then the atomic masses can be modified by hand in the DDB file, the new dynamical matrices need to be diagonalised, and thus the effect of the isotopes on the phonon dispersion can be observed. In this example, the mass of standard Cl (35.4527) is replaced with the masses of its two major isotopes: ^{35}Cl and ^{37}Cl.

The phonon frequencies, ω, of the modes, i, come into play in this equation via the u factors, which are defined as $u_i = hc\omega_i/k_B T$ for the light isotope and $u_i = hc\omega'_i/k_B T$ for the heavy isotopes. The other symbols are h for the Planck constant, c for the speed of light, and k_B for the Boltzmann constant. The temperature T can vary and gives the temperature dependence of the isotope partitioning.

Consequently, once the determination of the interatomic force constants is done, the atomic masses can be changed by hand in the dynamical matrices, like in Figure 15.5, and the phonon frequencies recalculated. The presence of heavier isotopes shifts the phonon frequencies to lower values (Figure 15.6).

The shift is proportional to the reduced mass of the isotope system:

$$\sqrt{\frac{mm'}{m+m'}} \tag{15.20}$$

The most intense effect is for hydrogen and deuterium, as deuterium has twice the mass of hydrogen. For example, in brucite, $Mg(OH)_2$, the OH stretching mode is shifted from 3,686 cm^{-1} in $Mg(OH)_2$ down to 2,684 cm^{-1} in $Mg(OD)_2$ [216].

The mass difference between isotopes also affects the boundaries between different polymorphs. The effect is most visible for hydrogen. As the phase transition sequence is the same for the two isotopes, the boundaries are shifted towards lower pressures for the deuterium system. Consequently, performing experiments on heavy water, or more precisely

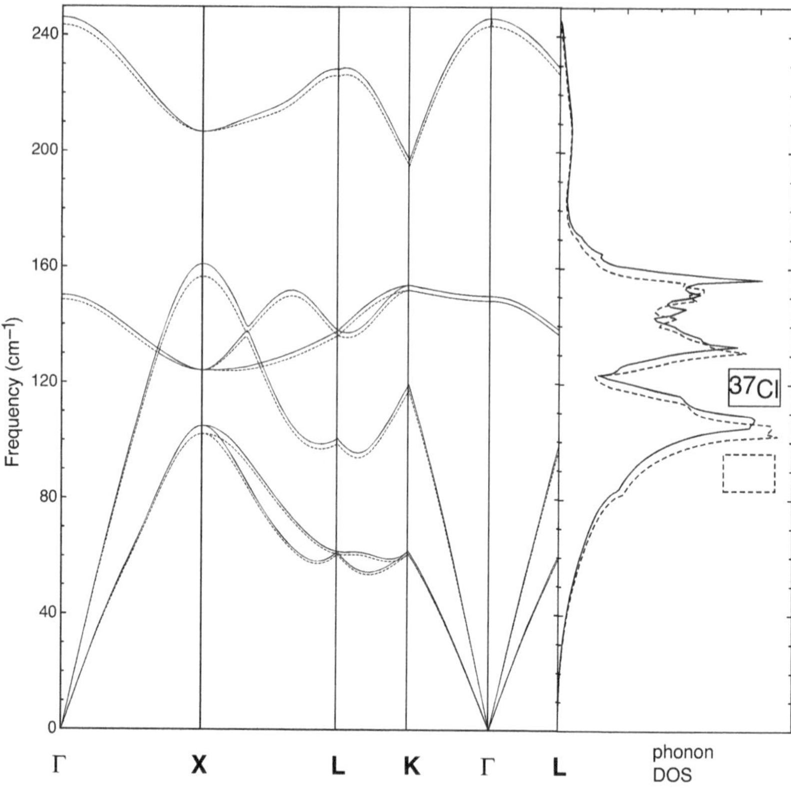

Fig. 15.6 The change of the mass of Cl, from the standard value to the ^{35}Cl or to the ^{37}Cl value, has a distinct influence on the dispersion of the phonon bands. The phonon bands in the Na^{35}Cl system (solid lines) have higher frequencies than the Na^{37}Cl system (dashed lines), as heavier isotopes vibrate more *sluggishly* than the lighter ones. The phonon density of states shows clearly the shift. Na has only one isotope.

on heavy ice, allows experimentalists to sample an extensive phase diagram with a smaller effort [113].

The shift in phonon band frequencies is proportional to the participation of the various isotopes in those particular vibrations. Figure 15.7 shows the phonon band dispersion in zincite, ZnO, when exchanging the isotope mass of Zn, as ^{64}Zn, ^{66}Zn, or ^{68}Zn, and of O, as ^{18}O. Depending on the atomic participation, the bands are shifted by different amounts in different parts of the Brillouin zone. The partial phonon density of states (Figure 15.8) reveals clear shifts for both the Zn and the O isotopes in different spectral regions: Zn, being heavier, dominates the low-frequency modes, and O, being lighter, dominates the high-frequency region.

The temperature dependence of the isotope partitioning stems from the same Equation 15.19. Once the ratios are performed, the figures are typically reported as $1{,}000\ ln\alpha_{X/X'}(A/B)$ as a function of $1/T^6$. A second axis can show the normal temperature in degrees to correlate partitioning with

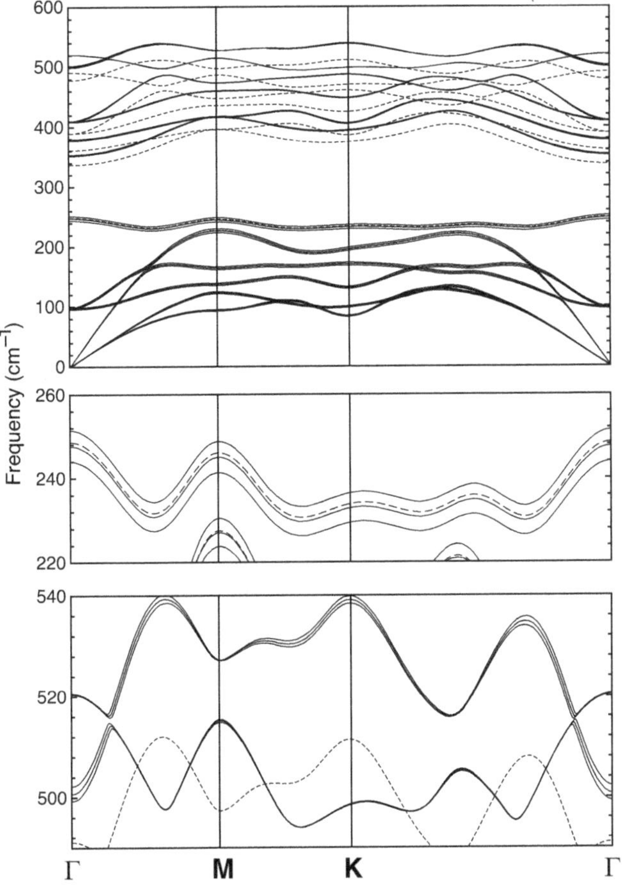

Fig. 15.7 The effect of the change of isotope mass on the phonon band dispersion is visible for zincite, ZnO. The two lower panels show a magnified view of two regions of the frequency range. The bands computed with the ^{18}O isotope are shown with dashed lines. There is a clear shift in phonon frequency with the heavier oxygen isotope. The high-frequency modes, dominated by the O movements, are more affected than the lower-frequency modes, dominated by Zn. The three solid lines are close to each other, and they correspond in decreasing frequency to the ^{68}Zn, ^{66}Zn, and ^{64}Zn isotopes. As their mass is very similar, the change in frequency is small.

actual temperatures. The theoretical results, expressed in *per mil*, are in the same range as measurements are reported.

In the quasi-harmonic approximation, the phonons are computed at T = 0 K and the thermodynamics are extrapolated to higher temperatures under isobaric conditions. While this is reasonable at relatively low temperatures, the anharmonic effects become essential for the phonon frequencies as the temperature approaches the melting point. Consequently, there is a limit up to which this approach can be reliably used, though this limit is not sharp. On the other hand, the partitioning is less and less effective as the temperature increases, and the mass effects are overwhelmed by the entropic terms. This is perfectly similar to observations from experiments and natural systems. However, at high temperatures, phase changes,

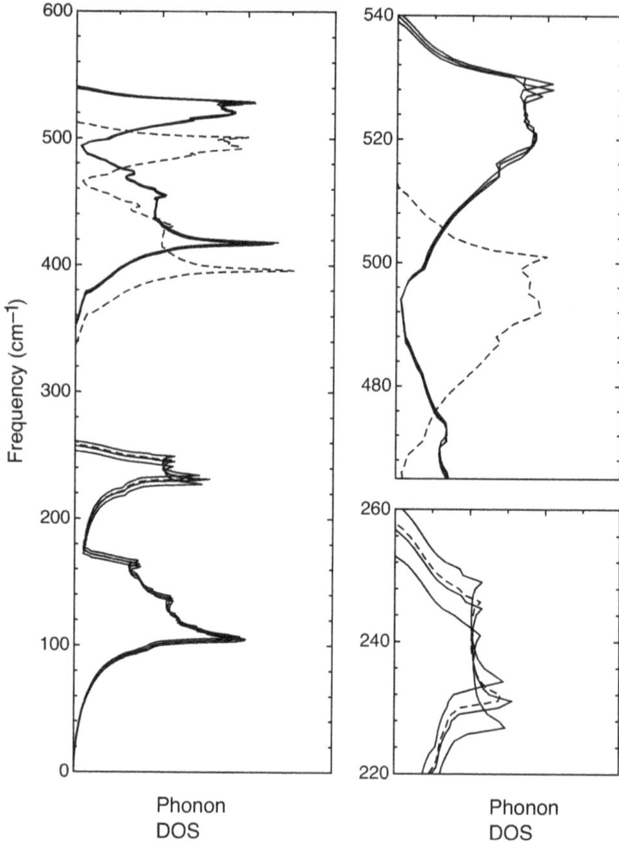

Fig. 15.8   Phonon density of states of zincite, ZnO, corresponding to the phonon band dispersions from Figure 15.7. The shifts in DOS with the isotope mass are apparent. As in Figure 15.7, ^{18}O DOS is shown with dashed lines, and the ^{68}Zn, ^{66}Zn, and ^{64}Zn isotopes with the solid lines.

like vaporisation or melting, can affect the isotope partitioning because of kinetic effects during that particular phase change. These need to be studied using molecular dynamics techniques (see Chapter 16).

This procedure is valid for studying mass-dependent fractionation only. But for some particular isotopes, the addition of neutrons changes the shape of the nuclei. The changes are large enough to modify the electrostatic field induced by the nuclei in their surroundings. As a result, the electronic clouds are slightly deformed from one isotope to another. The result is isotope partitioning that is independent of isotope masses. This can be studied with ab initio simulations only by generating adequate pseudopotentials for each isotope. These isotope-dependent pseudopotentials must specifically take into account the induced changes in the angle-dependent components. The entire simulation will need to be redone with the isotope-dependent pseudopotentials that consider these different electronic distortions.

15.4 Anharmonicities: Beyond the Quasi-Harmonic Approximation

Starting with Section 9.3, the derivation terms higher than second order were ignored from Equation 9.3. This truncation was defined as the quasi-harmonic approximation. The justification relies on the infinitesimal character of the perturbations. The result is Equation 9.5, in which only second-order derivatives are present.

In terms of energy, the quasi-harmonic approximation signifies symmetric potential wells. However, a careful look at the energy versus Zn-S bond distance in sphalerite, illustrated in Figure 15.9, shows that this curve is not symmetric. Similarly, the energy curves as a function of bonding distance in diatomic molecules, discussed in Section 9.1 and plotted in Figure 9.2, were not symmetric either. The discrepancy between the second-order polynomial fitting and the actual energy values increases with increasing atomic displacement, that is, with increasing energy difference from the equilibrium state. These errors are propagated to the vibrational frequencies: Because the phonon frequencies reflect the curvature of the energy line around the equilibrium, the computed phonons would, of course, be slightly off with respect to experimental measurements.

However, if the energy curve is fitted with a higher-order polynomial instead of a second-order polynomial, a much better result can be obtained. This analysis can be quickly done on the Zn-S potential well in sphalerite (Figure 15.9). In this image, a second-order polynomial fitted only on the few closest points to the equilibrium yields an acceptable fit with an R-value of 0.97. The fit is reasonable only if the perturbation remains infinitesimal. But if more points are added to the fit, with increasing perturbation amplitude, the fit deteriorates, and the R decreases towards 0.9. At large atomic displacements, due, for example, to high temperature, the quasi-harmonic approximation fails. However, a third-order polynomial fits the data points with an impressive R-value of 0.9954, even for perturbations with large amplitudes, as can be seen in Figure 15.9. And, of course, the fits become better if higher-order terms are added in the Taylor expansion.

The interatomic forces are treated as ideal elastic oscillators in the quasi-harmonic approximation. There is no dissipation, no interaction between them. Any amount of vibrational energy that is added to the system, for example, by heating, is conserved. Heat is conducted without loss. This idealised model works well for treating low-temperature phenomena and as a first approximation of the vibrations. But in many applications and materials, the quasi-harmonic approximation fails to describe the physics accurately. Thus, many physical problems require an improvement of the quasi-harmonic approximation.

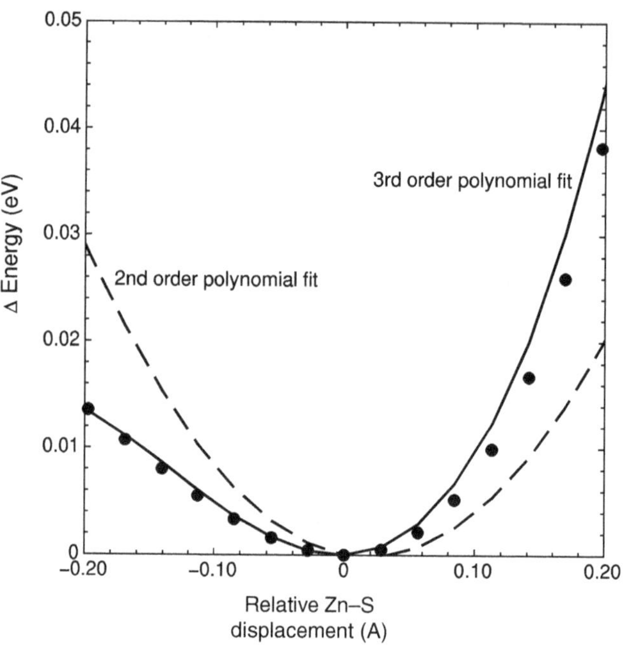

Fig. 15.9 The potential wells of the Zn-S bond distance as computed by point-by-point calculations in wurtzite, ZnS. Solid circles show the actual calculated points. The harmonic fit, *the second*-order polynomial, represented by the dashed line, fails to reproduce the energy variation with respect to the atomic displacements. As a result, the frequencies determined in the quasi-harmonic approximation are considerably shifted from the experimental values. The anharmonic fit, involving third-order terms, represented with the solid line, reproduces the actual computed variation much better. However, obtaining the frequencies from the third energy derivative is not yet routinely available in any standard post-processing code.

What needs to be achieved is to describe the interactions between phonons correctly. The first step is to treat processes between two phonons, with wavevectors in $\mathbf{q_1}$ and $\mathbf{q_2}$. Their interaction results in the creation of another phonon with wavevector $\mathbf{q_1 + q_2 + q_3}$. If $\mathbf{q_3}$ is in the same Brillouin zone as the first two wavevectors, the process is called Normal. If $\mathbf{q_3}$ lies in a neighbouring zone, the process is called Umklapp (from the German word *fold over*). The third wavevector must be folded back to the original zone. Both the N and the U processes play a role in heat transport.

Describing the interaction between phonons implies the explicit consideration of the higher-order derivative terms in the energy expansion. For the corrections of the pure interatomic forces, we need the derivative of the energy as a function of the displacements κ, κ', and κ'' of the three atoms α, β, and γ:

$$E^{\kappa\kappa'\kappa''} = \sum_{\alpha,\beta,\gamma} \frac{\partial^3 E_{el}}{\partial \tau_\alpha^\kappa \partial \tau_\beta^{\kappa'} \partial \tau_\gamma^{\kappa''}} \tag{15.21}$$

Because of the coupling between different perturbations, this represents the last term of the full expression of the energy derived to the third order

(Equation 9.12) with respect to electric fields and atomic displacements. In this expansion for insulators, in all, there are four third-order terms:

$$E^{(3)} = \sum_{\kappa,\kappa',\kappa''} \frac{\partial^3 E_{el}}{\partial \mathcal{E}^\kappa \partial \mathcal{E}^{\kappa'} \partial \mathcal{E}^{\kappa''}}$$

non-linear optical properties

$$+ \sum_{\kappa,\kappa',\alpha,\kappa''} \frac{\partial^3 E_{el}}{\partial \mathcal{E}^\kappa \partial \mathcal{E}^{\kappa'} \partial \tau_\alpha^{\kappa''}}$$

Raman tensors $\qquad\qquad$ (15.22)

$$+ \sum_{\kappa,\alpha,\kappa',\beta,\kappa''} \frac{\partial^3 E_{el}}{\partial \mathcal{E}^\kappa \partial \tau_\alpha^{\kappa'} \partial \tau_\beta^{\kappa''}}$$

$$+ \sum_{\alpha,\kappa,\beta,\kappa',\gamma,\kappa''} \frac{\partial^3 E_{el}}{\partial \tau_\alpha^\kappa \partial \tau_\beta^{\kappa'} \partial \tau_\gamma^{\kappa''}}$$

anharmonic interatomic force constants

where the sums run over all atoms and all directions.

The world of anharmonic interatomic forces is less studied than the harmonic one and is mainly centred on describing damped oscillations, or overtones [51, 73, 22]. The terms from Equation 15.22 are used in computing N-type processes and thus allow a first determination of the thermal conductivity (for applications for the Earth's deep mantle se, for example, [81, 126, 103]). They can also offer a first approximation to the bandwidth of the Raman and infrared peaks and correct the temperature-dependent phonons [201, 22, 247, 40].

Further terms not restricted by symmetry and the fourth-order derivatives are needed to obtain the full thermal conductivity. Several DFT packages have started to offer implementations of these terms, for example, quantum-espresso, abinit, castep, and vasp, though within some restrictions. However, their implementation within DFPT is a very tedious task, and a faster solution, for the time being, might be to use finite differences over dynamical matrices.

## 16

Molecular Dynamics

There are several major methods to sample the configurational space of a dynamical system. We have already seen the perturbation theory in Chapters 10–15, where the changes in energy that occur due to perturbations are related to the physical properties of lattices. This approach provides an elegant and highly efficient way to calculate, for example, the vibrational state of the system. An alternative to sampling the configurational space is the molecular dynamics (MD) method.

16.1 The Essentials of Molecular Dynamics

In MD, the configurational space is explored along a stochastic deterministic trajectory. The simulation starts at a point in the configurational space and from there, the atoms *evolve* over time, visiting their neighbouring configurations. The atoms move according to the Newtonian laws of mechanics under the action of interatomic forces. As such, MD calculations are very well suited for simulations of fluids. The quality of the computed interatomic forces is a major factor in defining the quality of the simulations. But the calculation of the interatomic forces is entirely independent of the MD simulation itself.

In classical MD, the forces between the atoms are determined using interatomic potentials. Because of the low computational cost of potentials, this method is highly efficient in studying large systems, which can go up to millions of atoms. Historically, this was the first type of MD method in use. Any atomic potential (see Chapter 2) can be used in MD simulations.

In *ab initio* molecular dynamics (AIMD), or first-principles molecular dynamics (FPMD), the forces between atoms are computed using first-principles methods. Density functional theory (DFT), of course, is the preferred method. While this method provides accurate details and thus an accurate description of the interactions between atoms, because of its high computational cost, it is limited to small system sizes, typically tens to a few hundred atoms.

In an ideal case, the atoms explore all the relevant configurational space over a finite amount of time. But in reality, things are a bit different. If the simulations are too short, then the configurational sampling is limited.

Sometimes, even if the simulations are long, the system can be trapped in a deep local minimum of the potential. In these cases, the system cannot leave that minimum. It needs some energetic help that would *push* the atoms away from that particular configuration.

A plethora of information stemming from the simulations can be analysed. First, various variables can be directly related to thermodynamic parameters, both intensive, like temperature and pressure, and extensive, like energy or volume of the simulation cell. Then the trajectories of the atoms contain a tremendous amount of information, which, once processed, offer an atomistic view of very complex processes. The geometric relations between the atoms reveal their chemical behaviour, and the time-dependent evolution of their trajectories yields the transport properties. However, MD remains a relatively straightforward technique to sample atomic environments at finite temperatures.

16.2　Atomic Displacements

Figures 16.1–16.7 summarise the different stages of an MD calculation. To illustrate it, we use a model molecule, similar to a passivated silica tetrahedron, like SiO_4H_4.

The simulations can start from any arbitrary atomic configuration, and the atoms do not necessarily lie at the bottom of their potential wells. Figure 16.1 shows the position $r_\alpha(t_i)$ of the atoms α, at an arbitrary step t_i.

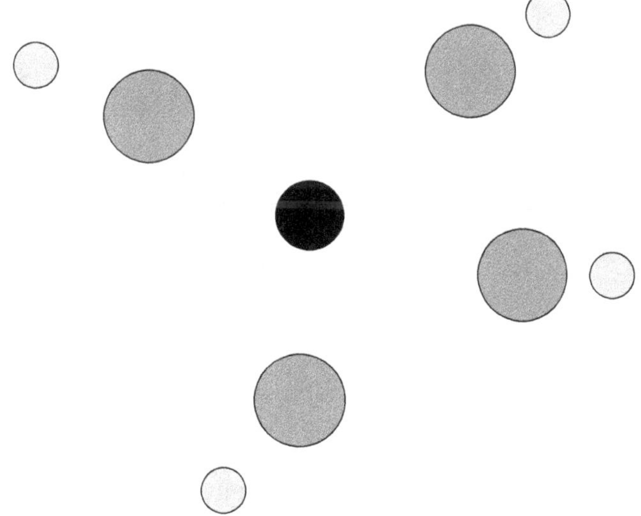

Fig. 16.1 The molecular dynamics simulations start with a collection of atoms placed in space. They can describe a periodic lattice or a molecule.

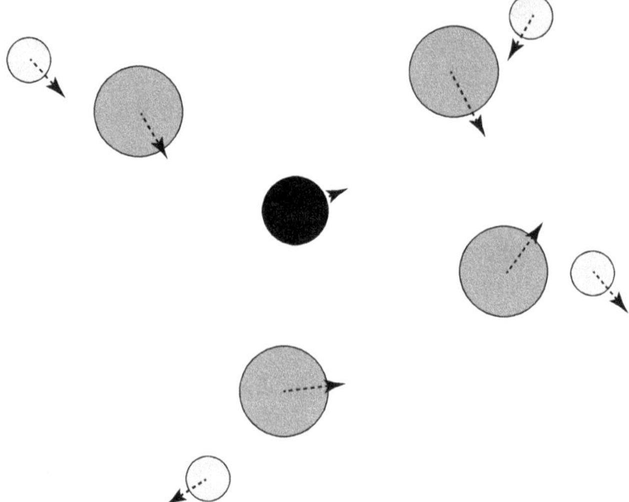

Fig. 16.2　As the simulations are done at finite temperature, the atoms have an initial velocity, represented by the dashed arrows. The average kinetic energy of the atoms defines the temperature of the system. The electrons are assumed to be adiabatic, that is, to follow and relax instantaneously around the nuclei. The electronic temperature is taken into account via the Fermi–Dirac distribution.

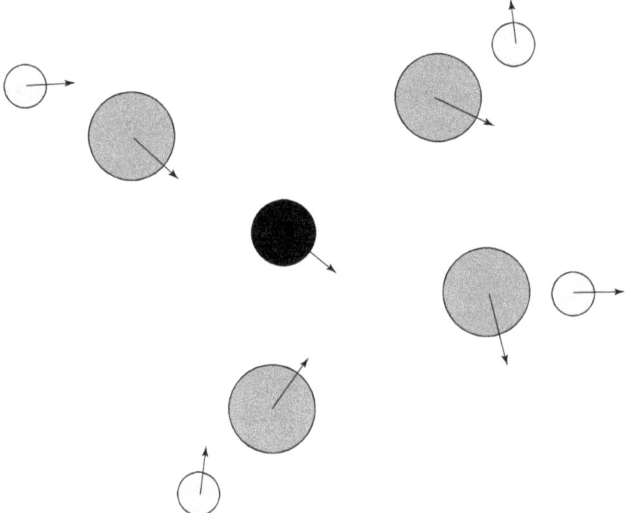

Fig. 16.3　As the atoms interact, they act with forces on each other, represented by the solid arrows. The resulting interatomic forces are the ones that rule the dynamics of the system. In ab initio simulations, the interatomic forces are computed as energy derivatives with respect to atomic displacements.

The MD simulations are realised at finite temperatures. Because of the temperature, the atoms are moving. Hence, at every step, they carry a specific kinetic energy. Figure 16.2 shows the velocity $v_\alpha(t_i)$ of the atoms at the same step t_i. The system temperature is measured by the kinetic energy of the nuclei. In the Born–Oppenheimer approximation, also called adiabatic approximation, we assume that the electrons follow the nuclei.

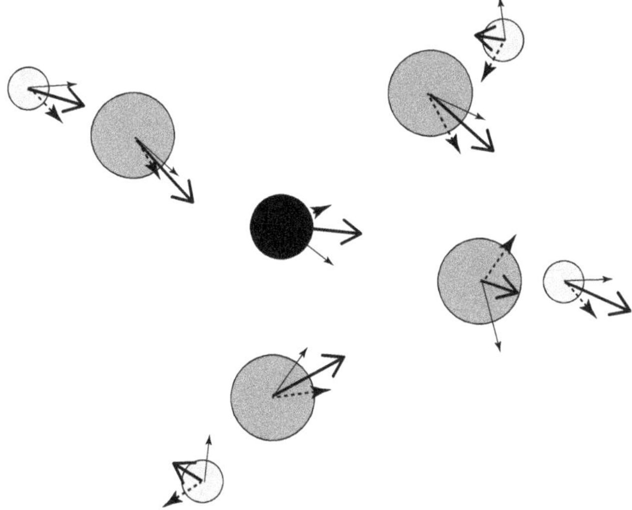

The initial positions, velocities (solid arrows), and interatomic forces (dashed arrows) are the ingredients necessary to write the Newtonian equations of motion.

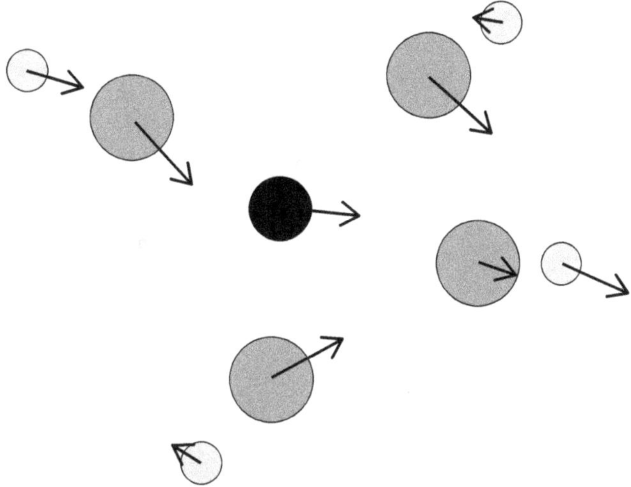

The equations of motion determine the trajectories of the atoms (thick solid arrows).

Thus, electronic relaxation is instantaneous. The electrons have a pure electronic temperature, which is ensured by the Fermi–Dirac distribution (Equation 4.8). Still, they do not carry a specific kinetic energy other than that defined by the electronic Hamiltonian.

After the atoms are in place, the interatomic forces are computed for all the atoms of the system (Figure 16.3). They can be calculated in various ways. What makes the whole interest of performing AIMD simulations is that the forces are highly accurate, as they are computed using ab initio methods, usually DFT. But this is also the most time-consuming step of the simulation.

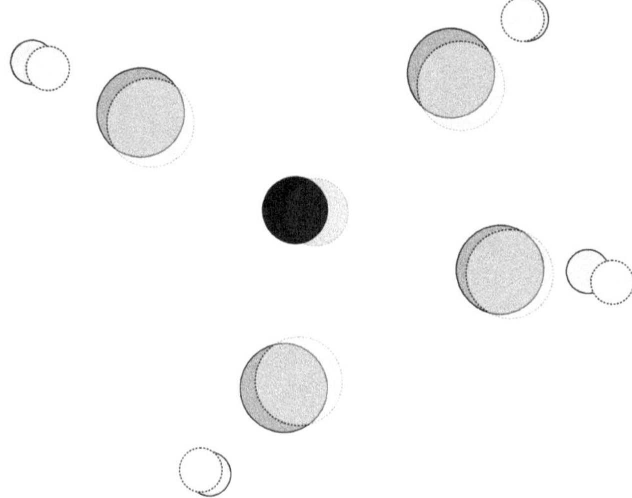

Fig. 16.6 The atoms are then displaced by a small and discrete amount along the computed trajectories.

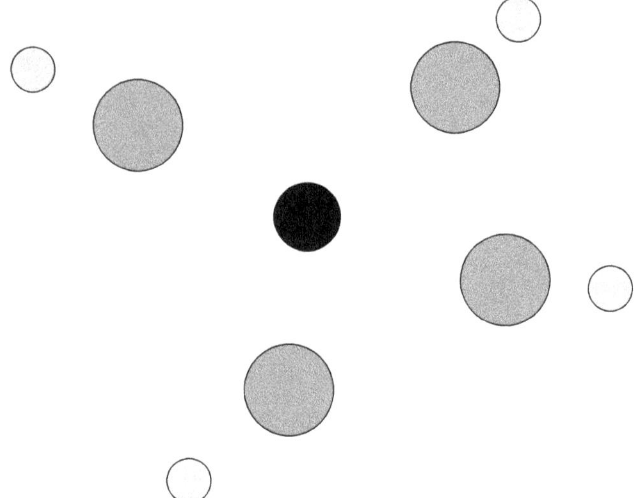

Fig. 16.7 In the new atomic configuration, a new MD step may begin. It is a matter of statistics to decide when a simulation has run enough time for the atoms to sample a representative part of the configurational space.

At this point, all the necessary ingredients to make a move are known (Figure 16.4): the positions and the velocity of the atoms and the interatomic forces. The Newtonian equations of motion for classical particles under the action of forces determine the stochastic behaviour of the system:

$$r_\alpha(t_{i+1}) = r_\alpha(t_i) + v_\alpha(t_i)\tau + a_\alpha(t_i)\frac{\tau^2}{2} \tag{16.1}$$

where τ defines the time step as:

$$t_{i+1} = t_i + \tau \tag{16.2}$$

The accelerations in Equation 16.1 are obtained from the interatomic forces, which are computed in DFT, and the atomic masses, which are known:

$$F_\alpha(t_0) = m_\alpha a_\alpha(t_0) \tag{16.3}$$

The atoms move according to Equation (Figure 16.5), with the moves determined by their interatomic forces and the previous state of the system. The moves are discrete, with time steps τ. During this time, τ, the atoms are assumed to move according to linear trajectories (Figure 16.6). This approximation holds if the time steps of the moves are small, that is, typically on the order of the femtosecond. The length of the time steps should allow for a complete description of the atomic vibrations whose durations are a few dozen femtoseconds or less.

Once the move is effectuated, the atoms have new positions in the configurational space (Figure 16.7), and the procedure can restart (Figure 16.1).

At the new positions, $r_\alpha(t_{i+1})$, the atoms have new initial velocities, also computed according to the Newtonian equations of motion:

$$v_\alpha(t_i + \tau) = v_\alpha(t_i) + a_\alpha(t_i)\tau \tag{16.4}$$

The interatomic forces are re-calculated, and a new MD step may begin.

16.3 Trajectories

One of the most reliable and widely used algorithms to determine atomic displacements is the Verlet algorithm [259]. It has the advantage that it does not specifically require the value of the velocity and makes use only of the consecutive atomic positions and the interatomic forces:

$$r_\alpha(t_i + \tau) = 2r_\alpha(t_i) - r_\alpha(t_i - \tau) + a_\alpha(t_i)\frac{\tau^2}{2} \tag{16.5}$$

However, the atomic positions are needed over several consecutive steps to estimate the next positions. The velocities, even if they do not participate directly to Equation 16.5, need to be calculated to estimate the temperature:

$$v(t_0) = \frac{r_\alpha(t_0 + \tau) - r_\alpha(t_0 - \tau)}{2\tau} \tag{16.6}$$

Several schemes have been proposed to reduce some numerical errors of the Verlet algorithm and provide more accurate and faster ways of computing the velocities. The LeapFrog algorithm [122] uses the positions and the velocities at the middle positions along the jumps to estimate the next positions and velocities. This algorithm is better for conserving the energy of the simulated system.

Because the equations of motion (Equation 16.1) are discrete, the atomic trajectories are made of a (long) succession of step movements. There is a delicate balance between the time step, the accuracy of the properties that need to be investigated, and the duration of the simulation. If the time steps are short, the trajectories and the physical properties are better described. Moreover, the DFT reaches convergence fast because the electronic density is not highly perturbed between consecutive steps. But the simulations need to contain more MD steps to ensure ergodicity.

On the other hand, longer time steps allow for longer trajectories to be sampled. This enhances ergodicity, as the system can sample more atomic configurations during a simulation. There can be some computing time loss at the DFT cycle level if the electronic density around the atoms is too perturbed. Also, too long time steps might lead to a loss of accuracy, as the simulation might miss some interatomic bonds with short lifetimes.

Finally, the value of the time step is also dependent on the mass of the atoms, as lighter atoms travel faster than heavier atoms. Consequently, the typical time step for AIMD simulations of silicate melts at Earth's mantle conditions, that is, 0–150 GPa and 3,000–5,000 K, is 1 fs. The time step can be longer at lower temperatures. The time step should be at least halved as soon as hydrogen is present in the system. The same is true for much higher temperatures. There is no absolute rule, but these values can be used as guidance.

16.4 The Simulation Box

The MD simulations can model finite systems, like molecules and clusters of atoms, and infinite systems, like liquids and gases. In the case of finite systems, the simulation contains all the atoms in the system, and there is no periodicity. For AIMD simulations, the codes that use localised basis sets are ideal for studying such systems.

For liquids and gases, the strategy of the simulations is different. These systems are approximated as crystals. The *unit cell* is replaced with a *simulation box*, which is periodically repeated along the three directions of the space. The periodicity means that when an atom exits the simulation box on one side, it enters the neighbouring box. This represents the essence of the periodic boundary conditions (Box 16.1). This way, the number of

Box 16.1   **How to set up the simulation box for molecular dynamics?**

Because fluids lack periodicity and the simulations have strong limitations in terms of the number of particles/atoms, we cannot directly simulate the structure of fluids. Instead, we need to use periodic approximants (Box Figure 5). They make the simulations tractable, but they impose artificial periodicity on the fluid structure:

Box Figure 5. Fluids are treated using periodic boundary conditions.

As always, you need to check the influence of the box size on the results you are looking for. Some of the processes investigated are highly size-dependent. Figure 2.2 is an excellent example of such a size-dependent process: The resulting texture of an aggregate of Ta crystallising from liquid changes drastically with the number of atoms in the periodic simulation box. Similarly, the large rings and long polymers revealed by Raman spectroscopy in silica glass [133] cannot be captured unless the simulation box is sufficiently large to contain these structures.

During the simulations, because of the periodic boundary conditions, when one atom exits the simulation box on one side, its periodic identical image enters from the opposite side. The white arrows in the example later show a few such atoms that traverse the cell boundaries (Box Figure 6):

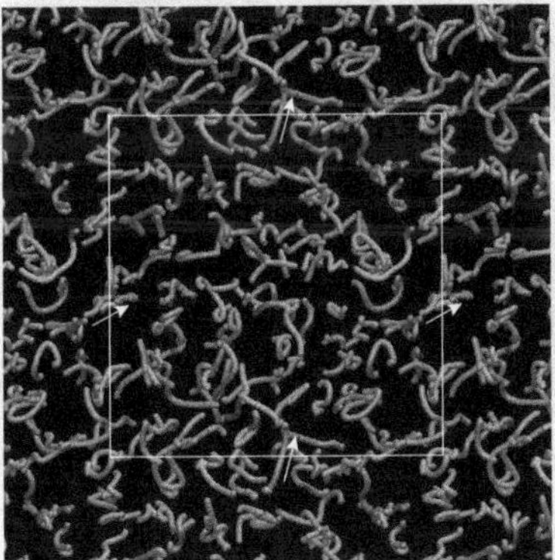

Box Figure 6. Periodicity ensures the preservation of the number of particles.

This procedure also preserves the number of particles in the system. By running long simulations, the configurational space can be sampled sufficiently well. In a way, this approach is similar to sampling different parts of the fluid. There are limitations related to the number of particles: Any structure containing more atoms than in the simulation box is not observed.

atoms in the simulation cell is preserved. Then the standard DFT codes with planewaves and pseudopotentials can work just as in the case of regular crystalline solids.

16.5 Ergodicity

Figure 16.8 shows examples of atomic trajectories obtained from an actual MD simulation of brucite, $Mg(OH)_2$ at high temperature. At 500 K, brucite is solid, so even a simulation short of only 500 fs samples about

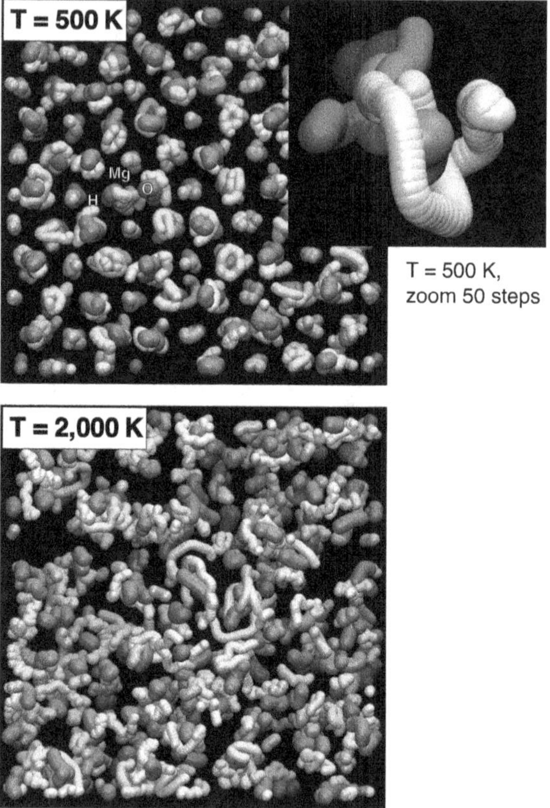

Fig. 16.8 The atomic trajectories in a brucite MD simulation. The upper panel shows the trajectories over 500 fs at T = 500 K. At this temperature, brucite is solid. During even a relatively short simulation, the atoms reproduce the thermal agitation, which is the sum of the atomic vibrations. The inset shows a zoom on the trajectory of only a few atoms over 50 steps, where you can follow their discrete displacements. The lower panel shows the trajectories over 50 fs, but at T = 2,000 K. At this temperature, brucite is liquid, and the structure loses any short-range ordering. The actual simulation needs to be considerably longer to ensure that all representative configurations are sampled.

two dozen OH stretching modes for each OH present in the simulation box, which is a representative part of the configurational space. At 2,000 K, brucite becomes liquid. At this temperature, the simulation must be longer to allow the atoms to sample all the relevant configurations in the liquid. This requirement defines the *ergodicity* of the simulation, that is, the quality of the configurational sampling. Of course, we can never be 100% sure that the configurational space has been sufficiently sampled. Comparison to experiment can help, as well as some other empirical criteria, which we will discuss later. Nevertheless, the limited ergodicity is why AIMD simulations have seen limited applications in the study of gases, in which the frequency of atomic interactions is considerably shorter than in the case of liquids.

It depends on the final purpose of the simulation – the ergodicity of the simulation scales with the length of the simulations. Ensuring good ergodicity for various magmas at mantle conditions would typically require tens or even hundreds of picoseconds. An alternative is to realise several shorter simulations, for example, on the order of 20–30 picoseconds, but starting with different independent configurations, all running at the same thermodynamic conditions. Eventually, the statistical description of the system can be obtained as the average between these independent simulations.

The ergodicity can also be improved by increasing the simulation box size. The simulation box for fluids should obviously be as large as possible. The limits are the number of bands a DFT code can solve on an available computer in a reasonable amount of time. Most of the simulations performed today are realised with simulation boxes containing up to a few hundred atoms or less than 2,000 electronic bands. For example, when studying liquid iron that preserves some remnant magnetisation, a typical simulation box contains less than 120 atoms. When studying molten silicates, this can go up to a few hundred atoms, while for simple oxides like MgO or H_2O simulations containing more than 500 atoms are standard.

16.6 Ensembles and Averaged Properties

There are always three thermodynamic parameters that are fixed in an FPMD simulation. These are the independent variables. These can be the number of atoms in the simulation box (N), the volume of the unit cell (V), and the temperature (T). In this case, the chemical potential μ, the pressure (P), and the internal energy (E) vary and become dependent variables. Alternatively, μ, V, and T are fixed, and N, P, and the Hill energy ($L = E - \sum \mu_i N_i$, where i covers all the chemistry in the system) vary.

Table 16.1 The different possible thermodynamic ensembles are obtained by fixing or controlling three thermodynamic parameters (after [147]). Another three parameters are then variable or dependent. The abbreviations correspond to the chemical potential μ_i, of all the chemical species i, the pressure (P), the volume of the simulation box (V), the temperature (T), the enthalpy ($H = E + PV$), the Hill energy ($L = E - \sum \mu_i N_i$), and the Ray enthalpy ($R = L + PV$). Each ensemble has a definite name. The name *canonical* in Latin means 'according to the rules', and was first used by Jacobi in 1837 to define coordinates respecting the Hamiltonian rule. For detailed discussions of each ensemble, see [147] and the references therein.

Controlled	Dependent	Name of the ensemble
NVE	μPT	Microcanonical
NVT	μPE	Canonical
NPH	μVT	Isoenthalpic-isobaric
NPT	μVH	Isothermal-isobaric
μVL	NPT	Grand-microcanonical
μVT	NPL	Grand-canonical
μPR	NVT	Grand-isothermal-isobaric
μPT	NVR	Generalised

The different combinations of fixed and variable thermodynamic parameters yield eight possibilities for the simulations. They are all covered in Table 16.1.

The control of the thermodynamic parameters is done in several ways. In the simulations where the number of particles is fixed, the periodic boundary conditions ensure that no atoms are *lost*. In practice, when one atom leaves the simulation box, its periodical image re-enters the box from the opposite side (see Box 8.1 and the discussion earlier).

The shape and volume of the simulation box are straightforward to fix. The temperature and pressure are controlled using thermostats and barostats, respectively. They are detailed in Section 16.7. Finally, when the chemical potential is controlled, the number of particles can vary to adjust that potential.

16.7 Thermostats and Barostats

Many of the first MD simulations were realised in the micro-canonical ensemble (the NVE ensemble) in which the volume and the energy of the system were conserved, while the temperature and the pressure were obtained from the simulation [3, 250, 47]. From a purely statistical point of view, they ensure the best ergodicity and the correct statistical distribution of the atomistic system. However, the exploration of large phase spaces soon made clear that specific properties were more accessible and faster described if the system temperature or pressure could be set rather than the energy.

Thus, much effort was put into finding efficient schemes for controlling the temperature and the pressures. Today, these are the most widely used controls in MD simulations.

The temperature is controlled with thermostats (see [147] for an excellent discussion about thermostats). The thermostat regulates the kinetic energy of the atoms. If the temperature is too low, it pumps kinetic energy into the system by increasing the velocity of the atoms. If the temperature is too high, it pumps kinetic energy out of the system by decreasing the velocity of the atoms. The thermostat acts on all the atomic velocities or forces. But each atom will have a different velocity, though their distribution is described by a Gaussian centred on the value of the corresponding target temperature. A good thermostat needs not only to preserve the temperature of the system but also to ensure the ergodicity of the system.

The temperature is expressed via the atomic velocities, v_i, as:

$$K = \frac{1}{2}\sum_{i=1}^{N} m_i v_i^2 = \frac{3}{2}N_A k_B T \tag{16.7}$$

Here k_B is the Boltzmann constant, N is the number of atoms, N_A is Avogadro's number, m_i is the mass of the atoms i, and T is the temperature.

A straightforward thermostat is the rescaling of all the atomic velocities with a factor λ, which is defined as [47]:

$$\lambda = \sqrt{T_{target}/T} \tag{16.8}$$

In the Barendsen thermostat [29], the system temperature is readjusted to asymptotically reach the target temperature after a periodic interval. This interval is a parameter set by the user. In the Andersen thermostat [6], the adjustment of the atomic velocities is made at every step. Consequently, the resulting simulation is also called iso-kinetical because the average kinetic energy of the atomic system is always fixed to the same temperature. The iso-kinetical ensemble provides the same numerical results as the proper NVT ensemble [66, 135], though from a purely statistical point of view, the description of the time-dependent properties, like diffusion, is not correct. Consequently, using a proper thermostat, like the Barendsen thermostat, is safer and more accurate.

The alternative to velocity rescaling is to alter the Newtonian motion equations directly. Instead of considering the interatomic force as mass times acceleration, various damping and friction terms are added to equations of motion. The basic simplified form for the resulting force acting on atom i is:

$$F_i(t) = F_i^{DFT} - \gamma(t)m_i v_i(t) \tag{16.9}$$

The coefficient γ is invariant for all the atoms and is the equivalent of a friction coefficient. This formulation is a simplification of the Langevin dynamics but has the advantage of requiring only one parameter [212].

The Nose–Hoover thermostat is probably the widest in use today [144, 193]. It puts in contact the atomistic system with an external thermal bath. The coupling between the atoms and the thermal bath is realised via a mass-like parameter, which defines the thermostat. This is the actual numerical parameter that needs to be set when performing the MD simulation. The additional term appears in the definition of the Hamiltonian (the total mechanical energy), which in this case becomes:

$$H^{Nose} = E_{kin} + E_{pot} + \frac{p^2}{2Q} + (g+1)k_B T ln(s) \qquad (16.10)$$

In this formulation, Q is the mass of the thermostat, g is the number of degrees of freedom, p is the momentum of the thermostat, and s is the *position* of the thermostat. The explicit presence of the p and s variables is necessary to ensure the coupling. In practice, the atomic velocities are scaled using the p and the s variables [158]. But in practice, only Q, the mass of the thermostat, is needed, the MD code setting up the other variables. This is one of the most important parameters of the MD simulation, and its setting determines if the dynamics is stable or divergent. Box 16.2 exemplifies setting up the mass of the Nose–Hoover thermostat. In general, the Q value should ensure that the energy fluctuations are on the same order of magnitude as the mean frequency of the atomic vibrations.

Pressure is controlled via barostats. As for thermostats, there are several formulations, each with its characteristics, but all follow the same idea.

The Barendsen barostat [29] follows the idea of the thermostat proposed in the same work. The pressure is readjusted to the target value, and the volume is rescaled accordingly. A thermal compressibility is used, whose value comes either from experiments, if known, or it is approximated. The volume scaling can be done either homogeneously or by assuming different distortions along different strain directions.

The Andersen barostat [6] couples the system with an external variable, Q, which is the volume of the system. This variable intervenes in the kinetic and potential energy, rescaling the volume of the system and the positions of the particles.

One of the most used barostats is the Parinello–Rahman [198, 199] and its further generalisation [263]. The idea is the same: to use a mass for the cell as a factor for the barostat to rescale the simulation box. For liquids, the rescaling is homogeneous, just a simple dilation or contraction, but for solids, this is done using a proper tensor.

The value of the thermostat should be such that there is an efficient coupling between the atomic system and the external thermal bath, whichever form you choose it. This is realised when the periodicity of the atomic vibrations is in phase with the oscillations of the exchange with the thermal bath. Because there are different vibrations with different frequencies and wavelengths, it is impossible to obtain a perfect in-phase coupling. But a few tens of picoseconds of thermal fluctuation should generally be fine. Box Figure 7 shows the temperature fluctuations in a liquid feldspar for several values of the mass of the Nose–Hoover thermostat [159]. The calculations were realised with VASP v5.

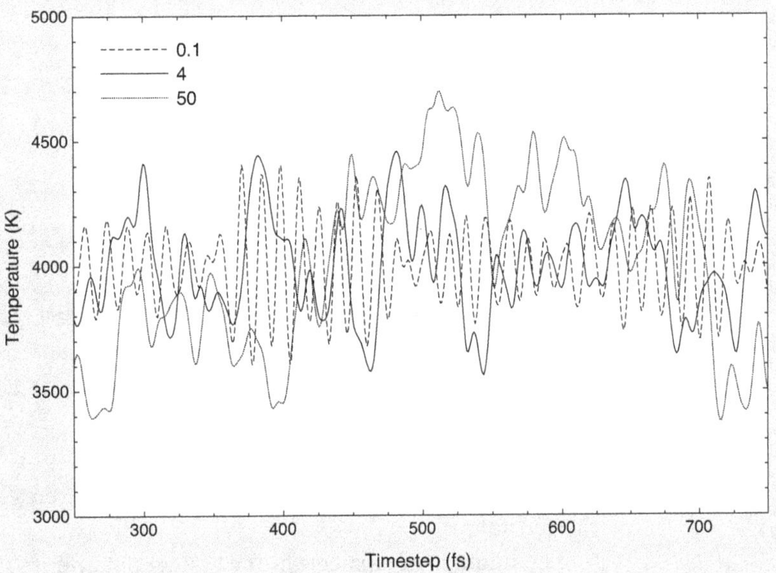

Box Figure 7. Temperature fluctuations depending on the number of particles and the mass of the thermostat.

The standard deviation of the thermal fluctuations is the same (see Section 16.8), but the periodicity of the fluctuations changes. A mass thermostat in the range of 1–10 gives a good coupling and thermal description of this system.

16.8 Other Practical Aspects of the Machinery Behind Molecular Dynamics Simulations

Figure 16.9 shows the temperature fluctuations along a $\tilde{2}7$ ps long simulation of molten albite, $NaAlSi_3O_8$ [159]. The simulation had 16 molecules of albite in a cubic cell, that is, 208 atoms in total. The temperature fluctu-

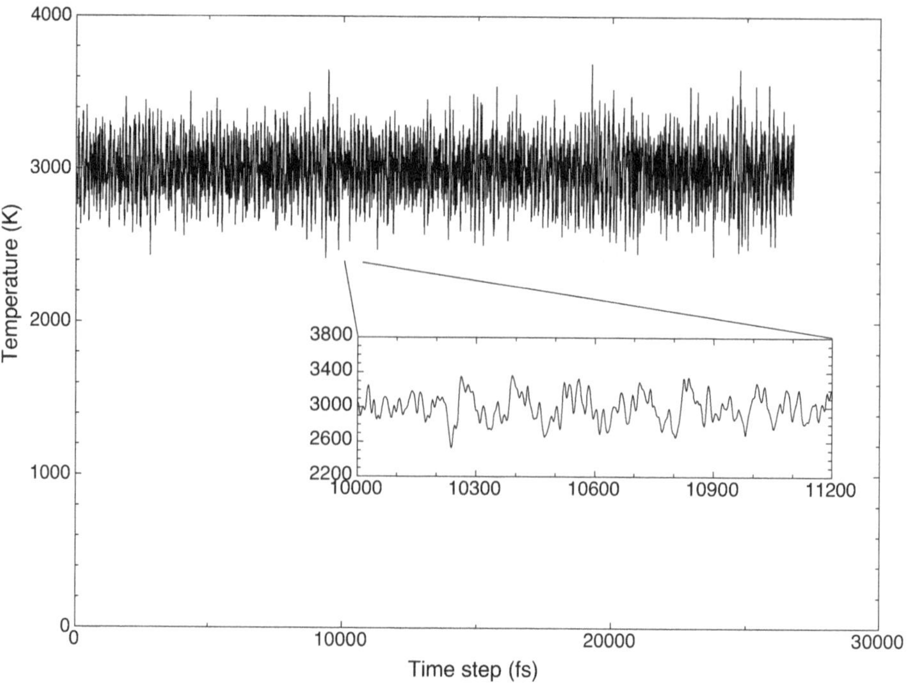

Fig. 16.9 Temperature fluctuations along an MD simulation of Na-feldspar, $NaAlSi_3O_8$ at 3,000 K. Note the length of the simulations, which allows for collecting very good statistics for the different structural and thermodynamic properties of the melt (courtesy of A. Kobsch, after [159]).

ates within a few hundred degrees around the average value, fixed by the thermostat.

The number of particles in the system determines the amplitude of the fluctuations in an NVT simulation. This comes from the Gaussian distribution of the atomic velocities [5, 147] and the size effects. The standard deviation of the temperature fluctuations is given by:

$$\sigma_T^2 = <T_i^2> - <T_i>^2 = 2\frac{T^2}{g} \tag{16.11}$$

The brackets in the second relation indicate the average over the temperature values at all steps i. This relation shows just one other measure of the importance of the size in MD simulations. Increasing the number of particles decreases the temperature fluctuations, as these are damped by the long-range interactions between independent particles. For small cell sizes, the fluctuations can quickly become extremely large. Figure 16.10 illustrates the effect of cell size on the amplitude of the temperature oscillations for an MD simulation of MgO at 5,000 K [48]. The amplitude decreases from about 1,000 K for a cell with 32 formula units to about 500 K for a cell with 108 formula units and down to only about 325 K for the largest cell with 256 formula units.

The internal energy variations along the MD simulation of albite [159] are shown in Figure 16.11. If the thermostat is well set and the simulation

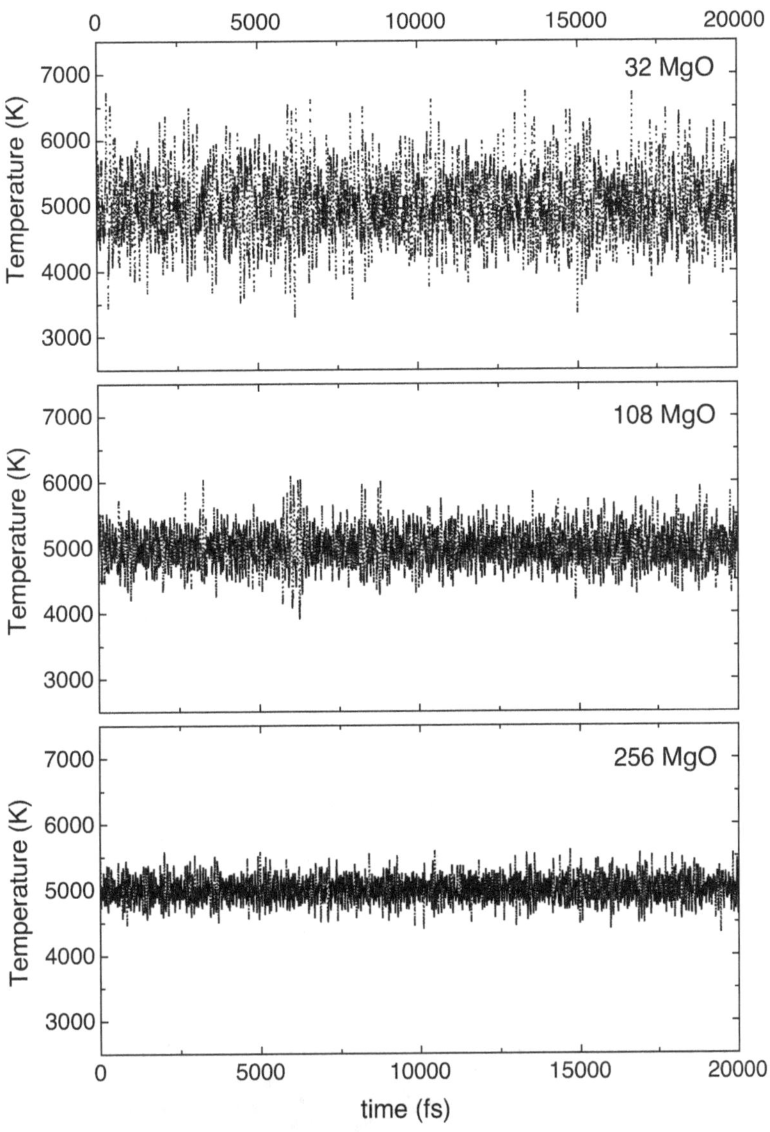

Fig. 16.10   Size dependence of the temperature fluctuations for an MD simulation of liquid MgO at 5,000 K and 2.63 g/cm^3 (courtesy of Tim Bögels; after [48]).

is well thermalised, there should be no energy drift throughout the simulation. Figure 16.12 shows the size dependence of the energy fluctuations for liquid MgO at 5,000 K [48]. At 32 MgO formula units, the amplitude is about 0.6 eV/molecule, decreasing to about 0.3 eV/molecule for the system with 108 MgO and about 0.15 eV/molecule for the system with 256 MgO.

The fluctuations of the energy in an NVT simulation can be related to the heat capacity, c_v, as [5]:

$$\sigma_E^2 = <E_i^2> - <E_i>^2 = k_B T^2 c_v \tag{16.12}$$

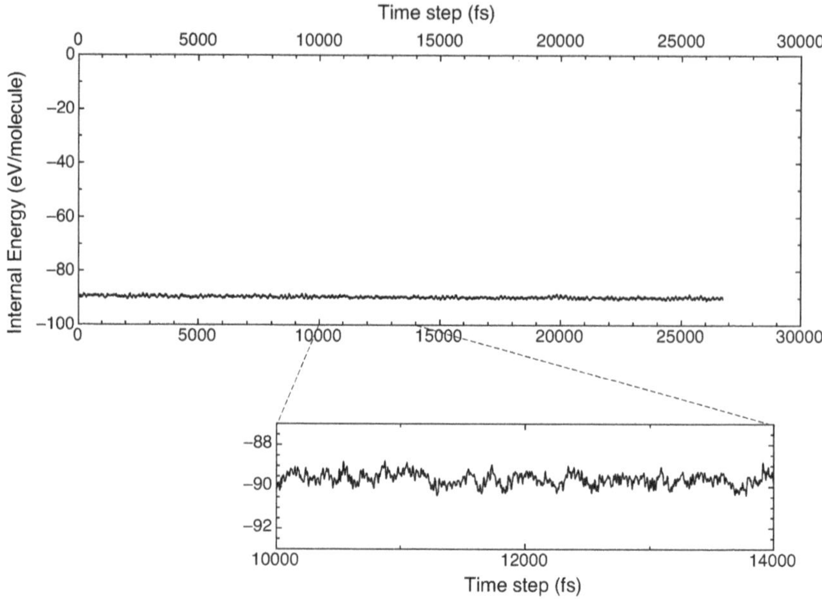

Fig. 16.11 Energy fluctuations along the same MD simulation of Na-feldspar, $NaAlSi_3O_8$ at 3,000 K, as in Figure 16.9 (courtesy of A. Kobsch, after [159]).

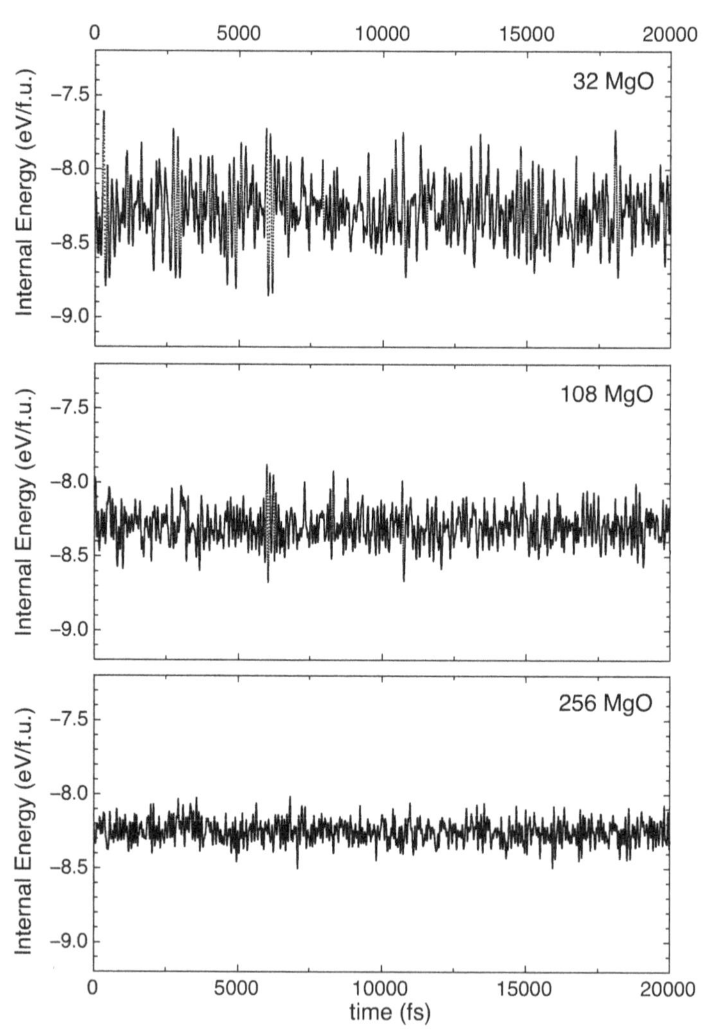

Fig. 16.12 Size dependence of the energy fluctuations for an MD simulation of liquid MgO at 5,000 K and 2.63 g/cm^3. (courtesy of Tim Bögels; after [48]).

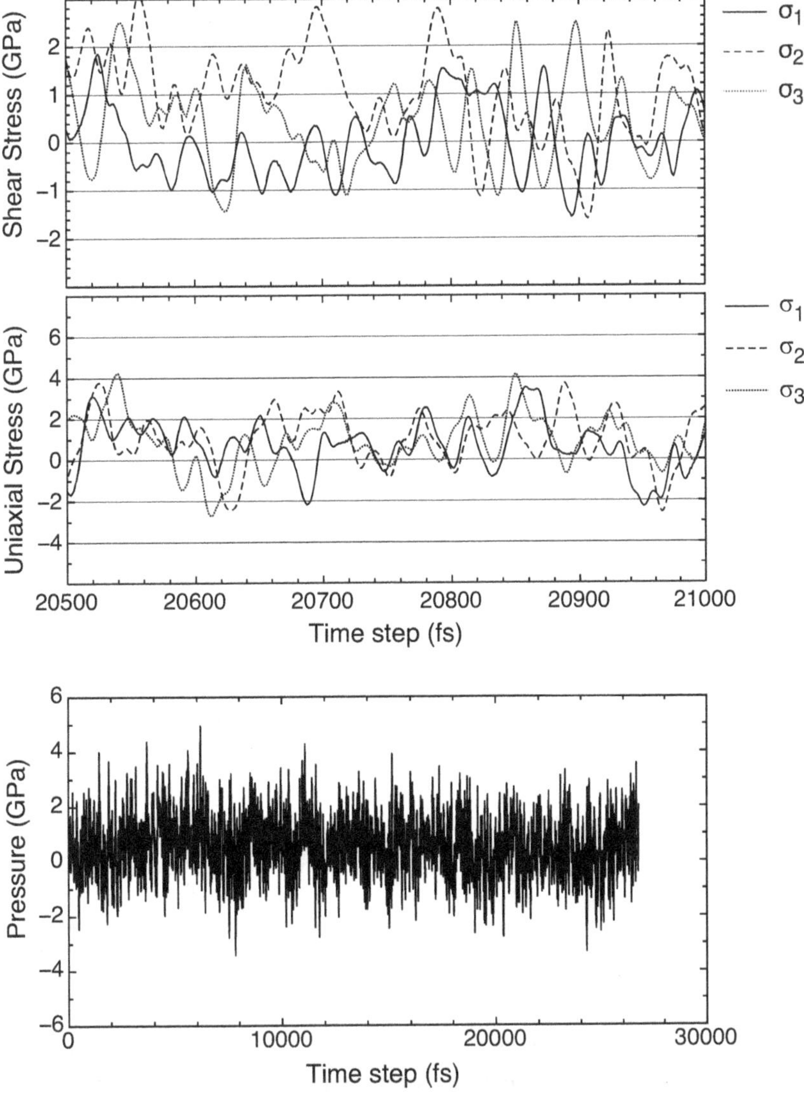

Fig. 16.13   Stress fluctuations during the MD simulations of molten albite, $NaAlSi_2O_8$. The two upper panels show the variations of the stress components during 500 MD steps. The lower panel shows the hydrostatic pressure variation throughout the entire simulation (courtesy of A. Kobsch, after Ref. [159]).

The brackets in the second relation indicate the average over the energy values at all steps i. It is advisable to have large cell sizes and long simulations performed at conditions far from structural transitions to obtain reliable values for the heat capacity.

Finally, Figure 16.13 shows the fluctuations of all six components of the stress tensor along the same MD simulation of molten Na-feldspar [159]. The three components of the uniaxial stresses show similar fluctuations around a similar average value of about half a gigapascal. This is expected for a well-homogenised fluid. The three off-diagonal components fluctuate

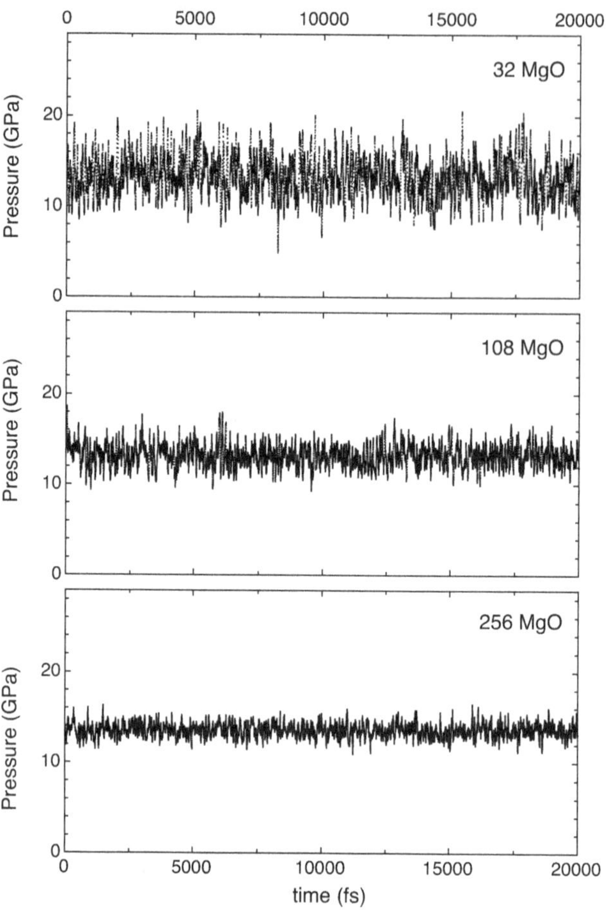

Fig. 16.14 Size dependence of the total pressure fluctuations for an MD simulation of liquid MgO at 5,000 K and 2.63 g/cm^3 (courtesy of Tim Bögels; after [48]).

around zero, with a slight tendency for positive values. The two upper panels show the stress tensor fluctuations over half a picosecond, that is, 500 MD steps. When the entire simulation is analysed, the average of the off-diagonal components is zero. The trace of the stress tensor, which gives the hydrostatic pressure, is shown in the lower panel. Figure 16.14 shows the effect of the system size on the pressure fluctuations for the same liquid MgO. The amplitude of the fluctuations decreases from about 6 GPa for 32 MgO to 3 GPa for 108 MgO, and less than 2 in the system with 256 MgO formula units.

In the NPT ensemble, the fluctuations of the volume can be related to the isothermal compressibility, β_T, at that specific temperature [5, 147]:

$$\sigma_V^2 = <V_i^2> - <V_i>^2 = Vk_BT\beta_T \tag{16.13}$$

The fluctuations of the enthalpy, $H = E + PV$, are related to the isobaric heat capacity:

$$\sigma_H^2 = <H_i^2> - <H_i>^2 = k_BT^2c_P \tag{16.14}$$

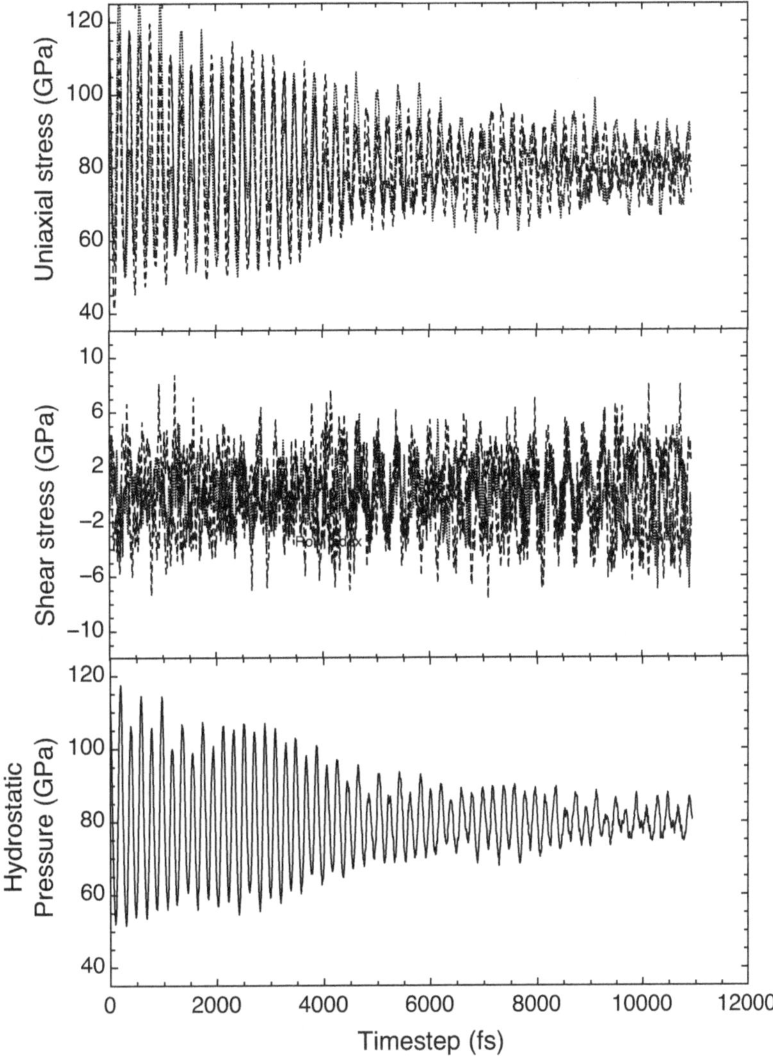

Fig. 16.15   Fluctuations of the six components of the stress tensor (same symbols as in Figure 16.13) and the resulting hydrostatic pressure of bridgmanite, $MgSiO_3$ at 2,000 K. The equilibrium pressure is 80 GPa. It is reached after a long equilibration and thermalisation period of more than 9,000 steps, that is, 9 ps.

Figures 16.15–16.17 show respectively the fluctuations of the stresses and pressure, of the energy and temperature, and of the unit cell parameters along an NPT simulation of bridgmanite, $MgSiO_3$, at 2,000 K and 80 GPa. Under these conditions, bridgmanite is solid. The initial configuration is almost at a 20 GPa higher pressure (about 100 GPa) than the target value of the pressure (80 GPa) (Figure 16.15). Consequently, imposing such a different target hydrostatic pressure necessitates a long time of thermalisation and equilibration. In this case, the system needed about 9,000 steps to equilibrate (Figure 16.16). The three unit cell parameters tend asymptotically to the average values characteristic at those P

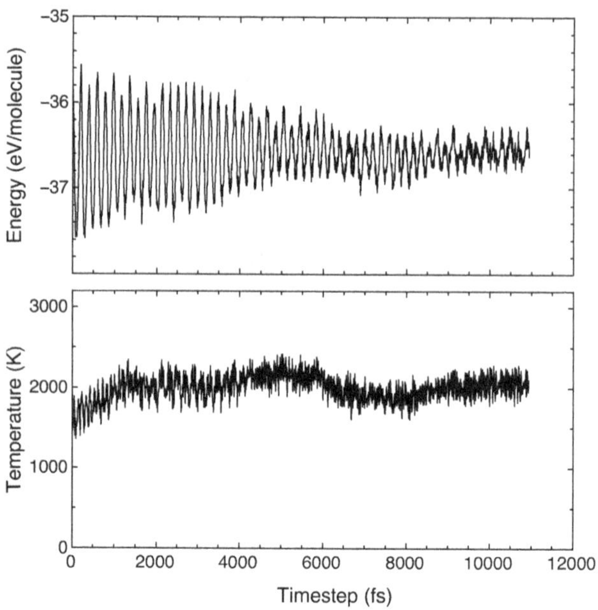

Fig. 16.16   Instantaneous energy and temperature along the NPT simulation of bridgmanite, MgSiO$_3$ at 2,000 K and 80 GPa. The thermalisation period lasts about 9,000 steps, that is, 9 ps.

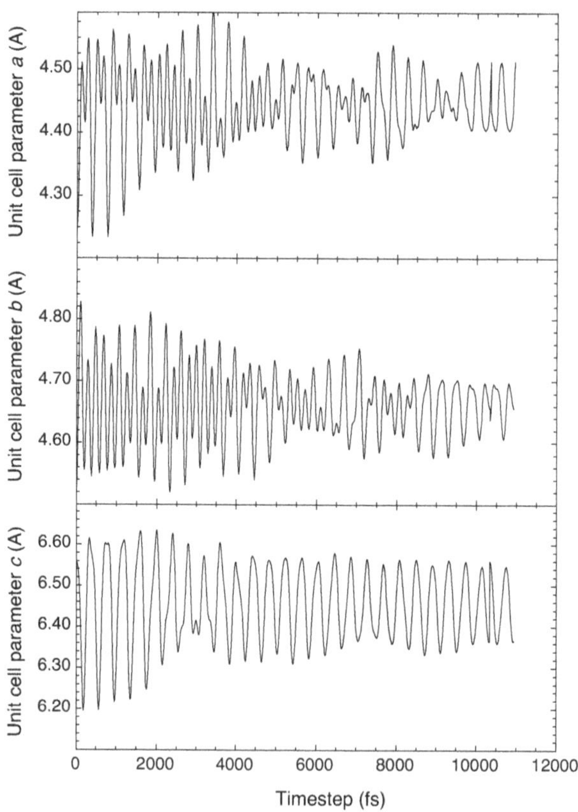

Fig. 16.17   Variation of the unit cell parameters of the orthorhombic structure of bridgmanite, MgSiO$_3$ at 2,000 K and 80 GPa. After the long equilibration and thermalisation period, the oscillations stabilise around equilibrium values. Such calculations, performed on a grid of pressure and temperature points, can be used to compute the thermal dilatation of a system. Beware, however, of their heavy computational toll.

and T conditions (Figure 16.17). The equilibration period is characterised by large fluctuations of all structural and thermodynamic variables. After about 9,000 steps, the system starts to get stabilised. In the last two picoseconds of the simulation, all the parameters of bridgmanite exhibit the usual oscillation behaviour around equilibrium values. This NPT simulation exemplifies a straightforward way to obtain the thermal dilatation tensor for any given solid, though it is highly demanding computationally.

Analysis of the Molecular Dynamics Simulations

17.1 General Considerations

An enormous amount of information gets accumulated by the end of each molecular dynamics (MD) simulation. Besides the average values of the energy, pressure, and temperature, the MD simulations contain the atomic trajectories and the relative relations between all the atoms. This is a real trove of information that requires specific analysis.

There are several options and packages for performing MD post-processing, each with its advantages and limitations. Some of the visualisation packages, like VMD [148], Ovito [270] or Travis [43], perform some analysis, like calculating interatomic bonding and diffusion. For the present purpose, we will employ the UMD post-processing tool [67] (Figure 17.1), which is available as a unified collection of open-source Python scripts on the GitHub platform (https://github.com/rcaracas/UMD_package). Further dedicated analysis scripts can be built using the atomic simulation environment (ASE), a large collection of tools released as a library written in Python [166].

To use the UMD post-processing tool, the output of the *ab initio* MD simulation must first be translated into a specific ASCII format, which is inspired by the XML/CML mark-up format rules. The resulting UMD files start with a header containing all the information kept fixed throughout the simulation, like the number and type of atoms, their masses, and so on. Then for each time step of the simulation, the file summarises the thermodynamics and structural state of the system, including the position and dynamical characteristics of the atoms (velocities, forces, etc.). The analysis is done at the command line.

17.2 Pair Distribution Function

Similar to defining the atomic charges in various ways (see Section 4.3), defining the chemical bonds between the atoms involves a certain amount of arbitrariness. The electronic density can be used as a criterion. All the atoms inside a given iso-density surface are considered bonded; they can

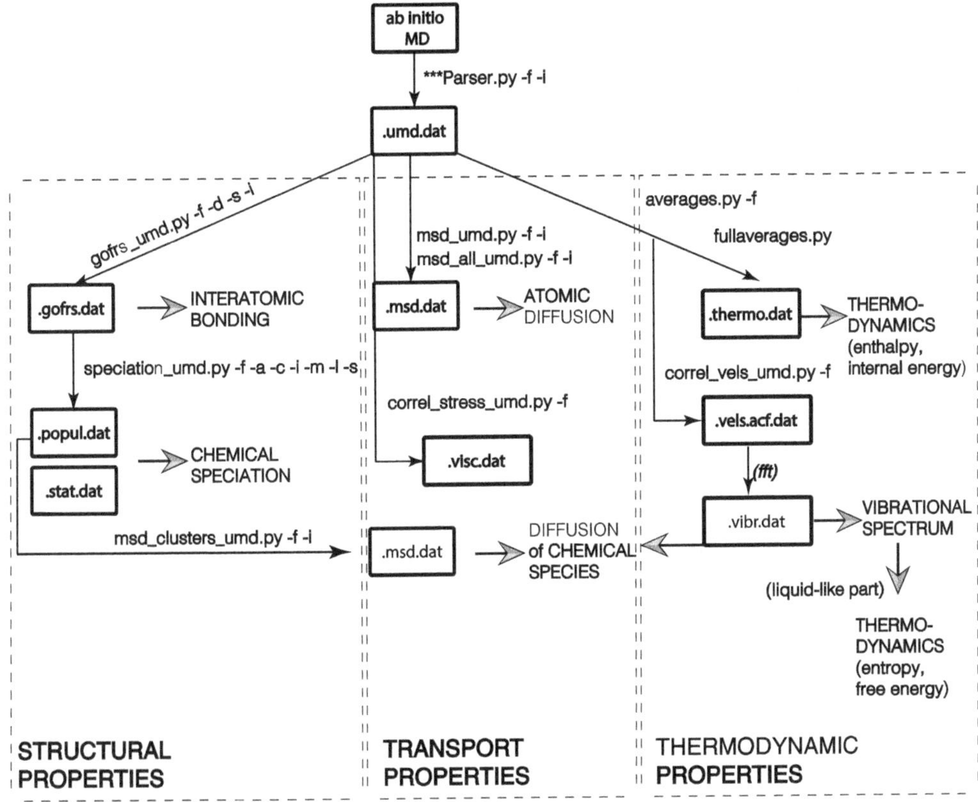

Fig. 17.1 The post-processing of *ab initio* molecular dynamics simulations is just as important as the simulation itself. While there are several software packages on the market, the UMD code is a highly versatile open-source software (https://github.com/rcaracas/UMD_package) that allows for the extraction of a large suite of properties, including structural, transport, and thermodynamic. The general flowchart of the package is presented here. Each Python script (*.py) is run at the command line in a terminal. Specific flags select between the various available options of each script. (General scheme after [67]).

even form a branching polymer. This approach has a valid physical meaning, but the electronic density value is arbitrary and depends on the atoms involved in the bonding. Moreover, this approach is time-consuming, requiring analysing the electronic density at each time step.

The alternative is to restrict the analysis to a purely geometrical one, where only the distances between pairs of atoms are computed and analysed statistically. This is a straightforward approach, which requires only the position of atoms. The analysis is based on the pair distribution function (PDF), also called the radial distribution function (RDF), which represents a counter of the number of neighbouring ligands around a central atom as a function of radial distance. The counting is averaged over all atoms of the ligand type and all atoms of the central type and normalised to the crystal density. For an MD simulation, the average is also done over the entire simulation duration. The PDF as a function of the distance r is abbreviated as $g(r)$ and calculated as follows:

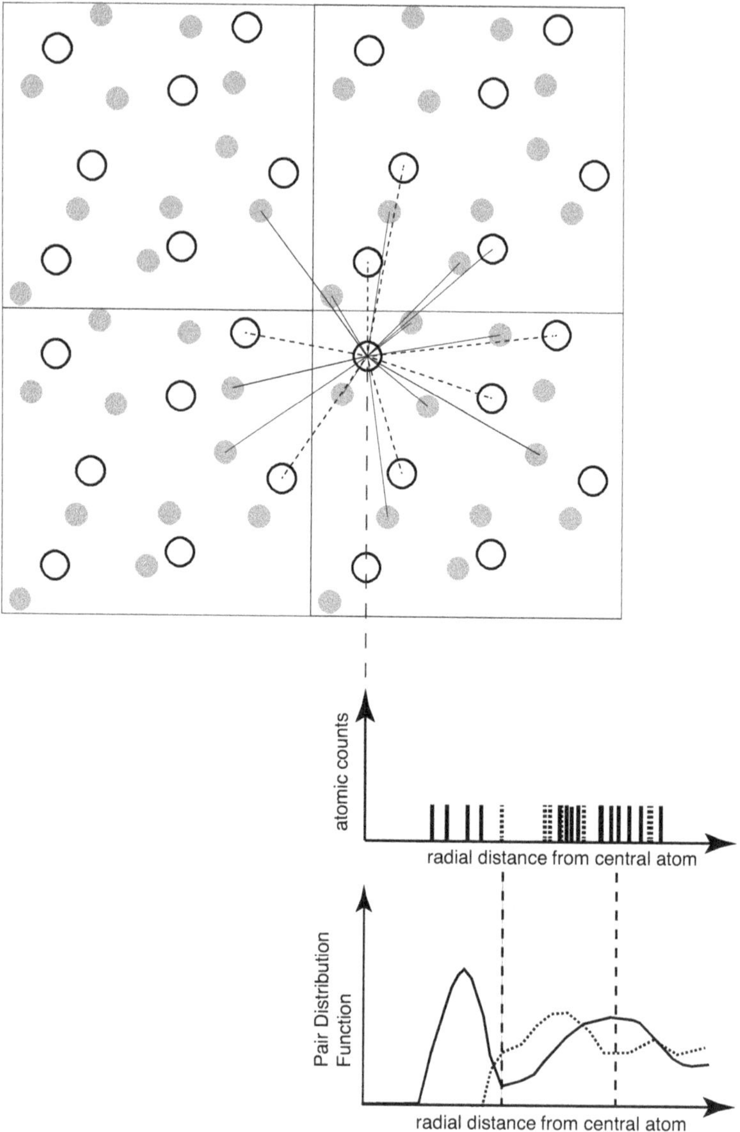

Fig. 17.2 The construction of the pair distribution function, $g_{AB}(r)$, between atoms A and B, or the radial distribution function, necessitates the counting of all the interatomic distances between all the atoms A and B. The counting is weighed over atomic density and is integrated over the entire duration of the molecular dynamics simulation. Graphical scheme after [67].

$$g_{AB}(r) = \frac{N_A - 1}{4\pi r^2} < \delta(r - r_{AB}) >_B \qquad (17.1)$$

In Equation 17.1, A is the type of central atom and B is the type of ligand. The average runs over all atoms of type A and all atoms of type B. The delta function $\delta(r - r_{AB})$ is equal to 1 if the two atoms (central and ligand) lie at a distance between r and $r + dr$, and 0 otherwise. The $1/4\pi r^2$ prefactor is for normalisation.

Figure 17.2 illustrates the steps taken for computing the PDF. The structure of the fluid is approximated with periodically repeated simulation boxes. All the distances between all the atoms of types A and B are calculated. The periodic images of the B atoms from the neighbouring cells are also considered. However, the calculation of the PDFs is restricted to a distance equivalent to half the size of the cubic simulation box to avoid multiple counting of the periodic images. This is the equivalent of computing the geometric relations between all the atoms of one unit cell.

The atoms A and B can be of any of the atomic types present in the simulation, including the same atomic type. The sum of the Fourier transformations of all the $g_{AB}(r)$ of all the A, B pairs gives the structure factor, which is directly related to the powder diffraction of the structure. To obtain a meaningful diffraction pattern for fluids, the $g(r)$ functions need to be computed up to large r values to include the long-range patterns that might be present in the fluid; otherwise, the quality of the theoretical predictions is very poor. But large r values require large simulation cells, which quickly imply untractable simulations. Consequently, this analysis, even if straightforward, should be employed with care.

For long MD simulations, the PDFs are smooth as more configurational space is sampled. The quality of the PDFs also increases with the number of atoms of each type, A and B, present in the simulation.

The PDFs always start at 0 at the origin, as the atoms A and B cannot be superposed. The PDFs remain at 0 up to a radius that defines the interatomic exclusion zone. Then the PDFs start to increase as more and more atoms B can be found at that distance around atoms A. The increase culminates with a maximum corresponding to the most likely A-B bond distance. The peak of the PDFs represents the bond distance with the highest probability but is usually wrongly associated with the bond distance.

The PDFs then start to decrease until they reach the first minimum. This first minimum corresponds to the radius of the first coordination sphere. Then as the distance r_{AB} increases, the pattern repeats itself with a second maximum, followed by a second minimum. They correspond, respectively, to the most likely bond distance in the second coordination sphere and the radius of the second coordination sphere.

The PDF values tend to 1 at large distances, corresponding to a 100% probability of finding an atom of type B at any large-enough distance r from an atom of type A.

The integral of the PDFs up to the radius of the first coordination sphere corresponds to the coordination number. The PDFs exhibit very narrow peaks for solids, which are δ functions in the limit of static structure at 0 K where atoms do not move. They are straightforward to interpret and transform in terms of coordination numbers. For this reason, a coordination number is always an integer number in solids.

For liquids, the peaks of the PDFs are smoothed out because of the variation in atomic environments. For molecular fluids, like uncompressed water or CO_2, the first coordination sphere of O by H in water and C by

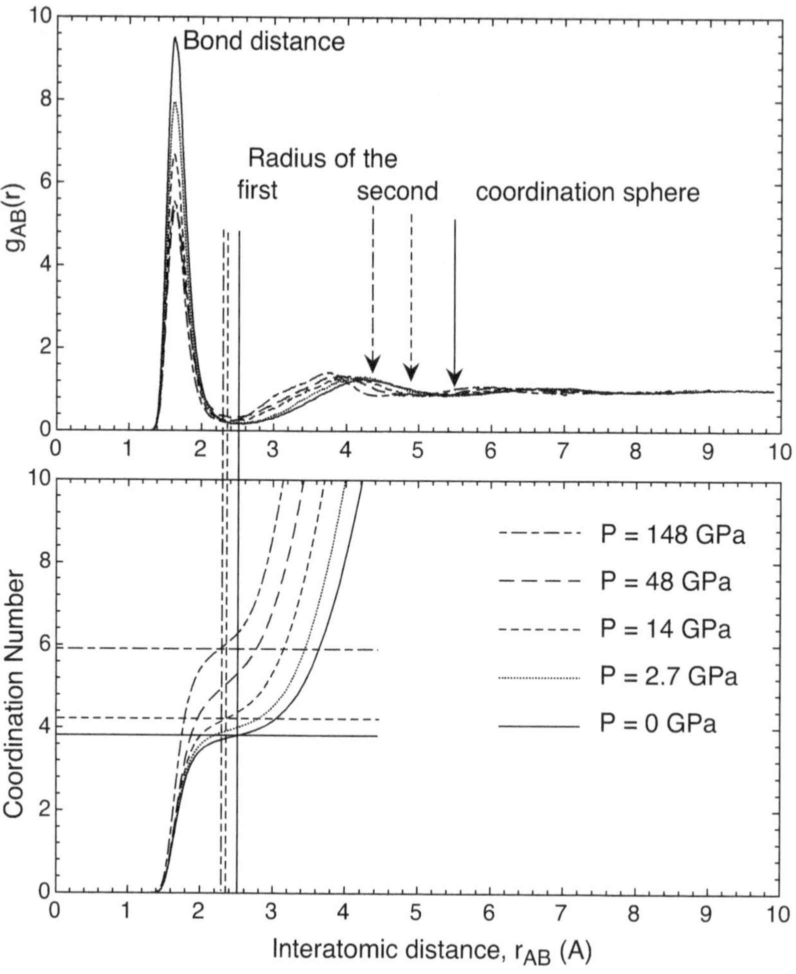

Fig. 17.3 The radial distribution functions, $g_{SiO}(r)$, calculated in MD simulations of pyrolite along the 4,000 K isotherm at several pressure points (upper panel) [227]. Thanks to the good statistics over long simulation times and many Si and O atoms, the PDFs are smooth. The local maxima (peaks) and minima of the PDFs correspond to the most likely bond distances and the radii of the coordination spheres. The integral of the PDFs gives the coordination of atoms A by atoms B (lower panel). The integral up to the first minimum corresponds to the coordination number.

O in liquid CO_2 provide coordination numbers very close to the nominal 2. For non-molecular fluids, the coordination numbers obtained from the integral of the PDFs may show deviations from the nominal values and are generally fractional numbers. The coordination number of Si by O in a pyrolitic melt at low pressure and low temperature is about 3.8, which is close to but not exactly 4, as SiO_4 tetrahedra dominate the melt. The deviation from the ideal SiO_4 comes from SiO_3 groups present in minor amounts in the melt. At higher pressures, the structure becomes more compact, and the coordination number increases to about six or even more at conditions close to the Earth's core-mantle boundary. Figure 17.3 shows the $g_{Si-O}(r)$ functions in pyrolite at several pressures and their corresponding integrals that provide the coordination numbers [227]. The transition from about 4 to about 6 is gradual with non-integral values at intermediate pressures and is common in most silicate melts [235, 161, 89, 66, 229].

17.3 Chemical Speciation

The radii of the first coordination sphere can be used to estimate coordination and polymerisation. Two atoms are considered bonded if the distance that separates them is smaller than the radius of the first coordination sphere, that is, the first minimum of the PDF. This is a purely geometric bonding approach, as stated at the beginning of this chapter. With this convention, a connectivity matrix is built, whose entries are the bond lengths between the atoms in the unit cell. This matrix can be transformed into a sparse matrix, whose entries are 1 if the distance between the two atoms is shorter than the radius of the first coordination sphere, that is, the atoms are bonded, or 0 otherwise. The coordination number obtained from the PDF is the weighted average of all the coordination polyhedra (Table 17.1).

Connecting all the atoms that are thus *bonded* results in several polymers, as in Figure 17.4. In natural silicate melts, by analogy with glasses, we can define a network of coordination polyhedra of silica, SiO_x, and of alumina, AlO_y, which form the backbone of the melt. In this case, Si and Al are the network-forming cations. The other cations, like Na, Ca, or Mg, diffuse in between this backbone and tend to break the polymer(s). These cations are called network modifier. Several short polymers can coexist

Table 17.1 Statistical analysis of the coordination polyhedra around Si and Al. The average coordination polyhedra are $AlO_{4.13}$ and $SiO_{4.02}$.

Polyhedron	Population (%)
AlO_2	2.5E-02
AlO_3	6.4
AlO_4	8.4
AlO_5	16.3
AlO_6	1.4
SiO_2	2.5E-03
SiO_3	1.2
SiO_4	95.6
SiO_5	3.1
SiO_6	2.6E-02

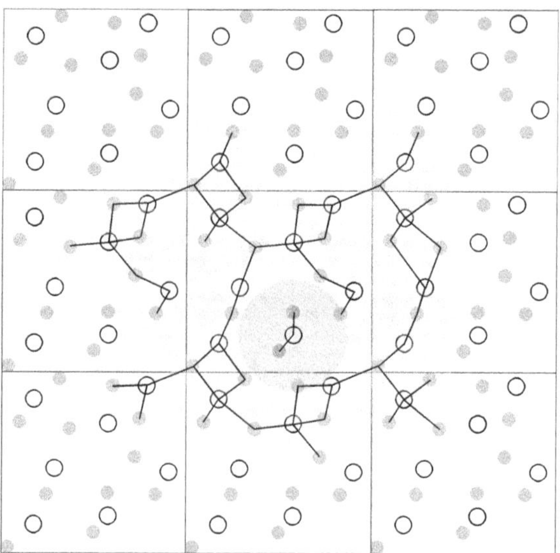

Fig. 17.4 Connecting all the atoms that lie within the first coordination sphere yields the polymerisation state of the melt. The unconnected central molecule can be assimilated into a volatile phase or another individual polymer. Graphical scheme after [67].

at low pressures in a melt, and even sometimes, volatile molecules can get detached from the melt and form a gas phase [228]. At higher pressures, the connectivity increases and eventually, the entire silicate melt forms an infinite polymer [161, 227]. The coordination polyhedra and the polymerisation state yield information about the chemical speciation in a fluid.

Building the connectivity matrix during the MD simulations allows us to estimate the lifetimes of various chemical bonds, coordination polyhedra, and polymerised structures. The lifetimes are obtained by counting how much time passed between the moment a given cluster formed and dissociated. After a cluster dissociates, the atoms often do not diffuse far but instead return and reform the same cluster. In the population analysis of the melt, this should count as separate entities, each with its own lifetime.

The lifetime analysis, straightforward in atomistic simulations, gives information about the population of the chemical species present in a fluid. Species with short lifetimes, on the order of up to a few dozen femtoseconds, would hardly be visible in a macroscopic experimental study. But their presence is captured in simulations, where their trajectories can be closely monitored. These species may be important as intermediate steps in chemical reactions and isotope exchanges. Figure 17.5 shows the lifetimes of various silica clusters and polymers encountered in a simulation of molten pyrolite at 2,000 K and 3.1 GPa, conditions very similar to an ultramafic melt at subcrustal conditions.

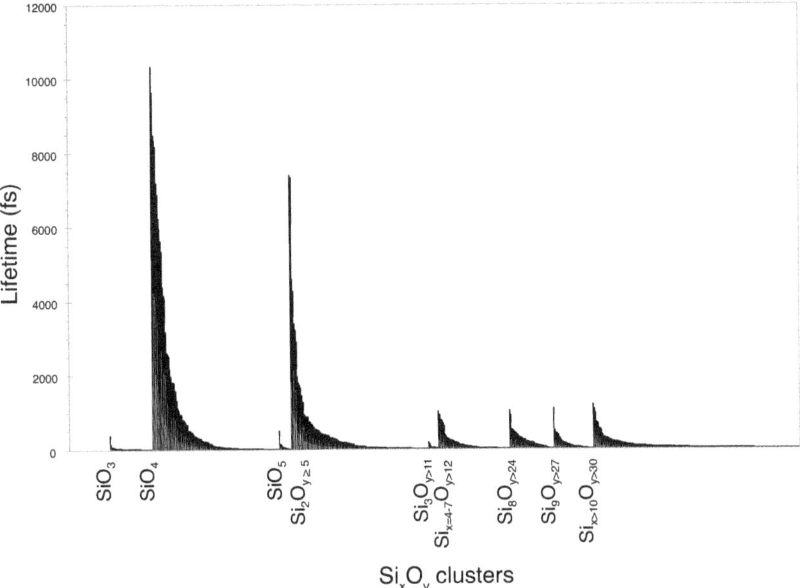

Fig. 17.5 Lifetimes of silica coordination polyhedra (upper panel) and polymers (lower panel) in molten pyrolite at 2,000 K and 3.1 GPa. Every vertical bar represents one Si_xO_y cluster. The height of the bar gives the lifetime of the cluster. At shallow Earth conditions, such ultramafic melts are dominated by isolated SiO_4 tetrahedra. Two Si and O atoms are bonded if their distance is shorter than the average radius of the first coordination sphere (Figure 17.3).

17.4 Atomic Diffusion in a Fluid

Monitoring the length of the atomic trajectories over time yields diffusion coefficients. The procedure consists of measuring the distance travelled by particles during various time intervals (Figure 17.6). The distances are averaged and then squared. The length of the time intervals is increased regularly. The maximum time length needs to be half the time of the simulation. Consequently, the sampling of the trajectory is restricted to the first half of the trajectory, meaning that the last starting point of the measurement can never be beyond half of the trajectory. In contrast, the endpoint of the last segment of the sampling can be the last snapshot of the trajectory.

The mean square displacements (MSD) of the atoms α are calculated as:

$$MSD_\alpha(\tau) = \frac{1}{N_\alpha} \frac{1}{N_{init}} \sum_{1}^{N_\alpha} \sum_{t}^{\tau/2} [r_\alpha(t+\tau) - r_\alpha(t)]^2 \qquad (17.2)$$

where N_{init} is the number of initial positions, which is the number of total segments, τ is the time interval, and $r_\alpha(t)$ is the position of the atom α at the time t. The MSD can be averaged over all the atoms of a given

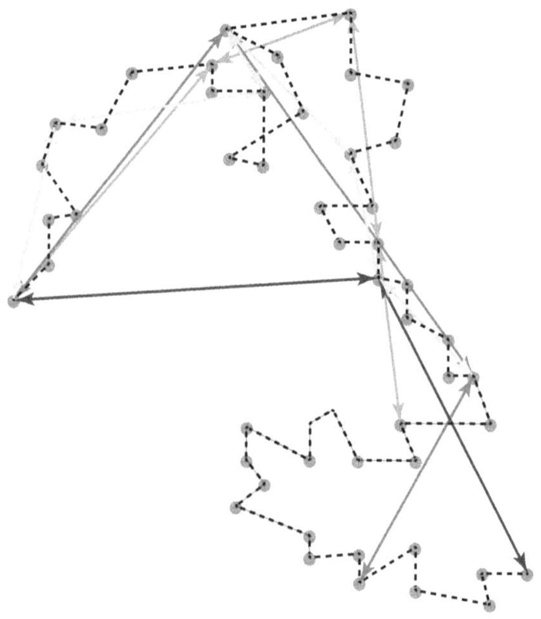

Fig. 17.6 The mean square displacement (MSD) of the atoms is calculated by sampling the trajectory of the atoms throughout the simulation. The distance travelled over a given time is represented by arrows of various grey shades, corresponding to a different number of time steps. Eventually, the distances are averaged and then squared to yield the MSDs. With increasingly large time intervals, the MSD offers a dynamic measure of atomic diffusivity.

type, over all the atoms belonging to a given atomic cluster/molecule, or it can be computed atom by atom. The time intervals can be defined as multiples of τ, and each segment starts at the next regular step along the trajectory.

The UMD package offers a faster way of computing the MSD, where not all the steps are sampled. This is illustrated in Figure 17.7. If the i-th interval starts at time $t_0(i)$, the (i+1)-th window may begin at time $t_0(i) + \tau + v$. Similarly, the window width is increased in discrete steps defined by the user: $\tau(i+1) = \tau(i) + z$. The values of both v and z are user-defined. If they are set to 0, then the formula for MSD from Equation 17.2 is fully recovered.

Box 17.1 explains how you can compute the MSDs in various situations for a fluid. The UMD package allows you to compute not only the diffusion coefficients for a given atomic type, like Si and O in a silicate melt, but also to trace the diffusion of individual atoms, and even of separate clusters of atoms, like the SiO_4 tetrahedra in a silicate melt, or the H_2O molecules in water or solution.

There are two regimes of the mass movement of atoms in an MD simulation: ballistic and diffusive. The first part of the trajectories represents the ballistic regime, in which the free particle motion dominates the movement of the atoms. In the second part, in the diffusive regime, the particles

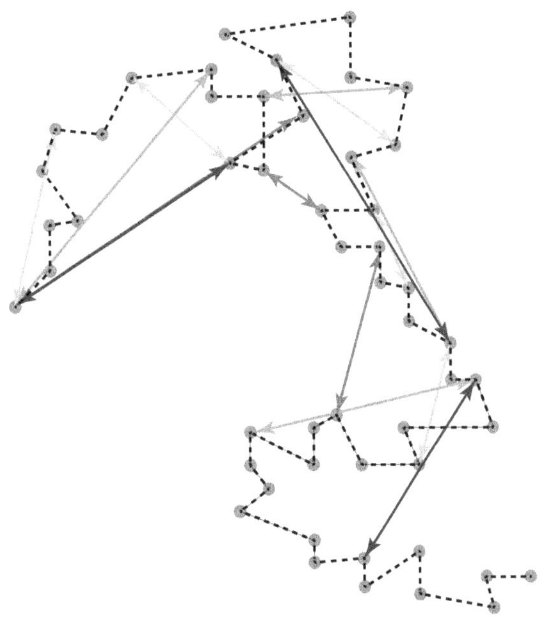

Fig. 17.7 A sparse calculation of the mean square displacements, as proposed in [67], allows a more rarefied sampling of the trajectories. With fewer distances to compute, this procedure shows a considerable increase in speed compared to the *classical* way of computing the mean square displacements described in Figure 17.6. If the diffusion coefficients are not affected by the sparse sampling, with this sparse procedure, you might miss capturing the ballistic region of the diffusion.

have the time to undergo several collisions, which thermalise their velocities. The definition of the two regimes is a statistical quantity applied to all the moments of the trajectories. In a way, the change between the two regimes represents a measure of the inertia of the system. The difference between the two regimes is marked by a slope change in a log-log plot of the MSDs as a function of time (Figure 17.8).

The variation of the MSDs as a function of time in the diffusive part of the trajectories yields the diffusion coefficient:

$$D_\alpha = \frac{MSD_\alpha}{2Z\tau} \qquad (17.3)$$

where Z is the number of degrees of freedom, 2 for diffusion in plane, and 3 for regular diffusion in space.

For long trajectories, in simulations with many atoms, the MSDs exhibit nice linear dependences with time (Figure 17.8). But short simulations or poor representation of atoms in a fluid lead to more scattered MSDs over time. The calculation of the diffusion coefficients should be done only in the linear part of the MSDs.

The diffusion coefficients can be determined for atomic types in a melt and individual atoms and atomic clusters.

The main characteristic of a fluid, for example, liquid, gas, or supercritical, is the continuous movement of atoms over time. The diffusion coefficient measures this amount of displacement. It is defined as the mean square displacement (MSD) the atoms travel divided by the time. Its units are cm^2/s, or m^2/s. The *umd* code allows for calculating various types of MSDs: average displacements for every atomic type, for individual atoms, and for chemical clusters. The different scripts and their requirements are summarised in Box Figure 8:

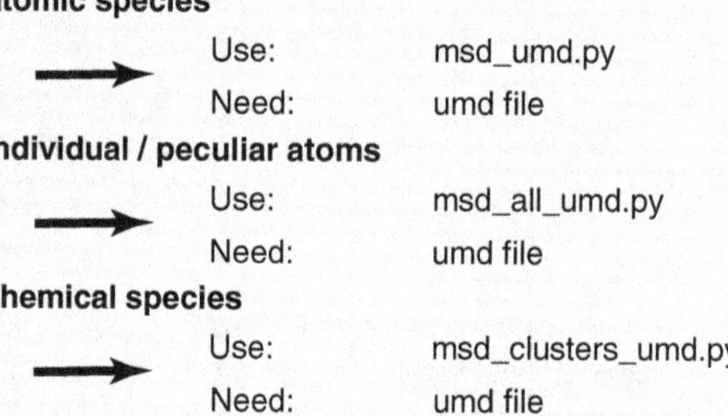

Box Figure 8. Different types of diffusion of atoms, chemical species, average, etc.

Then, for each case, you can calculate the diffusion coefficient as 1/6 MSD/time.

17.5 Vibrational Spectra

The time-dependent phenomena in a fluid are extracted from the self-correlation functions (SCFs) analysis. The SCFs of a time-dependent property $a(t)$ are defined as the average of the values of a at time t and $t + \tau$. The averages are normalised by the time t:

$$SCF(a(\tau)) = \frac{<a(t)a(t+\tau)>}{<a(t)a(t)>} \tag{17.4}$$

The SCFs give another measure of the inertia of the system. Large values of SCFs correspond to highly correlated time snapshots, while lower values correspond to independent configurations. As the SCFs capture periodic phenomena present in the simulations, the Fourier transform of the SCFs can extract these periodicities.

If the functions $a(t)$ are the atomic velocities, the SCFs become the velocity autocorrelation functions (VAFs), which are calculated for each atomic type, i, as:

$$VAF_i(\tau) = \frac{<v_i(t)v_i(t+\tau)>}{<v_i(t)v_i(t)>} \tag{17.5}$$

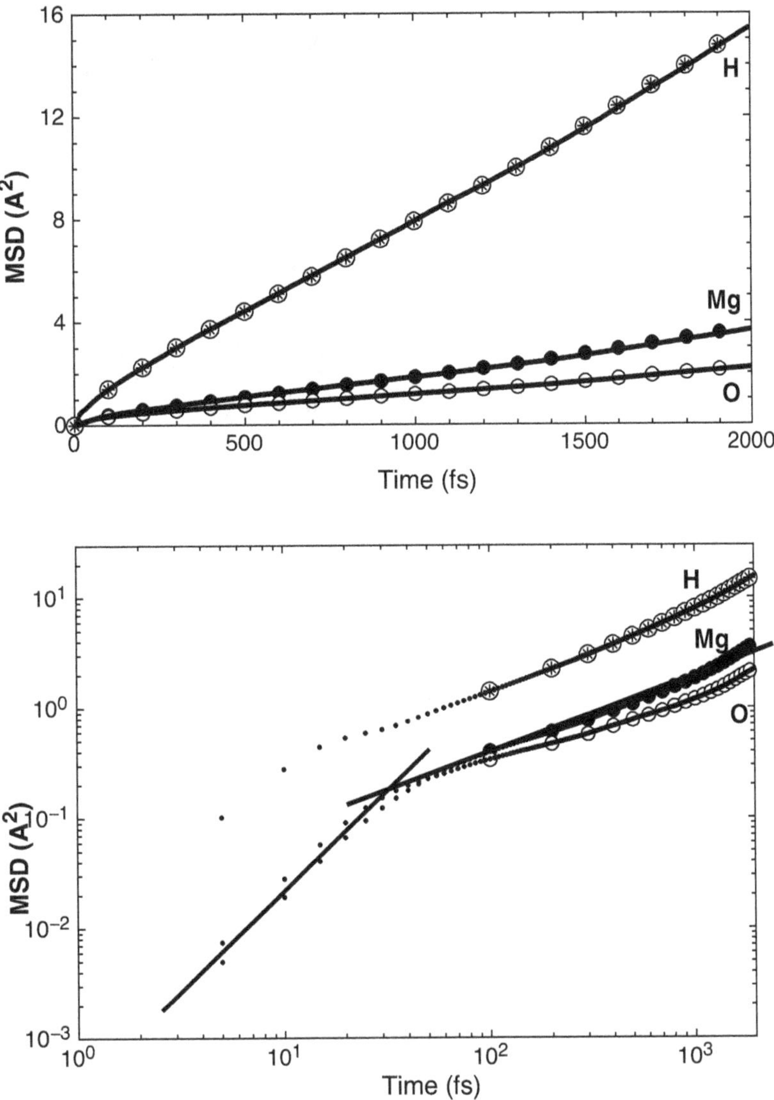

Fig. 17.8 The linear fit of the MSDs versus time in the diffusive regime gives the diffusion coefficients (upper panel). H is the most diffusive species in liquid brucite and O is the slowest one. The variation of the MSDs shows the two regimes, ballistic at the beginning of the trajectories and diffusive in the second part of the trajectories. The transition between the two is marked by a change in slope, clearly visible in the log-log representation (lower panel).

Then the Fourier transform of the VAFs captures the periodic patterns of the atomic velocities and implicitly of atomic displacements. These periodicities correspond to atomic vibrations. Thus, the SCF analysis of the atomic velocities yields the vibrational spectrum.

Figure 17.9(a) shows the atomic trajectories for brucite, $Mg(OH)_2$, at 3 GPa and 500 K. As the atoms describe oscillatory movements centred on their equilibrium positions, the resulting averaged image corresponds to the thermal ellipsoids. The SCF analysis shows clear periodicities at

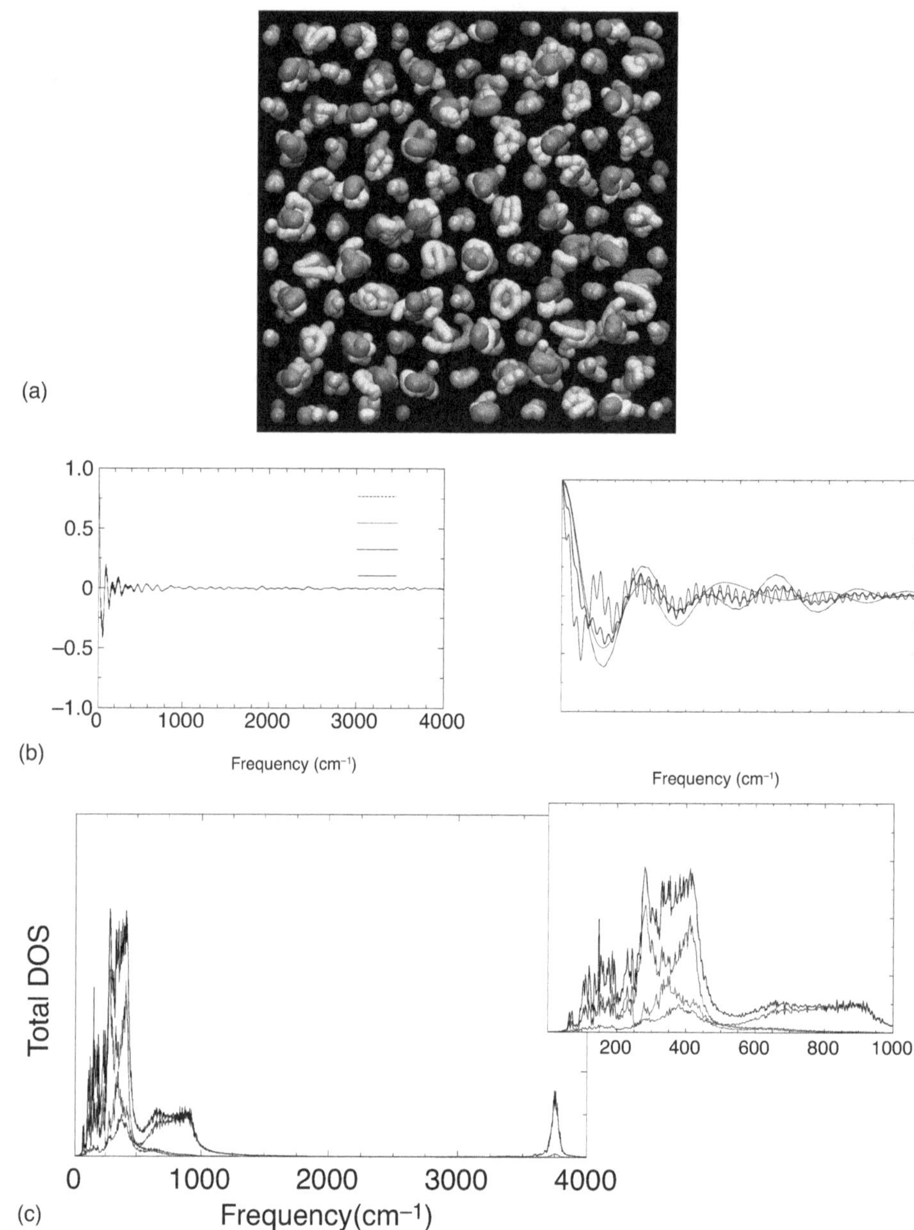

(a)

(b)

(c)

Fig. 17.9 (a) The atomic trajectories in solid brucite, $Mg(OH)_2$ at 500 K, over 2 picoseconds. The movements correspond to thermal agitation. (b) The velocity self-correlation functions. (c) The vibrational spectrum of brucite at the same conditions, as obtained from the Fourier transform of the velocity self-correlations.

low frequency (Figure 17.9(b)), which the Fourier transform helps to deconvolute as the vibrational spectrum, with several high-intensity peaks (Figure 17.9(c)).

Under heating, the amplitude of the atomic displacements increases. At 2,000 K and the same volume, the trajectory of the atoms in brucite shows clear net displacements (Figure 17.10(a)). The SCF analysis of the atomic

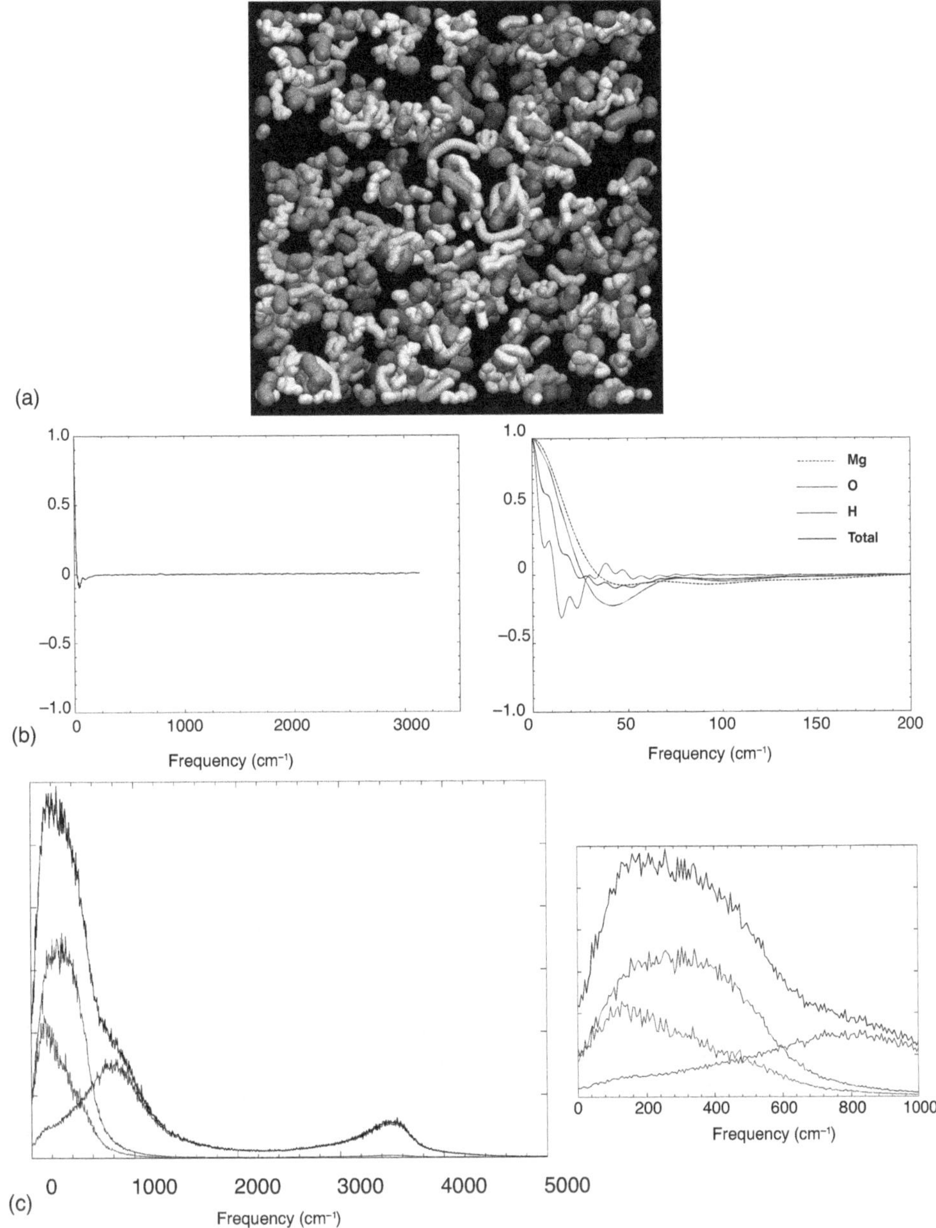

Fig. 17.10 (a) The atomic trajectories in brucite, $Mg(OH)_2$ at 2,000 K, over 2 picoseconds. There is a clear net movement on the atoms, corresponding to the thermal agitation. (b) The middle panel shows the velocity self-correlation functions (SCFs). (c) The vibrational spectrum of brucite at the same conditions, as obtained from the Fourier transform of the velocity SCFs. The spectra are broader and have fewer features than at low temperatures (Figure 17.9). Due to atomic diffusion, there is a non-zero net component at zero frequency. This is characteristic of fluids.

velocities shows a smaller correlation with fewer features (Figure 17.10(b)). The Fourier transform of the velocity SCF reveals a vibrational spectrum with broader peaks and fewer features.

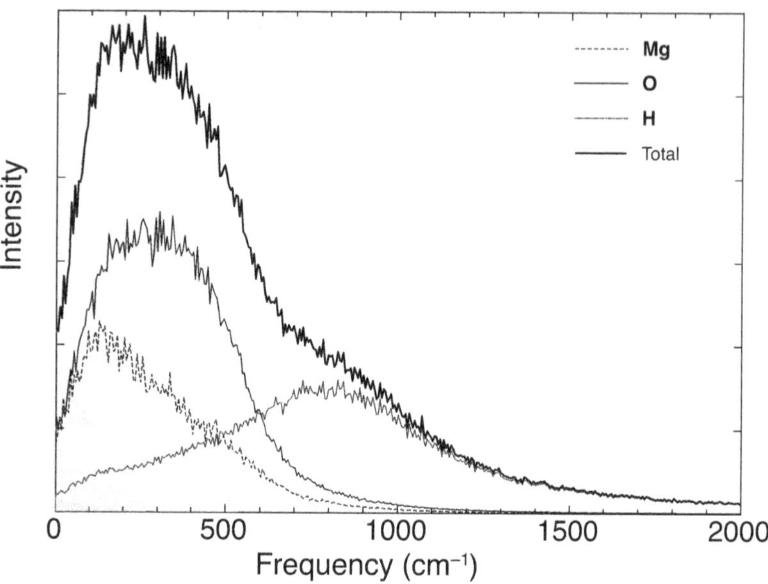

Fig. 17.11 The vibrational spectrum of a fluid, here exemplified by liquid brucite, $Mg(OH)_2$, can be decomposed into a gas-like part and a solid-like part. The gas-like part shown with a grey shadow at the bottom left is diffusive. If the gas-like part can be extracted from the total density of states using 2TP-MS theory [172, 173, 200], the thermodynamic properties can be obtained by integration of the remaining solid-like part. However, the results need to be taken with caution because of large errors due to anharmonicities and the approximations induced by estimating the diffusive dispersion.

In fluids, the movement of the atoms comprises a diffusion and a vibrational component. Consequently, the density of states of the vibrational spectrum at zero frequency has a finite non-zero value, which depends on the diffusion coefficient (Figure 17.10(c)). This diffusive part is extended and decays over the low frequencies of the spectrum. The contribution of each atomic species to the zero component is thus atom-dependent, and the thermodynamic parameters cannot be obtained by simply integrating the spectrum. Instead, the diffusive part needs to be first eliminated from the spectrum (Figure 17.11).

The diffusive part can be extracted according to the two-phase thermodynamic formalism with hard spheres [172, 173, 200]. The total vibrational spectrum is split into a gas-like part, which represents diffusion, and a solid-like part, which represents vibration. Assuming a hard sphere model, the diffusion contribution depends on the temperature, mass, and size of the spheres and the diffusion coefficients of the spheres. For the frequency-dependence decay of the gas-like diffusive component, Lorentzian or Gaussian models can be used as a start and later improved with memory and correlation functions [172, 96, 200, 137].

The hard-sphere model needs atomic volumes to extract the diffusive part of the spectrum. While obtaining the atomic volume is straightforward for monoatomic systems, it quickly becomes impossible to estimate for stoichiometric multi-atomic systems. For example, to determine the atomic volume of Fe in molten iron, it is enough to divide the volume of the simulation box by the number of Fe atoms in the box [170]: $V_{Fe} = V_{box}/N_{Fe}$. But for a system like molten MgO, the volume of the MgO molecule is $V_{MgO} = V_{box}/N_{MgO} = V_{Mg} + V_O$. Adding one Mg or one O atom to the system might provide their partial atomic volumes, but the resulting system, $N_{MgO} + 1\,Mg$ or $N_{MgO} + 1\,O$, does not correspond to the *molten* MgO from the initial simulations. The chemistry has changed with this addition. The same is the case for multicomponent systems, like $MgSiO_3$ or $CaSiO_3$, where adding an oxide component, like MgO, SiO_2, or CaO, only provides the partial volume of that component and not of the atoms.

The vibrational spectra of fluids exhibit considerably fewer details than the spectra of solids. At temperatures immediately above the melting point, the vibrational pattern of the fluid preserves certain features similar to the ones of the solid. With increasing temperature, the remaining features are lost. Eventually, the vibrational spectrum of a fluid resembles that of a hot, non-interacting ideal gas.

17.6　Viscosity of a Fluid

The viscosity of a fluid is a measure of the resistance to flow. It is the dominant parameter affecting the laminar flow.

The viscosity of the natural silicate liquids is highly dependent on the content of volatiles and the degree of polymerisation of the silica units. The latter is directly related to the amount of silica present in the melt [153]. Various basaltic liquids have viscosity values on the order of 0.1–10 Pa.s at 1 bar. Molten peridotite at upper mantle conditions is one order of magnitude less viscous [107, 154, 239, 89, 38]. Liquid iron at outer core conditions has a viscosity about 2–3 orders of magnitude smaller than the liquid silicates [264, 4, 169].

The viscosity of a melt can be easily computed with the SCFs of the stresses, η_{ij}. The self-correlation is:

$$SCF(\sigma_{ij}(\tau)) = \frac{<\sigma_{ij}(t)\sigma_{ij}(t+\tau)>}{<\sigma_{ij}(t)\sigma_{ij}(t)>} \qquad (17.6)$$

The viscosity is obtained from the SCF of the stresses according to the Green–Kubo formalism:

$$\eta_{ij} = \frac{V}{3k_B T}SCF(\sigma_{ij}) = \frac{V}{3k_B T}\int_0^\infty <\eta_{ij}(t)\eta_{ij}(t+\tau)>_t d\tau \qquad (17.7)$$

In this formulation, the viscosity represents the inertia of the stresses. Only the shear stresses are used, resulting in the shear viscosity, typically reported in the literature.

In general, in simulations, the convergence of the SCFs is obtained after longer simulation times than other properties. For silicate melts at mantle conditions, the MD simulations need to be on the order of hundreds of picoseconds, if not more. The simulations can be accelerated by taking longer simulation steps. However, too-long steps can easily mislead the decaying behaviour of the stress SCFs. An alternative is to use several independent starting configurations at the same volume and temperature and then average the results.

The quality of the SCF calculation can be improved by considering not only the self-correlation of the three shear stresses, xy, yz, and zx, but also the cross-correlations:

$$SCF(\tau) = \frac{<\sigma_{ij}(t)\sigma_{jk}(t+\tau)>}{<\sigma_{ij}(t)\sigma_{jk}(t)>} \tag{17.8}$$

where i, j, and k can alternatively be x, y, and z, as well as the tetragonal shear:

$$\eta_{ij} = \frac{V}{3k_BT}SCF(\sigma_{ij}) = \frac{V}{3k_BT}\int_0^\infty <\eta_{ii}(t)\eta_{jj}(t+\tau)>_t d\tau \tag{17.9}$$

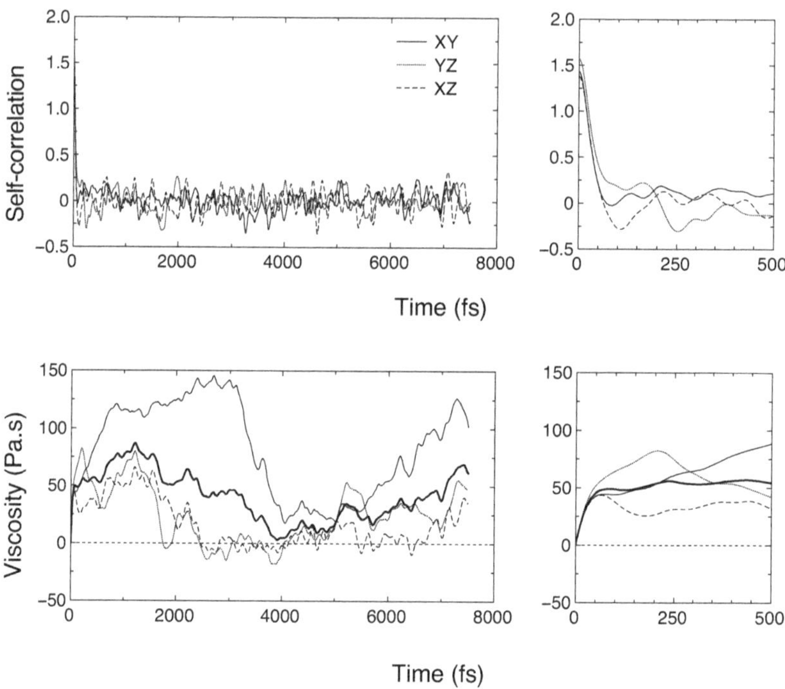

Fig. 17.12 (a) Self-correlation of the shear stresses is represented as a function of correlation time. After a slight decrease, the correlation oscillates around or slightly above 0 for long correlation times. The inset to the right shows that the 0 correlation is reached after slightly less than 100 steps – this represents the decorrelation time. The three shear stresses do not decorrelate at the same time. Longer simulation times can improve this. (b) The integral of the self-correlation, via a prefactor dependent on density and temperature, yields the viscosity of the fluid. The inset to the right shows a convergence reached for correlation times between 100 and at least 500 steps. This gives a good estimation of the viscosity of fluid iron at 16 kbars and 4,000 K.

where $i \neq j$. Only two of the three shear combinations can be used, as the last ones are always a linear combination of the first two. Then the best numerical result is obtained by averaging over all the different combinations of self-correlations.

Figure 17.12(a) shows the SCFs of the shear stresses as a function of time as computed for molten iron at 16 kbars and 4,000 K. The time range when the SCFs go first below zero gives the scale of the decorrelation length of the stresses. This time interval can be interpreted as the inertia of the system – the time the system needs to *forget* previous states. The integrals of the SCFs, and even better, their averages, also tend to hit a first plateau around that time length. The total time of the simulation needs to be sufficiently long so that enough entirely uncorrelated cycles are sampled in the SCFs. This value should typically be tens to hundreds of cycles.

Figure 17.12(b) illustrates the resulting value of the viscosity. The decorrelation time is about 100 fs or less. The three shear self-correlations do not hit the 0 value simultaneously. Differences are most likely due to the short simulation time. Averaging the corresponding integrals smoothes out the irregularities and shows a plateau around 50 Pa.s that lasts long.

The pages of this book tried to cover some of the most fundamental aspects of the density functional theory (DFT). The hope is that at the end of this book, the DFT language is not as incomprehensible and mysterious as it was at its beginnings. However, much more lies ahead.

Over the last decades, we have witnessed the exponential growth of computational power. Together with remarkable theory developments and algorithm improvements, these advancements brought computational physics to an equally important standing as experimental and theoretical physics. Mineralogists fully embraced the developments in condensed matter physics. This was primarily favoured by the increased availability of software packages, which made calculations based on DFT available to the entire community. Availability of extensive computing resources and access to supercomputer centres also became much easier over time.

Today access to supercomputers can be done at several levels: local, national, and international. Typically, mid-size supercomputers are hosted by local computing centres at universities. Access can be granted based on an internal agreement, via a fee or free of charge. National supercomputers can range anywhere from mid-size to exascale facilities. Their access is typically regulated via a common platform. Annual or multi-annual calls for computing time are organised, and committees decide on allocating the available hours. The most common computational projects may require anywhere between 1 and several tens of millions of CPU hours. For larger allocations, the international platforms centralise exascale resources, ranging in several hundreds of millions of CPU hours annually or more. These resources are available for the academic world. In most cases, the associated costs are covered by various research and education institutions and ministers.

In most cases, the available computing time of a supercomputer is measured in CPU hours, representing one core running for one hour dedicated to a given job. This functioning ensures that the resources, both computing and active memory, are used most efficiently.

When submitting an application for access to computing time, it is necessary to prove that the software works on that specific machine and demonstrate its good scaling. That implies finding the ideal number of processors a particular code should use to answer the scientific question. The scaling is rarely linear. Especially in DFT, the diagonalisation of the large matrices containing the weights of the planewaves can be only partially distributed to several processors. This implies considerable communication time between the processors and the nodes, which, beyond a particular value, becomes prohibitive. The scaling tests ensure that the calculations are done before reaching this threshold.

But once you have access to computational resources, a world of possibilities opens. As our understanding of the surrounding world evolves and improves, the necessity for performing more accurate simulations on specific properties for realistic materials increases. In the nineties, the simulations would typically cover ideal structures with only a handful of atoms in a unit cell, for which we could, at best, compute the vibrational spectrum and the electronic properties. Today, we can afford DFT-based molecular dynamics simulations or complex phonon calculations with hundreds or even thousands of atoms, for which we can determine not only equations of state but also transport properties [249]. And we can explore thermodynamic conditions that characterise the interior of giant exoplanets and even stars [182]. We can assist experiments, help interpret experimental results and observations, and predict and suggest future directions of investigations.

Beyond DFT, the machine-learning techniques based on ab initio data are exponentially developing. Only during the writing of this book were several packages developed for generating interatomic potentials based on machine learning. They allow us to perform accurate simulations with millions of atoms lasting for nanoseconds. These are clear steps that, together, machine learning and density functional theory can bring our simulations to the next level in the near future.

References

[1] Aguado, A, and Madden, P A. 2003. Oxide potentials from ab initio molecular dynamics: An assessment of their transferability. *The Journal of Chemical Physics*, **118**(13), 5718–5728.

[2] Alboussiere, T, Deguen R, and Melzani, M. 2010. Melting-induced stratification above the Earth's inner core due to convective translation. *Nature*, **466**(7307), 744–747. DOI:10.1038/nature09257.

[3] Alder, B J, and Wainwright, T E. 1959. Studies in molecular dynamics. I. General method. *The Journal of Chemical Physics*, **31**(2), 459–466.

[4] Alfè, D, Kresse, G, and Gillan, M J. 2000. Structure and dynamics of liquid iron under Earth's core conditions. *Physical Review B*, **61**(1), 132–142.

[5] Allen, M P, and Tildesley, D J. 1989. *Computer Simulation of Liquids* (Oxford: Oxford University Press).

[6] Hans, C A. 1980. Molecular dynamics simulations at constant pressure and/or temperature. *Journal of Chemical Physics*, **72**(4), 2384–2393.

[7] Anderson, D G. 1965. Iterative Procedures for nonlinear integral equations. *Journal of the ACM*, **12**(4), 547–560.

[8] Angel, R J, Alvaro, M, and Gonzalez-Platas, J. 2014. EosFit7c and a Fortran module (library) for equation of state calculations. *Zeitschrift für Kristallographie – Crystalline Materials*, **229**(5), 1–15.

[9] Anisimov, V I, Poteryaev, A I, Korotin, M A, Anokhin, A O, and Kotliar, G. 1997. First-principles calculations of the electronic structure and spectra of strongly correlated systems: Dynamical mean-field theory. *Journal of Physics: Condensed Matter*, **9**(35), 7359–7367.

[10] Aroyo, M I, Kirov, A, Capillas, C, Perez-Mato, J M, and Wondratschek, H. 2006. Bilbao Crystallographic Server. II. Representations of crystallographic point groups and space groups. *Acta Crystallographica Section A Foundations of Crystallography*, **62**(Pt 2), 115–128.

[11] Aryanpour, M, van Duin, A C T, and Kubicki, J D. 2010. Development of a reactive force field for iron – oxyhydroxide systems. *The Journal of Physical Chemistry A*, **114**(21), 6298–6307.

[12] Ashcroft, N W, and Mermin, N D. 1976. *Solid State Physics*. Holt-Saunders.

[13] Audouze, C, Jollet, F, Torrent, M, and Gonze, X. 2006. Projector augmented-wave approach to density-functional perturbation theory. *Physics Review B*, **73**, 235101.

[14] Austin, B M, Zubarev, D Y, and William, A L Jr. 2011. Quantum Monte Carlo and related approaches. *Chemical Reviews*, **112**(1), 263–288.

[15] Bader, R F W, Anderson, S G, and Duke, A J. 1979. Quantum topology of molecular charge distributions. 1. *Journal of the American Chemical Society*, **101**(6), 1389–1395.

[16] Badinski, A, Haynes, P D, Trail, J R, and Needs, R J. 2010. Methods for calculating forces within quantum Monte Carlo simulations. *Journal of Physics: Condensed Matter*, **22**(7), 074202–9.

[17] Badro, J, Siebert, J, and Nimmo, F. 2016. An early geodynamo driven by exsolution of mantle components from Earth's core. *Nature*, **536**(7616), 326–328.

[18] Balan, E, Refson, K, Blanchard, M, Delattre, S, Lazzeri, M, Ingrin, J, Mauri, F, Wright, K, and Winkler, B. 2008. Theoretical infrared absorption coefficient of OH groups in minerals. *American Mineralogist*, **93**(5–6), 950–953.

[19] Baldereschi, A. 1973. Mean-value point in the Brillouin zone. *Physics Review B*, **7**(June), 5212–5215.

[20] Balyakin, I A, Rempel, S V, Ryltsev, R E, and Rempel, A A. 2020. Deep machine learning interatomic potential for liquid silica. *Physical Review E*, **102**(5), 052125.

[21] Baláž, P, Achimovičová, M, Baláž, M, Billik, P, Cherkezova-Zheleva, Z, Criado, J M, Delogu, F, Dutková, E, Gaffet, E, Gotor, F J, Kumar, R, Mitov, I, Rojac, T, Senna, M, Streletskii, A, and Wieczorek-Ciurowa, K. 2013. Hallmarks of mechanochemistry: From nanoparticles to technology. *Chemical Society Reviews*, **42**(18), 7571–7567.

[22] Barone, V, Biczysko, M, and Bloino, J. 2014. Fully anharmonic IR and Raman spectra of medium-size molecular systems: Accuracy and interpretation. *Physical Chemistry Chemical Physics*, **16**(5), 1759–1787.

[23] Baroni, S, Giannozzi, P, and Testa, A. 1987. Green's-function approach to linear response in solids. *Physical Review Letters*, **58**(18), 1861–1864.

[24] Baroni, S, de Gironcoli, S, Corso, A D, and Giannozzi, P. 2001. Phonons and related crystal properties from density-functional perturbation theory. *Reviews of Modern Physics*, **73**(2), 515–562.

[25] Bartók, A P, and Csányi, G. 2016. Gaussian approximation potentials: A brief tutorial introduction (vol 115, pg 1051, 2015). *International Journal of Quantum Chemistry*, **116**(13), 1049–1049.

[26] Bass, J D. 1995. Elasticity of minerals, glasses, and melts. In Ahrens, T J, ed., *Mineral Physics & Crystallography: A Handbook of Physical Constants*, (Hoboken, New Jersey: American Geophysical Union), 2 vols., pp. 1–19.

[27] Becke, A D. 1988. Density-functional exchange-energy approximation with correct asymptotic behavior. *Physical Review. A, General Physics*, **38**(6), 3098–3100.

[28] Behler, J. 2016. Perspective: Machine learning potentials for atomistic simulations. *The Journal of Chemical Physics*, **145**(17), 170901–170910.

[29] Berendsen, H J C, Postma, J P M, van Gunsteren, W F, DiNola, A, and Haak, J R. 1998. Molecular dynamics with coupling to an external bath. *The Journal of Chemical Physics*, **81**(8), 3684–3690.

[30] Bernu, B, and Cepeley, M D. 2002. Quantum simulations of complex many-body systems: From theory to algorithms, Eds. J. Grotendorst, D. Marx, A. Muramatsu. *Lecture Notes on Path Integral Monte Carlo*, ipps. 51–62.

[31] Bigeleisen, J, and Mayer, M G. 1947. Calculation of equilibrium constants for isotopic exchange reactions. *The Journal of Chemical Physics*, **15**(5), 261–267.

[32] Bindi, L, Steinhardt, P J, Yao, N, and Lu, P J. 2009. Natural Quasicrystals. *Science*, **324**(5932), 1306–1309.

[33] Birch, F. 1947. Finite elastic strain of cubic crystals. *Physics Review*, **71**(June), 809–824.

[34] Blaha, P, Schwarz, K, Tran, F, Laskowski, R, Madsen, G K H, and Marks, L D. 2020. WIEN2k: An APW+lo program for calculating the properties of solids. *The Journal of Chemical Physics*, **152**(7), 1–31.

[35] Blöchl, P E. 1994. Projector augmented-wave method. *Physics Review B*, **50**(Dec), 17953–17979.

[36] Blöchl, P E, Jepsen, O, and Andersen, O K. 1994. Improved tetrahedron method for Brillouin-zone integrations. *Physics Review B*, **49**(June), 16223–16233.

[37] Blöchl, P E, Först, C J, and Schimpl, J. 2003. Projector augmented wave method: Ab initio molecular dynamics with full wave functions. *Bulletin of Materials Science*, **26**(1), 33–41.

[38] Bonechi, B, Stagno, V, Kono, Y, Hrubiak, R, Ziberna, L, Andreozzi, G B, Perinelli, C, and Gaeta, M. 2022. Experimental measurements of the viscosity and melt structure of alkali basalts at high pressure and temperature. *Scientific Reports*, **12**(1), 2599.

[39] Borlido, P, Schmidt, J, Huran, A W, Tran, F, Marques, M A L, and Botti, S. 2020. Exchange-correlation functionals for band gaps of solids: Benchmark, reparametrization and machine learning. *npj Computational Materials*, **6**(1), 96.

[40] Bouchet, J, Bottin, F, Recoules, V, Remus, F, Morard, G, Bolis, R M, and Benuzzi-Mounaix, A. 2019. Ab initio calculations of the B1-B2 phase transition in MgO. *Physical Review B*, **99**(9), 094113.

[41] Bouckaert, L P, Smoluchowski, R, and Wigner, E. 1936. Theory of Brillouin zones and symmetry properties of wave functions in crystals. *Physical Review B*, **50**(00), 58–67.

[42] Bradley, CJ, and Cracknell, A P. 1972. *The Mathematical Theory of Symmetry in Solids: Representation Theory for Point Groups and Space Groups* (Oxford: Clarendon Press).

[43] Brehm, M, Thomas, M, Gehrke, S, and Kirchner, B. 2020. TRAVIS—A free analyzer for trajectories from molecular simulation. *The Journal of Chemical Physics*, **152**(16), 164105.

[44] Buckingham, R A. 1938. The classical equation of state of gaseous helium, neon and argon. *Proceedings of the Royal Society of London. Series A, Mathematical and Physical Sciences*, **168**(933), 264–283.

[45] Bulatov, V, and Cai, W. 2006. *Computer Simulations of Dislocations* (Oxford; New York: Oxford University Press on Demand).

[46] Burns, R G. 1993. *Mineralogical applications of crystal field theory*. 2nd ed. Cambridge Topics in Mineral Physics and Chemistry. (New York: Cambridge University Press).

[47] Bussi, G, Donadio, D, and Parrinello, M. 2007. Canonical sampling through velocity rescaling. *The Journal of Chemical Physics*, **126**(1), 014101.

[48] Bögels, T F J, and Caracas, R. 2022. Critical point and supercritical regime of MgO. *Physical Review B*, **105**(6), 064105.

[49] Cady, W G, 1874-1974 (viaf)110373708. 1946. *Piezoelectricity: An Introduction to the Theory and Applications of Electromechanical Phenomena in Crystals*. 1st ed. (New York: McGraw-Hill).

[50] Campbell, A J. 2016. Phase diagrams and thermodynamics of core materials. In Terasaki, H, and Fischer, R A, eds., *Physics and Chemistry of the Lower Mantle and Core*, Geophysical Monograph **217**, pp. 191–199 (Hoboken, New Jersey: American Geophysical Union and John Wiley and Sons, Inc.).

[51] Cappelli, C, Lipparini, F, Bloino, J, and Barone, V. 2011. Towards an accurate description of anharmonic infrared spectra in solution within the polarizable continuum model: Reaction field, cavity field and nonequilibrium effects. *The Journal of Chemical Physics*, **135**(10), 104505–104516.

[52] Caracas, R. 2002. A database of incommensurate phases. *Journal of Applied Crystallography*, **35**(2), 120–121.

[53] Caracas, R, and Bobocioiu, E. 2011. The WURM project – a freely available web-based repository of computed physical data for minerals. *American Mineralogist*, **96**(2–3), 437–443.

[54] Caracas, R, and Boffa-Ballaran, T. 2010. Elasticity of (K,Na)AlSi3O8 hollandite from lattice dynamics calculations. *Physics of the Earth and Planetary Interiors*, **181**(1–2), 21–26.

[55] Caracas, R, and Gonze, X. 2004. High-pressure isosymmetrical phase transition in calaverite. *Physics and Chemistry of Minerals*, **31**(8), 553–558.

[56] Caracas, R, and Wentzcovitch, R M. 2004. Equation of state and elasticity of FeSi. *Geophysical Research Letters*, **31**(20), 784.

[57] Caracas, R, and Wentzcovitch, R M. 2006. Theoretical determination of the structures of CaSiO3 perovskites. *Acta Crystallographica Section B: Structural Science*, **62**(6), 1025–1030.

[58] Caracas, R, Wentzcovitch, R, Price, G D, and Brodholt, J. 2005. CaSiO3 perovskite at lower mantle pressures. *Geophysical Research Letters*, **32**(6), L06306.

[59] Caracas, R. 2008. Dynamical instabilities of ice X. *Physical Review Letters*, **101**(8), 085502–085504.

[60] Caracas, R. 2010. Spin and structural transitions in AlFeO3 and FeAlO3 perovskite and post-perovskite. *Physics of the Earth and Planetary Interiors*, **182**(1–2), 10–17.

[61] Caracas, R. 2016. Deep earth: Physics and chemistry of the lower mantle and core. *Geophysical Monograph Series*, 55–68.

[62] Caracas, R. 2017. The influence of carbon on the seismic properties of solid iron. *Geophysical Research Letters*, **44**(1), 128–134.

[63] Caracas, R, and Cohen, R E. 2006. Theoretical determination of the Raman spectra of MgSiO 3perovskite and post-perovskite at high pressure. *Geophysical Research Letters*, **33**(12), 5969.

[64] Caracas, R, and Gonze, X. 2003. Ab initio determination of the ground-state properties of Ca2MgSi2O7 åkermanite. *Physical Review B, Condensed Matter*, **68**(18), 425.

[65] Caracas, R, and Hemley, R J. 2015. Ferroelectricity in high-density H2O ice. *The Journal of Chemical Physics*, **142**(13), 134501.

[66] Caracas, R, Hirose, K, Nomura, R, and Ballmer, M D. 2019. Melt–crystal density crossover in a deep magma ocean. *Earth and Planetary Science Letters*, **516**(6), 202–211.

[67] Caracas, R, Kobsch, A, Solomatova, N V, Li, Z, Soubiran, F, and Hernandez, J-A. 2021. Analyzing melts and fluids from ab initio molecular dynamics simulations with the UMD package. *Journal of Visualized Experiments*, (175), e61534, doi:10.3791/61534.

[68] Carrez, P, and Cordier, P. 2010. Modeling dislocations and plasticity of deep earth materials. *Reviews in Mineralogy and Geochemistry*, **71**(1), 225–251.

[69] Ceperley, D M. 2010. An Overview of Quantum Monte Carlo Methods. *Reviews in Mineralogy and Geochemistry*, **71**(1), 129–135.

[70] Ceperley, D M, and Alder, B J. 1980. Ground state of the electron gas by a stochastic method. *Physical Review Letters*, **45**(Aug), 566–569.

[71] Ceriotti, M. 2019. Unsupervised machine learning in atomistic simulations, between predictions and understanding. *The Journal of Chemical Physics*, **150**(15), 150901–150911.

[72] Chadi, D J, and Cohen, M L. 1973. Special points in the Brillouin zone. *Physical Review B*, **8**(Dec), 5747–5753.

[73] Cheng, X, and Steele, R P. 2014. Efficient anharmonic vibrational spectroscopy for large molecules using local-mode coordinates. *The Journal of Chemical Physics*, **141**(10), 104105–104117.

[74] Coccocioni, M. 2002 (04). *A LDA+U study of selected iron compounds*. PhD thesis.

[75] Cococcioni, M, and de Gironcoli, S. 2005. Linear response approach to the calculation of the effective interaction parameters in the LDA + U method. *Physical Review B*, **71**(Jan), 035105.

[76] Cummins, H Z. 1990. Experimental studies of structurally incommensurate crystal phases. *Physics Reports*, **185**(5–6), 211–409.

[77] Cygan, R T, Liang, J-J, and Kalinichev, A G. 2004. Molecular models of hydroxide, oxyhydroxide, and clay phases and the development of a general force field. *The Journal of Physical Chemistry B*, **108**(4), 1255–1266.

[78] Dal Corso, A. 2010. Projector augmented-wave method: Application to relativistic spin-density functional theory. *Physical Review B*, **82**(Aug), 075116.

[79] Darlington, C N W. 2002. Normal-mode analysis of the structures of perovskites with tilted octahedra. *Acta Crystallographica Section A Foundations of Crystallography*, **58**(Pt 1), 66–71.

[80] Davidson, E R. 1975. Iterative calculation of a few of lowest eigenvalues and corresponding eigenvectors of large real-symmetric matrices. *Journal of Computational Physics*, **17**(1), 87–94.

[81] Dekura, H, and Tsuchiya, T. 2019. Lattice thermal conductivity of MgSiO 3postperovskite under the lowermost mantle conditions from ab initio anharmonic lattice dynamics. *Geophysical Research Letters*, **46**(22), 12919–12926.

[82] Deuss, A. 2014. Heterogeneity and anisotropy of earth's inner core. *Annual Review of Earth and Planetary Sciences*, **42**(1), 103–126.

[83] Dick, H J B, MacLeod, C J, Blum, P, and Scientists, Expedition 360. 2016. *International Ocean Discovery Program Preliminary Report*. IODP Publications.

[84] Driver, K P, and Militzer, B. 2012. All-electron path integral Monte Carlo simulations of warm dense matter: Application to water and carbon plasmas. *Physical Review Letters*, **108**(11), 185–194.

[85] Driver, K P, Cohen, R E, Wu, Z, Militzer, B, Rios, P L, Towler, M D, Needs, R J, and Wilkins, J W. 2010. Quantum Monte Carlo computations of phase stability, equations of state, and elasticity of high-pressure silica. *Proceedings of the National Academy of Sciences*, **107**(21), 9519–9524.

[86] Duan, W, Paiva, G, Wentzcovitch, R M, and Fazzio, A. 1998. Optical transitions in ruby across the corundum to Rh2O3 (II) phase transformation. *Physical Review Letters*, **81**(15), 3267–3270.

[87] Dubrovinskaia, N, Dubrovinsky, L, Langenhorst, F, Jacobsen, S, and Liebske, C. 2005. Nanocrystalline diamond synthesized from C60. *Diamond and Related Materials*, **14**(1), 16–22.

[88] Duff, A I, Finnis, M W, Maugis, P, Thijsse, B J, and Sluiter, M H F. 2015. MEAMfit: A reference-free modified embedded atom method (RF-MEAM) energy and force-fitting code. *Computer Physics Communications*, **196**(11), 439–445.

[89] Dufils, T, Sator, N, and Guillot, B. 2018. Properties of planetary silicate melts by molecular dynamics simulation. *Chemical Geology*, **493**(8), 298–315.

[90] Duin, A C T van, Strachan, A, Stewman, S, Zhang, Q, Xu, X, and Goddard, W A. 2003. ReaxFF SIOReactive force field for silicon and silicon oxide systems. *The Journal of Physical Chemistry A*, **107**(19), 3803–3811.

[91] Dutta, R, Greenberg, E, Prakapenka, V B, and Duffy, T S. 2019. Phase transitions beyond post-perovskite in NaMgF3 to 160 GPa. *Proceedings of the National Academy of Sciences*, **116**(39), 19324–19329.

[92] Dutta, R, Tracy, S J, Cohen, R E, Miozzi, F, Luo, K, Yang, J, Burnley, P C, Smith, D, Meng, Y, Chariton, S, Prakapenka, V B, and Duffy, T S. 2022. Ultrahigh-pressure disordered eight-coordinated phase of Mg2GeO4: Analogue for super-Earth mantles. *Proceedings of the National Academy of Sciences*, **119**(8), e2114424119.

[93] Dziewonski, A M, and Anderson, D L. 1981. Preliminary reference earth model. *Physics of the Earth and Planetary Interiors*, **25**(4), 297–356.

[94] Fischer, R A. 2016. Melting of Fe alloys and the thermal structure of the core. In Rebecca Fischer and Hidenori Terasaki (eds.), *Deep Earth*: Physics

and Chemistry of the Lower Mantle and the Core. Geophysical Monograph 217. (Hoboken, NJ: John Wiley & Sons, Inc.). pp. 1–12.

[95] Foulkes, W M C, Mitas, L, Needs, R J, and Rajagopal, G. 2001. Quantum Monte Carlo simulations of solids. *Reviews of Modern Physics*, **73**(1), 33–83.

[96] French, M, Desjarlais, M P, and Redmer, R. 2016. Ab initio calculation of thermodynamic potentials and entropies for superionic water. *Physical Review E*, **93**(Feb), 022140.

[97] Fuchs, M, and Scheffler, M. 1999. Ab initio pseudopotentials for electronic structure calculations of poly-atomic systems using density-functional theory. *Computer Physics Communications*, **119**(1), 67–98.

[98] Gabovich, A M, Voitenko, A I, and Ausloos, M. 2002. Charge- and spin-density waves in existing superconductors: Competition between Cooper pairing and Peierls or excitonic instabilities. *Physics Reports*, **367**(6), 583–709.

[99] Gale, J D. 2006. Empirical potential derivation for ionic materials. *Philosophical Magazine B*, **73**(1), 3–19.

[100] Gastegger, M, Schwiedrzik, L, Bittermann, M, Berzsenyi, F, and Marquetand, P. 2018. wACSF—Weighted atom-centered symmetry functions as descriptors in machine learning potentials. *The Journal of Chemical Physics*, **148**(24), 241709–241712.

[101] Geiger, P, and Dellago, C. 2013. Neural networks for local structure detection in polymorphic systems. *The Journal of Chemical Physics*, **139**(16), 164105.

[102] Georges, A, Kotliar, G, Krauth, W, and Rozenberg, M J. 1996. Dynamical mean-field theory of strongly correlated fermion systems and the limit of infinite dimensions. *Reviews of Modern Physics*, **68**(1), 13–125.

[103] Ghaderi, N, Zhang, D-B, Zhang, H, Xian, J, Wentzcovitch, R M, and Sun, T. 2017. Lattice thermal conductivity of MgSiO3 perovskite from first principles. *Scientific Reports*, **7**(1), 1–9.

[104] Ghosez, P, and Gonze, X. 2000. Band-by-band decompositions of the Born effective charges. *Journal of Physics: Condensed Matter*, **12**(43), 9179–9188.

[105] Giannozzi, P, Baroni, S, Bonini, N, Calandra, M, Car, R, Cavazzoni, C, Ceresoli, D, Chiarotti, G L, Cococcioni, M, Dabo, I, Corso, A D, Gironcoli, S de, Fabris, S, Fratesi, G, Gebauer, R, Gerstmann, U, Gougoussis, C, Kokalj, A, Lazzeri, M, Martin-Samos, L, Marzari, N, Mauri, F, Mazzarello, R, Paolini, S, Pasquarello, A, Paulatto, L, Sbraccia, C, Scandolo, S, Sclauzero, G, Seitsonen, A P, Smogunov, A, Umari, P, and Wentzcovitch, R M. 2009. QUANTUM ESPRESSO: A modular and open-source software project for quantum simulations of materials. *Journal of Physics: Condensed Matter*, **21**(39), 395502–395520.

[106] Gillan, M J, Alfè, D, and Michaelides, A. 2016. Perspective: How good is DFT for water? *The Journal of Chemical Physics*, **144**(13), 130901–130934.

[107] Giordano, D, Russell, J K, and Dingwell, D B. 2008. Viscosity of magmatic liquids: A model. *Earth and Planetary Science Letters*, **271**(1–4), 123–134.

[108] Gironcoli, S de. 1995. Lattice dynamics of metals from density-functional perturbation theory. *Physical Review B*, **51**(10), 6773–6776.

[109] Glazer, A M, and IUCr. 1972. The classification of tilted octahedra in perovskites. *Acta Crystallographica Section B: Structural Crystallography and Crystal Chemistry*, **28**(11), 3384–3392.

[110] Glielmo, A, Husic, B E, Rodriguez, A, Clementi, C, Noé, F, and Laio, A. 2021. Unsupervised learning methods for molecular simulation data. *Chemical Reviews*, **121**(16), 9722–9758.

[111] Goedecker, S, Teter, M, and Hutter, J. 1996. Separable dual-space Gaussian pseudopotentials. *Physical Review B*, **54**(3), 1703–1710.

[112] Gomi, H, Ohta, K, Hirose, K, Labrosse, S, Caracas, R, Verstraete, M J, and Hernlund, J W. 2013. The high conductivity of iron and thermal evolution of the Earth's core. *Physics of the Earth and Planetary Interiors*, **224**(C), 88–103.

[113] Goncharov, A F, and Hemley, R J. 2006. Probing hydrogen-rich molecular systems at high pressures and temperatures. *Chem Soc Rev*, **35**(10), 899–907.

[114] Gonze, X, and Vigneron, J P. 1989. Density-functional approach to nonlinear-response coefficients of solids. *Physical Review B*, **39**(18), 13120–13128.

[115] Gonze, X, Allan, D C, and Teter, M P. 1992. Dielectric tensor, effective charges, and phonons in alpha-quartz by variational density-functional perturbation theory. *Physical Review Letters*, **68**(24), 3603–3606.

[116] Gonze, X, Jollet, F, Abreu Araujo, F, Adams, D, Amadon, B, Applencourt, T, Audouze, C, Beuken, J-M, Bieder, J, Bokhanchuk, A, Bousquet, E, Bruneval, F, Caliste, D, Côté, M, Dahm, F, Da Pieve, F, Delaveau, M, Di Gennaro, M, Dorado, B, Espejo, C, Geneste, G, Genovese, L, Gerossier, A, Giantomassi, M, Gillet, Y, Hamann, D R, He, L, Jomard, G, Laflamme Janssen, J, Le Roux, S, Levitt, A, Lherbier, A, Liu, F, Lukačević, I, Martin, A, Martins, C, Oliveira, M J T, Poncé, S, Pouillon, Y, Rangel, T, Rignanese, G-M, Romero, A H, Rousseau, B, Rubel, O, Shukri, A A, Stankovski, M, Torrent, M, Van Setten, M J, Van Troeye, B, Verstraete, M J, Waroquiers, D, Wiktor, J, Xu, B, Zhou, A, and Zwanziger, J W. 2016. Recent developments in the ABINIT software package. *Communications in Computational Physics*, **205**(August), 106–131.

[117] Gonze, X. 1995. Adiabatic density-functional perturbation theory. *Physical Review A*, **52**(2), 1096–1114.

[118] Gonze, X. 1997. First-principles responses of solids to atomic displacements and homogeneous electric fields: Implementation of a conjugate-gradient algorithm. *Physical Review. B, Condensed Matter*, **55**(16), 10337–10354.

[119] Gonze, X, and Lee, C. 1997. Dynamical matrices, Born effective charges, dielectric permittivity tensors, and interatomic force constants from density-functional perturbation theory. *Physical Review. B, Condensed Matter*, **55**(16), 10355–10368.

[120] Gonze, X, Rignanese, G-M, and Caracas, R. 2005. First-principle studies of the lattice dynamics of crystals, and related properties. *Zeitschrift für Kristallographie – Crystalline Materials*, **220**(5/6), 13120.

[121] Gramsch, S A, Cohen, R E, and Savrasov, S Y. 2003. Structure, metal-insulator transitions, and magnetic properties of FeO at high pressures. *American Mineralogist*, **88**(2–3), 257–261.

[122] Gunsteren, W F Van, and Berendsen, H J C. 1988. A leap-frog algorithm for stochastic dynamics. *Molecular Simulation*, **1**(3), 173–185.

[123] Guren, M G, Sveinsson, H A, Malthe-Sørenssen, A, and Renard, F. 2022. Nanoscale modelling of dynamic rupture and damage production in alpha-quartz. Geophysical Research Letters, **49**, e2022GL100468. doi.org/10.1029/2022GL100468.

[124] Gurevich, V M, Osadchii, V O, Polyakov, V B, Gavrichev, K S, and Osadchii, E G. 2016. Heat capacity and thermodynamic functions of sphalerite: Implication to sulfide solid-state galvanic cell measurements. *Thermochimica Acta*, **641**(10), 14–23.

[125] Hahn, edited by Theo. 2002. *International tables for crystallography. Volume A, Space-group symmetry.* Fifth, revised edition. Dordrecht; London: Published for the International Union of Crystallography by Kluwer Academic Publishers, 2002. Fifth, revised edition. Dordrecht; London: Published for the International Union of Crystallography by Kluwer Academic Publishers, 2002. Includes bibliographical references and index.

[126] Haigis, V, Salanne, M, and Jahn, S. 2012. Thermal conductivity of MgO, MgSiO3 perovskite and post-perovskite in the Earth's deep mantle. *Earth and Planetary Science Letters*, **355–356**(C), 102–108.

[127] Hamann, D R, Schluter, M, and Chiang, C. 1979. Norm-conserving pseudopotentials. *Physical Review Letters*, **43**(20), 1494–1497.

[128] Hamann, D R, Wu, X, Rabe, K M, and Vanderbilt, D. 2005. Metric tensor formulation of strain in density-functional perturbation theory. *Physical Review B*, **71**(3), 10–13.

[129] Hammes-Schiffer, S. 2017. A conundrum for density functional theory. *Science*, **355**(6320), 28–29.

[130] Hazen, R M, Papineau, D, Bleeker, W, Downs, R T, Ferry, J M, McCoy, T J, Sverjensky, D A, and Yang, H. 2008. Mineral evolution. *American Mineralogist*, **93**(11–12), 1693–1720.

[131] Heaney, P J, and Veblen, D R. 1991. Observations of the alpha-beta-phase transition in quartz – a review of imaging and diffraction studies and some new results. *American Mineralogist*, **76**(5–6), 1018–1032.

[132] Helman, D S. 2016. Symmetry-based electricity in minerals and rocks: A summary of extant data, with examples of centrosymmetric minerals that exhibit pyro-and piezoelectricity. *Periodico di Mineralogia*, **85**(3), 201–248.

[133] Hemley, R J, Mao, H K, Bell, P M, and Mysen, B O. 1986. Raman spectroscopy of Si glass at high pressure. *Physical Review B*, **57**, 747.

[134] Hermet, P, Veithen, M, and Ghosez, P. 2007. First-principles calculations of the nonlinear optical susceptibilities and Raman scattering spectra of lithium niobate. *Journal of Physics: Condensed Matter*, **19**(45), 456202.

[135] Hernandez, J-A, and Caracas, R. 2018. Proton dynamics and the phase diagram of dense water ice. *The Journal of Chemical Physics*, **148**(21), 214501.

[136] Hernandez, J-A, and Caracas, R. 2016. Superionic-superionic phase transitions in body-centered cubic H_2O ice. *Physical Review Letters*, **117**(13), 135503.

[137] Hernandez, J-A, Caracas, R, and Labrosse, S. 2022. Stability of high-temperature salty ice suggests electrolyte permeability in water-rich exoplanet icy mantles. *Nature Communications*, **13**(1), 3303.

[138] Heyd, J, Scuseria, G E, and Ernzerhof, M. 2003. Hybrid functionals based on a screened Coulomb potential. *The Journal of Chemical Physics*, **118**(18), 8207.

[139] Hill, R. 1952. The elastic behaviour of a crystalline aggregate. *Proceedings of the Physical Society. Section A*, **65**(5), 349–354.

[140] Hirose, K, Morard, G, Sinmyo, R, Umemoto, K, Hernlund, J, Helffrich, G, and Labrosse, S. 2017. Crystallization of silicon dioxide and compositional evolution of the Earth's core. *Nature Publishing Group*, **543**(7643), 99–102.

[141] Hirshfeld, F L. 1977. Bonded-atom fragments for describing molecular charge-densities. *Theoretica Chimica Acta*, **44**(2), 129–138.

[142] Hohenberg, P, and Kohn, W. 1964. Inhomogeneous electron gas. *Physical Review B*, **136**(3B), B864.

[143] Holzwarth, N A W, Matthews, G E, Dunning, R B, Tackett, A R, and Zeng, Y. 1997. Comparison of the projector augmented-wave, pseudopotential, and linearized augmented-plane-wave formalisms for density-functional calculations of solids. *Physical Review B*, **55**(Jan), 2005–2017.

[144] HOOVER, W G. 1985. Canonical dynamics – equilibrium phase-space distributions. *Physical Review A*, **31**(3), 1695–1697.

[145] Hsu, H, Umemoto, K, Wu, Z, and Wentzcovitch, R M. 2010. Spin-state crossover of iron in lower-mantle minerals: Results of DFT+U investigations. *Reviews in Mineralogy and Geochemistry*, **71**(1), 169–199.

[146] Hsu, H, Blaha, P, Cococcioni, M, and Wentzcovitch, R M. 2011. Spin-state crossover and hyperfine interactions of ferric iron in Mg-SiO3perovskite. *Physical Review Letters*, **106**(11), 169.

[147] Huenenberger, P H. 2005. *Thermostat Algorithms for Molecular Dynamics Simulations* (Berlin, Heidelberg: Springer Berlin Heidelberg), pp. 105–149.

[148] Humphrey, W, Dalke, A, and Schulten, K. 1996. VMD: Visual molecular dynamics. *Journal of Molecular Graphics*, **14**(1), 33–38.

[149] Jahn, S, and Madden, P A. 2007. Modeling Earth materials from crustal to lower mantle conditions: A transferable set of interaction potentials for the CMAS system. *Physics of the Earth and Planetary Interiors*, **162**(1–2), 129–139.

[150] Jones, J E. 1924. On the determination of molecular fields. – II. From the equation of state of a gas. *Proceedings of the Royal Society of London A*, **106**, 463–477.

[151] Karato, S. 2008. *Deformation of Earth Materials: An Introduction to the Rheology of Solid Earth* (New York: Cambridge University Press).

[152] Karato, S. 2010. Rheology of the Earth's mantle: A historical review. *Gondwana Research*, **18**(1), 17–45.

[153] Karki, B B, and Stixrude, L. 2010. Viscosity of MgSiO3 liquid at Earth's mantle conditions: Implications for an early magma ocean. *Science*, **328**(5979), 740–742.

[154] Karki, B B, Zhang, J, and Stixrude, L. 2013. First principles viscosity and derived models for $MgO-SiO_2$ melt system at high temperature. *Geophysical Research Letters*, **40**(1), 94–99.

[155] Keefer, K D, and Brown, G D. 1978. Crystal structures and compositions of sanidine and high albite in cryptoperthitic intergrowths. *American Mineralogist*, **63**, 1264–1273.

[156] Khorshidi, A, and Peterson, A A. 2016. Amp: A modular approach to machine learning in atomistic simulations. *Computer Physics Communications*, **207**(10), 310–324.

[157] Kim, S, Kumar, N, Persson, P, Sofo, J, Duin, A C T van, and Kubicki, J D. 2013. Development of a ReaxFF reactive force field for titanium dioxide/water systems. *Langmuir*, **29**(25), 7838–7846.

[158] Kleinerman, D S, Czaplewski, C, Liwo, A, and Scheraga, H A. 2008. Implementations of Nosé-Hoover and Nosé-Poincaré thermostats in mesoscopic dynamic simulations with the united-residue model of a polypeptide chain. *The Journal of Chemical Physics*, **128**(24), 245103.

[159] Kobsch, A, and Caracas, R. 2020. The critical point and the supercritical state of alkali feldspars: Implications for the behavior of the crust during impacts. *Journal of Geophysical Research: Planets*, **125**(9), 85.

[160] Kohn, W, and Sham, L J. 1965. Self-consistent equations including exchange and correlation effects. *Physical Review*, **140**(Nov), A1133–A1138.

[161] Koker, N P de, Stixrude, L, and Karki, B B. 2008. Thermodynamics, structure, dynamics, and freezing of Mg2SiO4 liquid at high pressure. *Geochimica et Cosmochimica Acta*, **72**(5), 1427–1441.

[162] Kotliar, G, Savrasov, S Y, Haule, K, Oudovenko, V S, Parcollet, O, and Marianetti, C A. 2006. Electronic structure calculations with dynamical mean-field theory. *Reviews of Modern Physics*, **78**(3), 865–951.

[163] Kresse, G, and Joubert, D. 1999. From ultrasoft pseudopotentials to the projector augmented-wave method. *Physical Review B*, **59**(3), 1758–1775.

[164] Kurth, S, Perdew, J P, and Blaha, P. 1999. Molecular and solid-state tests of density functional approximations: LSD, GGAs, and meta-GGAs. *International Journal of Quantum Chemistry*, **75**(4–5), 889–909.

[165] Lang, N D, and Kohn, W. 1970. Theory of metal surfaces: Charge density and surface energy. *Physical Review B*, **1**(June), 4555–4568.

[166] Larsen, A H, Mortensen, J J, Blomqvist, J, Castelli, I E, Christensen, R, Dułak, M, Friis, J, Groves, M N, Hammer, B, Hargus, C, Hermes, E D, Jennings, P C, Jensen, P B, Kermode, J, Kitchin, J R, Kolsbjerg, E L, Kubal, J, Kaasbjerg, K, Lysgaard, S, Maronsson, J, Maxson, T, Olsen, T, Pastewka, L, Peterson, A, Rostgaard, C, Schiøtz, J, Schütt, O, Strange, M, Thygesen, K S, Vegge, T, Vilhelmsen, L, Walter, M, Zeng, Z, and Jacobsen, K W. 2017. The atomic simulation environment—a Python library for working with atoms. *Journal of Physics: Condensed Matter*, **29**(27), 273002.

[167] Lazicki, A, Goncharov, A F, Struzhkin, V V, Cohen, R E, Liu, Z, Gregoryanz, E, Guillaume, C, Mao, H K, and Hemley, R J. 2009. Anomalous optical and electronic properties of dense sodium. *Proceedings of the National Academy of Sciences*, **106**(16), 6525–6528.

[168] Li, X, and Jeanloz, R. 1987. Measurement of the B1-B2 transition pressure in NaCl at high temperatures. *Physical Review B*, **36**(1), 474.

[169] Li, Z, and Caracas, R. 2021. Thermophysical properties of hot fluid iron in the protolunar disk. *Physics of the Earth and Planetary Interiors*, **321**(12), 106806.

[170] Li, Z, Winisdoerffer, C, Soubiran, F, and Caracas, R. 2021. Ab initio Gibbs ensemble Monte Carlo simulations of the liquid-vapor equilibrium and the critical point of sodium. *Physical Chemistry Chemical Physics*, **23**(1), 311–319.

[171] Lichtenstein, A I, Katsnelson, M I, and Kotliar, G. 2001. Finite-temperature magnetism of transition metals: An ab initio dynamical mean-field theory, **87**(6), 85–94.

[172] Lin, S-T, Blanco, M, and Goddard, III W A. 2003. The two-phase model for calculating thermodynamic properties of liquids from molecular dynamics: Validation for the phase diagram of Lennard-Jones fluids. *The Journal of Chemical Physics*, **119**(22), 11792–11805.

[173] Lin, S-T, Maiti, P K, and Goddard, III W A. 2010. Two-phase thermodynamic model for efficient and accurate absolute entropy of water from molecular dynamics simulations. *The Journal of Physical Chemistry B*, **114**(24), 8191–8198.

[174] Lindsey, R K, Fried, L E, and Goldman, N. 2017. ChIMES: A force matched potential with explicit three-body interactions for molten carbon. *Journal of Chemical Theory and Computation*, **13**(12), 6222–6229.

[175] Manzano, H, Pellenq, R J M, Ulm, F-J, Buehler, M J, and Duin, A C T van. 2012. Hydration of calcium oxide surface predicted by reactive force field molecular dynamics. *Langmuir*, **28**(9), 4187–4197.

[176] Mao, H K, Xu, J, and Bell, P M. 1986. Calibration of the ruby pressure gauge to 800 kbar under quasi-hydrostatic conditions. *Journal of Geophysical Research: Solid Earth*, **91**(B5), 4673–4676.

[177] Martin, A, Torrent, M, and Caracas, R. 2019. Projector augmented-wave formulation of response to strain and electric-field perturbation within density functional perturbation theory. *Physical Review B*, **99**(9), 1–17.

[178] Martin, R M. 2004. *Electronic Structure: Basic Theory and Practical Methods* (New York: Cambridge University Press), pp. 624.

[179] Medvedev, M G, Bushmarinov, I S, Sun, J, Perdew, J P, and Lyssenko, K A. 2017. Density functional theory is straying from the path toward the exact functional. *Science*, **355**(6320), 49–52.

[180] Mermin, N D. 1965. Thermal properties of the inhomogeneous electron gas. *Phys Rev*, **137**(5A), A1441–A1443.

[181] Miguel, Y, Guillot, T, and Fayon, L. 2016. Jupiter internal structure: The effect of different equations of state. *Astronomy & Astrophysics*, **596**, A114.

[182] Militzer, B, González-Cataldo, F, Zhang, S, Driver, K P, and Soubiran, F. 2021. First-principles equation of state database for warm dense matter computation. *Physical Review E*, **103**(Jan), 013203.

[183] Monkhorst, H J, and Pack, J D. 1976. Special points for Brillouin-zone integrations. *Physical Review B*, **13**(June), 5188–5192.

[184] Morgan, D, and Jacobs, R. 2020. Opportunities and challenges for machine learning in materials science. *Annual Review of Materials Research*, **50**(1), annurev–matsci–070218–010015–33.

[185] Moroni, E G, Kresse, G, Hafner, J, and Furthmüller, J. 1997. Ultrasoft pseudopotentials applied to magnetic Fe, Co, and Ni: From atoms to solids. *Physical Review B*, **56**(Dec), 15629–15646.

[186] Moskovitz, N, and Gaidos, E. 2011. Differentiation of planetesimals and the thermal consequences of melt migration. *Meteoritics and Planetary Science*, **46**(6), 903–918.

[187] Mulliken, R S. 2004. Report on notation for the spectra of polyatomic molecules. *The Journal of Chemical Physics*, **23**(11), 1997.

[188] Murnaghan, F D. 1937. Finite deformations of an elastic solid. *American Journal of Mathematics*, **59**(2), 235–260.

[189] Musgrave, M J P. 1970. *Crystal Acoustics* (London: Holden-Day).

[190] Narang, P, Garcia, C A C, and Felser, C. 2021. The topology of electronic band structures. *Nature Materials*, **20**, 293–300.

[191] Narayanan, B, Duin, A C T van, Kappes, B B, Reimanis, I E, and Ciobanu, C V. 2011. A reactive force field for lithium–aluminum silicates with applications to eucryptite phases. *Modelling and Simulation in Materials Science and Engineering*, **20**(1), 015002.

[192] Nordstrom, L, and Singh, D J. 2006. *Planewaves, Pseudopotentials, and the LAPW Method*. Springer Science Business Media, Inc. Printed in the United States of America. pp. 142.

[193] Nose, S. 1984. A unified formulation of the constant temperature molecular-dynamics methods. *The Journal of Chemical Physics*, **81**(1), 511–519.

[194] Noé, F, Tkatchenko, A, Müller, K-R, and Clementi, C. 2020. Machine learning for molecular simulation. *Annual Review of Physical Chemistry*, **71**(1), 361–390.

[195] Oganov, A R., and Glass, C W. 2006. Crystal structure prediction using ab initio evolutionary techniques: Principles and applications. *The Journal of Chemical Physics*, **124**(24), 244704.

[196] Ohta, K, Cohen, R E, Hirose, K, Haule, K, Shimizu, K, and Ohishi, Y. 2012. Experimental and theoretical evidence for pressure-induced metallization in FeO with rocksalt-type structure. *Physical Review Letters*, **108**(2), 374–375.

[197] Parliński, K, Li, Z Q, and Kawazoe, Y. 1997. First-principles determination of the soft mode in cubic ZrO2. *Physical Review Letters*, **78**(21), 4063.

[198] Parrinello, M, and Rahman, A. 1980. Crystal structure and pair potentials: A molecular-dynamics study. *Physical Review Letters*, **45**(14), 1196.

[199] Parrinello, M, and Rahman, A. 1981. Polymorphic transitions in single-crystals – a new molecular-dynamics method. *Journal of Applied Physics*, **52**(12), 7182–7190.

[200] Pascal, T A, Lin, S-T, and Goddard, III W A. 2011. Thermodynamics of liquids: standard molar entropies and heat capacities of common solvents from 2PT molecular dynamics. *Physical Chemistry Chemical Physics*, **13**(1), 169–181.

[201] Paulatto, L, Mauri, F, and Lazzeri, M. 2013. Anharmonic properties from a generalized third-order ab initio approach: Theory and applications to graphite and graphene. *Physical Review B*, **87**(21), 214303.

[202] Payne, M C, Teter, M P, Allan, D C, Arias, T A, and Joannopoulos, J D. 1992. Iterative minimization techniques for ab initio total-energy calculations: molecular dynamics and conjugate gradients. *Reviews of Modern Physics*, **64**(Oct), 1045–1097.

[203] Pearson, D G, Brenker, F E, Nestola, F, McNeill, J, Nasdala, L, Hutchison, M T, Matveev, S, Mather, K, Silversmit, G, Schmitz, S, Vekemans, B, and Vincze, L. 2014. Hydrous mantle transition zone indicated by ringwoodite included within diamond. *Nature*, **507**(7491), 221–224.

[204] Perdew, J P, Burke, K, and Ernzerhof, M. 1996. Generalized gradient approximation made simple. *Physical Review Letters*, **77**(18), 3865–3868.

[205] Perdew, J P, Kurth, S, Zupan, A, and Blaha, P. 1999. Accurate density functional with correct formal properties: A step beyond the generalized gradient approximation. *Physical Review Letters*, **82**(12), 2544.

[206] Perdew, J P, Ruzsinszky, A, Csonka, G, Vydrov, O A, Scuseria, G E, Constantin, L A, Zhou, X, and Burke, K. 2008. Restoring the density-gradient expansion for exchange in solids and surfaces. *Physical Review Letters*, **100**(13), 136406.

[207] Peverati, R, and Truhlar, D G. 2014. Quest for a universal density functional: The accuracy of density functionals across a broad spectrum of databases in chemistry and physics. *Philosophical Transactions of the Royal Society A: Mathematical, Physical and Engineering Sciences*, **372**(2011), 20120476.

[208] Pickard, C J, and Needs, R J. 2011. Ab initio random structure searching. *Journal of Physics: Condensed Matter*, **23**(5), 053201.

[209] Pinheiro, C B, and Abakumov, A M. 2015. Superspace crystallography: A key to the chemistry and properties. *IUCrJ*, **2**(1), 137–154.

[210] Prescher, C, Dubrovinsky, L, Bykova, E, Kupenko, I, Glazyrin, K, Kantor, A, McCammon, C, Mookherjee, M, Nakajima, Y, Miyajima, N, Sinmyo, R, Cerantola, V, Dubrovinskaia, N, Prakapenka, V, Rüffer, R, Chumakov, A, and Hanfland, M. 2015. High Poisson's ratio of Earth's inner core explained by carbon alloying. *Nature Geoscience*, **8**(3), 220–223.

[211] Prosandeev, S A, Waghmare, U, Levin, I, and Maslar, J. 2005. First-order Raman spectra of A(B1,B2)O3 double perovskites. *Physical Review B*, **71**(21), 241.

[212] Quigley, D, and Probert, M I J. 2004. Langevin dynamics in constant pressure extended systems. *The Journal of Chemical Physics*, **120**(24), 11432.

[213] Rappe, A M, Rabe, K M, Kaxiras, E, and Joannopoulos, J D. 1990. Optimized pseudopotentials. *Physical Review B*, **41**(2), 1227–1230.

[214] Resta, R. 2003. Ab initio simulation of the properties of ferroelectric materials. *Modelling and Simulation in Materials Science and Engineering*, **11**(4), R69–R96.

[215] Reynard, B, Caracas, R, and McMillan, P F. 2015. Lattice vibrations and spectroscopy of mantle phases. In Schubert, G, ed., *Treatise on Geophysics*, 2nd ed (Amsterdam: Elsevier), pp. 203–231.

[216] Reynard, B, and Caracas, R. 2009. D/H isotopic fractionation between brucite Mg(OH)2 and water from first-principles vibrational modeling. *Chemical Geology*, **262**(3–4), 159–168.

[217] Rountree, C L, Kalia, R K, Lidorikis, E, Nakano, A, Brutzel, L V, and Vashishta, P. 2002. Atomistic aspects of crack propagation in brittle materials: Multimillion atom molecular dynamics simulations. *Annual Review of Materials Research*, **32**(1), 377–400.

[218] Ryabov, A, Akhatov, I, and Zhilyaev, P. 2020. Neural network interpolation of exchange-correlation functional. *Scientific Reports*, **10**(1), B864–B867.

[219] Savrasov, S Y. 1996. Linear-response theory and lattice dynamics: A muffin-tin-orbital approach. *Physical Review B*, **54**(Dec), 16470–16486.

[220] Schlegel, H B. 1982. Optimization of equilibrium geometries and transition structures. *Journal of Computational Chemistry*, **3**(2), 214–218.

[221] Setten, M J van, Giantomassi, M, Bousquet, E, Verstraete, M J, Hamann, D R, Gonze, X, and Rignanese, G M. 2018. The PseudoDojo: Training

and grading a 85 element optimized norm-conserving pseudopotential table. *Computer Physics Communications*, **226**(5), 39–54.

[222] Shankland, T J. 1968. Band gap of forsterite. *Science*, 161(3836), 51–53.

[223] Shechtman, D, Blech, I, Gratias, D, and Cahn, J W. 1984. Metallic phase with long-range orientational order and no translational symmetry, **53**(20), 1951–1953.

[224] Shephard, G E, Houser, C, Hernlund, J W, Valencia-Cardona, J J, Trønnes, R G, and Wentzcovitch, R M. 2021. Seismological expression of the iron spin crossover in ferropericlase in the Earth's lower mantle. *Nature Communications*, **12**(1), 5905.

[225] Slater, J C. 1937. Wave functions in a periodic potential. *Physical Review*, **51**(May), 846–851.

[226] Smaalen, S van. 2004. The Peierls transition in low-dimensional electronic crystals. *Acta Crystallographica Section A Foundations of Crystallography*, **61**(1), 51–61.

[227] Solomatova, N V, and Caracas, R. 2019. Pressure-induced coordination changes in a pyrolitic silicate melt from ab initio molecular dynamics simulations. *Journal of Geophysical Research: Solid Earth*, **124**(11), 11232–11250.

[228] Solomatova, N V, and Caracas, R. 2021a. Genesis of a CO2-rich and H2O-depleted atmosphere from Earth's early global magma ocean. *Science Advances*, **7**(41), eabj0406.

[229] Solomatova, N V, and Caracas, R. 2021b. Buoyancy and structure of volatile-rich silicate melts. *Journal of Geophysical Research: Solid Earth*, **126**(2), e2020JB021045.

[230] Spaldin, N A. 2020. Multiferroics beyond electric-field control of magnetism. *Proceedings of the Royal Society A*, **476**(2233), 20190542.

[231] Speziale, S, Marquardt, H, and Duffy, T S. 2014. Brillouin scattering and its application in geosciences. *Reviews in Mineralogy and Geochemistry*, **78**(01), 543–603.

[232] Sternheimer, R. M. 1954. Electronic polarizabilities of ions from the Hartree–Fock wave functions. *Physical Review*, **96**(Nov), 951–968.

[233] Stillinger, F H, and Weber, T A. 1984. Computer simulation of local order in condensed phases of silicon. *Physical Review B*, **31**(8), 5262–5271.

[234] Stixrude, L, Cohen, R E, Yu, R, and Krakauer, H. 1996. Prediction of phase transition in CaSiO3 perovskite and implications for lower mantle structure. *American Mineralogist*, **81**(9–10), 1293–1296.

[235] Stixrude, L, Koker, N de, Sun, N, Mookherjee, M, and Karki, B B. 2009. Thermodynamics of silicate liquids in the deep Earth. *Earth and Planetary Science Letters*, **278**(3–4), 226–232.

[236] Stoneham, A M, and Harding, J H. 1986. Interatomic potentials in solid state chemistry. *Annual Review of Physical Chemistry*, **37**(1), 53–80.

[237] Streitz, F H, Glosli, J N, and Patel, M V. 2006. Beyond finite-size scaling in solidification simulations. *Physical Review Letters*, **96**(22), 195–204.

[238] Sun, Y, Cococcioni, M, and Wentzcovitch, R M. 2020. LDA + Usc calculations of phase relations in FeO. *Physical Review Materials*, **4**(6), 063605.

[239] Takeuchi, S. 2011. Preeruptive magma viscosity: An important measure of magma eruptibility. *Journal of Geophysical Research: Solid Earth (1978–2012)*, **116**(B10).

[240] Tang, W, Sanville, E, and Henkelman, G. 2009. A grid-based Bader analysis algorithm without lattice bias. *Journal of Physics: Condensed Matter*, **21**(8), 084204–084207.

[241] Tao, J, Perdew, J P, Staroverov, V N, and Scuseria, G E. 2003. Climbing the density functional ladder: Nonempirical meta-generalized gradient approximation designed for molecules and solids. *Physical Review Letters*, **91**(14), 146401.

[242] Tersoff, J. 1988. Empirical interatomic potential for silicon with improved elastic properties. *Physical Review B*, **38**(14), 9902–9905.

[243] Tersoff, J. 1989. Modeling solid-state chemistry: Interatomic potentials for multicomponent systems. *Physical Review B*, **39**(8), 5566–5568.

[244] Togo, A, and Tanaka, I. 2015. First principles phonon calculations in materials science. *Scripta Materialia*, **108**(C), 1–5.

[245] Torrent, M, Jollet, F, Bottin, F, Zérah, G, and Gonze, X. 2008. Implementation of the projector augmented-wave method in the ABINIT code: Application to the study of iron under pressure. *Computational Materials Science*, **42**(2), 337–351.

[246] Torsvik, T H, Müller, R D, Voo, R Van der, Steinberger, Bernhard, and Gaina, Carmen. 2008. Global plate motion frames: Toward a unified model. *Reviews of Geophysics*, **46**(3), RG3004.

[247] Treviño, P, Garcia-Castro, A C, López-Moreno, S, Bautista-Hernández, A, Bobocioiu, E, Reynard, B, Caracas, R, and Romero, A H. 2018. Anharmonic contribution to the stabilization of $Mg(OH)_2$ from first principles. *Physical Chemistry Chemical Physics*, **20**(26), 17799–17808.

[248] Troullier, N, and Martins, J L. 1991. Efficient pseudopotentials for plane-wave calculations. *Physical Reviews B: Condensed Matter*, **43**(3), 1993–2006.

[249] Tsuchiya, T, Tsuchiya, J, Dekura, H, and Ritterbex, S. 2020. Ab initio study on the lower mantle minerals. *Annual Review of Earth and Planetary Sciences*, **48**(1), 99–119.

[250] Tuckerman, M E. 2002. Ab initio molecular dynamics: Basic concepts, current trends and novel applications. *Journal of Physics: Condensed Matter*, **14**(50), R1297–R1355.

[251] Umemoto, K, Wentzcovitch, R M, Yu, Y G, and Requist, R. 2008. Spin transition in $(Mg,Fe)SiO_3$ perovskite under pressure. *Earth and Planetary Science Letters*, **276**(1–2), 1–9.

[252] Umemoto, K, Wentzcovitch, R M., Wu, S, Ji, M, Wang, C-Z, and Ho, K-M. 2017. Phase transitions in $MgSiO_3$ post-perovskite in super-Earth mantles. *Earth and Planetary Science Letters*, **478**, 40–45.

[253] Unke, O T, Chmiela, S, Sauceda, H E, Gastegger, M, Poltavsky, I, Schütt, K T, Tkatchenko, A, and Müller, K-R. 2021. Machine learning force fields. *Chemical Reviews*, **121**(16), 10142–10186.

[254] Unterborn, C T, Dismukes, E E, and Panero, W R. 2016. Scaling the Earth: A sensitivity analysis of terrestrial exoplanetary interior models. *The Astrophysical Journal*, **819**(1), 1–8.

[255] Vanderbilt, D, and Kingsmith, R D. 1993. Electric polarization as a bulk quantity and its relation to surface charge. *Physical Review B*, **48**(7), 4442–4455.

[256] Vanderbilt, D. 1990. Soft self-consistent pseudopotentials in a generalized eigenvalue formalism. *Physical Review B*, **41**(11), 7892–7895.

[257] Vanderbilt, D. 2004. First-principles theory of polarization and electric fields in ferroelectrics. *Ferroelectrics*, **301**(1), 9–14.

[258] Veithen, M, Gonze, X, and Ghosez, P. 2005. Nonlinear optical susceptibilities, Raman efficiencies, and electro-optic tensors from first-principles density functional perturbation theory. *Physical Review B*, **71**(12), 12–14.

[259] Verlet, L. 1967. Computer 'Experiments' on Classical Fluids. I. Thermodynamical properties of Lennard-Jones molecules. *Physical Review*, **159**(1), 98–103.

[260] Walley, S M. 2013. Historical origins of indentation hardness testing. *Materials Science and Technology*, **28**(9–10), 1028–1044.

[261] Wang, J, Neaton, J B, Zheng, H, Nagarajan, V, Ogale, S B, Liu, B, Viehland, D, Vaithyanathan, V, Schlom, D G, Waghmare, U V, Spaldin, N A, Rabe, K M, Wuttig, M, and Ramesh, R. 2003. Epitaxial $BiFeO_3$ multiferroic thin film heterostructures. *Science*, **299**(5613), 1719–1722.

[262] Wang, Y, Lv, J, Zhu, L, and Ma, Y. 2010. Crystal structure prediction via particle-swarm optimization. *Physical Review B*, **82**(9), 094116.

[263] Wentzcovitch, R M. 1991. Invariant molecular-dynamics approach to structural phase-transitions. *Physical Reviews B: Condensed Matter*, **44**(5), 2358–2361.

[264] Wijs, G A de, Kresse, G, Vočadlo, L, Dobson, D, Alfè, D, Gillan, M J, and Price, G D. 1998. The viscosity of liquid iron at the physical conditions of the Earth's core. *Nature*, **392**(6678), 805–807.

[265] Witze, A. 2015. Drill ship targets Earth's mantle. *Nature*, **528**(7580), 16–17.

[266] Woollam, J A, and Somoano, R B. 1976. Superconducting critical fields of alkali and alkaline-earth intercalates of MoS_2. *Physical Review B*, **13**(May), 3843–3853.

[267] Wu, Z, and Cohen, R E. 2006. More accurate generalized gradient approximation for solids. *Physical Review B*, **73**(23), R8800.

[268] Yuan, N F Q, Mak, K F, and Law, K T. 2014. Possible topological superconducting phases of MoS2. *Physical Review Letters*, **113**(9), 097001.

[269] Zen, A, Brandenburg, J G, Klimeš, J, Tkatchenko, A, Alfè, D, and Michaelides, A. 2018. Fast and accurate quantum Monte Carlo for molecular crystals. *Proceedings of the National Academy of Sciences*, **115**(8), 1724–1729.

[270] Zepeda-Ruiz, L A, Stukowski, A, Oppelstrup, T, and Bulatov, V V. 2017. Probing the limits of metal plasticity with molecular dynamics simulations. *Nature*, **492**(7677), 492–495.

[271] Zhang, G-X, Reilly, A M, Tkatchenko, A, and Scheffler, M. 2018. Performance of various density-functional approximations for cohesive properties of 64 bulk solids. *New Journal of Physics*, **20**(6), 063020.

Index

For EU product safety concerns, contact us at Calle de José Abascal, 56–1°,
28003 Madrid, Spain or eugpsr@cambridge.org.